d Regenerative Biology

nal Biology

ratory

dical Center

ry
arch Institute

Organogenesis Group
Biology

generative Biology

AF778128

NO LONGER THE PROPERTY OF
PHILLIPS MEMORIAL LIBRARY

Series Editor

Paul M. Wassarman
Department of Developmental
Mount Sinai School of Medicine
New York, NY 10029-6574
USA

Editorial Board

Blanche Capel
Duke University Medical Center
Durham, NC, USA

B. Denis Duboule
Department of Zoology and An
NCCR 'Frontiers in Genetics'
Geneva, Switzerland

Anne Ephrussi
European Molecular Biology Lab
Heidelberg, Germany

Janet Heasman
Cincinnati Children's Hospital Me
Department of Pediatrics
Cincinnati, OH, USA

Julian Lewis
Vertebrate Development Labora
Cancer Research UK London Res
London WC2A 3PX, UK

Yoshiki Sasai
Director of the Neurogenesis an
RIKEN Center for Developmental
Chuo, Japan

Philippe Soriano
Department of Developmental R
Mount Sinai Medical School
New York, USA

Cliff Tabin
Harvard Medical School
Department of Genetics
Boston, MA, USA

Founding Editors

A. A. Moscona
Alberto Monroy

VOLUME ONE HUNDRED AND ONE

CURRENT TOPICS IN DEVELOPMENTAL BIOLOGY

Planar Cell Polarity During Development

Edited by

YINGZI YANG
Bethesda, MD, USA

ELSEVIER

AMSTERDAM • BOSTON • HEIDELBERG • LONDON
NEW YORK • OXFORD • PARIS • SAN DIEGO
SAN FRANCISCO • SINGAPORE • SYDNEY • TOKYO
Academic Press is an imprint of Elsevier

AP

QL
951
C81
V.101

Academic Press is an imprint of Elsevier
525 B Street, Suite 1900, San Diego, CA 92101-4495, USA
225 Wyman Street, Waltham, MA 02451, USA
The Boulevard, Langford Lane, Kidlington, Oxford, OX51GB, UK
32, Jamestown Road, London NW1 7BY, UK
Radarweg 29, PO Box 211, 1000 AE Amsterdam, The Netherlands

First edition 2012

Copyright © 2012 Elsevier Inc. All rights reserved.

No part of this publication may be reproduced, stored in a retrieval system or transmitted in any form or by any means electronic, mechanical, photocopying, recording or otherwise without the prior written permission of the publisher

Permissions may be sought directly from Elsevier's Science & Technology Rights Department in Oxford, UK: phone (+44) (0) 1865 843830; fax (+44) (0) 1865 853333; email: permissions@elsevier.com. Alternatively you can submit your request online by visiting the Elsevier web site at http://elsevier.com/locate/permissions, and selecting *Obtaining permission to use Elsevier material*

Notice
No responsibility is assumed by the publisher for any injury and/or damage to persons or property as a matter of products liability, negligence or otherwise, or from any use or operation of any methods, products, instructions or ideas contained in the material herein. Because of rapid advances in the medical sciences, in particular, independent verification of diagnoses and drug dosages should be made

For information on all Academic Press publications
visit our website at store.elsevier.com

ISBN: 978-0-12-394592-1
ISSN: 0070-2153

Printed and bound in USA
12 13 14 15 11 10 9 8 7 6 5 4 3 2 1

Working together to grow
libraries in developing countries

www.elsevier.com | www.bookaid.org | www.sabre.org

ELSEVIER BOOK AID International Sabre Foundation

CONTENTS

CONTRIBUTORS

Paul N. Adler
Cell Biology and Biology Department, University of Virginia, Charlottesville, Virginia, USA

Jeffrey D. Axelrod
Department of Pathology, Stanford University School of Medicine, Stanford, California, USA

Camille Boutin
Institute of Neuroscience, Developmental Neurobiology, Université Catholique de Louvain, Brussels, Belgium

Thomas J. Carroll
Department of Internal Medicine (Nephrology) and Molecular Biology, UT Southwestern Medical Center, Dallas, Texas, USA

Bo Gao
National Human Genome Research Institute, Bethesda, Maryland, USA

André M. Goffinet
Institute of Neuroscience, Developmental Neurobiology, Université Catholique de Louvain, Brussels, Belgium

Philippe Gros
Department of Biochemistry, and Complex Traits Group, McGill University, Montreal, Quebec, Canada

Alexandra Iliescu
Department of Biochemistry, and Complex Traits Group, McGill University, Montreal, Quebec, Canada

Masatake Kai
Department of Anatomy and Cell Biology, Graduate School of Medicine, Osaka City University, Abeno-ku, Osaka, Japan

Matthew W. Kelley
Laboratory of Cochlear Development, NIDCD, NIH, Bethesda, Maryland, USA

Helen May-Simera
Laboratory of Cochlear Development, NIDCD, NIH, Bethesda, Maryland, USA

Ying Peng
Department of Biochemistry and Molecular Biology, Mayo Clinic, Rochester, Minnesota, USA

Hitoshi Sawa
Multicellular Organization Laboratory, National Institute of Genetics, 1111 Yata, Mishima, Japan

Masazumi Tada
Department of Cell and Developmental Biology, University College London, London, United Kingdom

Fadel Tissir
Institute of Neuroscience, Developmental Neurobiology, Université Catholique de Louvain, Brussels, Belgium

Elena Torban
Department of Medicine, McGill University, Montreal, Quebec, Canada

Anthony Wynshaw-Boris
UCSF School of Medicine, San Francisco, California, USA

Jing Yu
Department of Cell Biology, University of Virginia, School of Medicine, Charlottesville, Virginia, USA

Yimin Zou
Neurobiology Section, Biological Sciences Division, University of California, San Diego, La Jolla, California, USA

PREFACE

During the morphogenesis of multicellular organisms, directional information has to be provided in order to form functional tissues and organs with specific organization and morphology. Planar cell polarity (PCP) is an evolutionarily conserved fundamental mechanism that provides such directional information in development and physiological functions. PCP refers to the uniform organization of cells in a multicellular structure within a plane, typically orthogonal to the apicobasal axis. PCP is required for morphogenesis in many contexts, including orienting mammalian hairs, Drosophila bristles, ciliary structures in the inner ear, the embryonic node, and the brain; setting up spindle orientation of dividing cells; and directing cell migration as well as axon path finding. PCP has been studied most extensively in Drosophila for more than three decades, which has laid down the framework for understanding PCP at the tissue, cellular, and molecular levels. More and more important functions of PCP in vertebrate development and physiology have been revealed during the past decade by studies cutting across several disciplines, which has made PCP a field of rapid growth. It is clear that PCP in vertebrate is a major branch of Wnt signaling and requires several new PCP regulators. PCP is also of increasing interest in the fields of stem cell biology and cancer. Not surprisingly, defective PCP has been found to cause several human disorders including ciliopathies, cardiac malformation, spina bifida, cystic kidney disease, and short-limb dwarfism.

Drosophila has been the key model system for studies on PCP. In Chapter 1, Adler described the genes required for PCP. These discoveries in Drosophila have contributed significantly to our current understanding of the cellular and molecular mechanisms underlying PCP. In Chapter 2, Peng and Axelrod further discussed the global and core modules that operate upstream of effector proteins to produce morphological PCP. They have provided insights in these two unresolved questions: the molecular and cell mechanisms underlying core module function in the generation and communication of subcellular asymmetry and the relationship between the global and core modules.

After PCP was described in Drosophila, it was found that the underlying molecular regulatory circuit might have been developed earlier in evolution. A hallmark of PCP is asymmetrical distribution of PCP regulators within a cell. In Chapter 3, Sawa discussed asymmetrical cell division in *C. elegans* controlled by a distinct Wnt/β-catenin asymmetry pathway, which appears to be a "hybrid" of the Wnt/β-catenin and PCP pathway.

In vertebrates, the functional spectrum of PCP in development and physiology is getting broader. Regulating polarized and asymmetrical tissue and cellular behaviors is critically important throughout vertebrate development. This is exemplified in several tissues and organs where PCP's function and mechanisms have been extensively studied. In Chapter 4, Tada and Kai described a critical function of PCP invertebrates, which is to control collective cell migration and cell intercalation in a variety of developmental processes. In Chapter 5, May-Simera and Kelley reviewed two crucial steps in the developing inner ear regulated by PCP, the orientation of stereociliary bundles, and the outgrowth of the cochlear duct. As the neural tissues contain polarized cells (neurons) and require directional behaviors such as axon path finding, it is not surprising that PCP plays critical roles in both neural development and physiology, and these are reviewed by Zou in Chapter 6 and Boutin et al. in Chapter 7. There is increasing evidence that PCP plays an essential role in kidney development and disease. However, the underlying cellular and molecular mechanisms still await further elucidation. This field was comprehensively reviewed by Carroll and Yu in Chapter 8. More recently, cartilage elongation is also found to be regulated by the Wnt/PCP pathway and this was discussed by Gao in Chapter 11.

As evolutionarily conserved regulators of PCP, Van gogh-like (Vangl) and Dishevelled proteins play a centrally important role in many vertebrate developmental processes. In Chapters 9 and 10, Wynshaw-Boris and Torban et al., respectively, highlighted the roles of Dishevelled and Vangl in both mammalian development and human diseases. Vangl proteins also appear to be critical mediators of cellular polarity in response to external cues, attractive candidates of which are Wnts. The question whether Wnts are indeed important components of the global modules of PCP is key to understanding PCP mechanism and function, and this was reviewed and discussed by Gao in Chapter 11.

Taken together, these chapters provide a multifaceted view of the fundamentally important PCP pathway in both morphogenesis and physiological functions. It must be pointed out here that the studies of PCP go beyond what has been covered here. The significance of the Wnt/PCP pathway in biology has been increasingly recognized, and this pathway has attracted attentions from different disciplines. However, the field still faces major conceptual and technical challenges that have to be addressed in the future. For instance, the relationship of Wnt/PCP and the Wnt/β-catenin pathways, the identity of the global cue(s) that breaks the initial tissue and cellular symmetry, and a sensitive and reliable *in vivo* and *in vitro* readout of PCP activity are not known yet. Much of these may be attributed, at least in part, to the nature of PCP, which is a uniform cell polarity in a group of cells within a three-dimensional tissue structure that is difficult to model *in vitro* in a traditional cell culture system. It is my hope that these chapters can serve as "food for thoughts" that will enlighten avenues for further investigation.

Yingzi Yang

CHAPTER ONE

The *frizzled/stan* Pathway and Planar Cell Polarity in the Drosophila Wing

Paul N. Adler[*,†,1]

[*]Biology Department, University of Virginia, Charlottesville, Virginia, USA
[†]Cell Biology Department, University of Virginia, Charlottesville, Virginia, USA
[1]Corresponding author: e-mail address: pna@virginia.edu

Contents

Current Topics in Developmental Biology, Volume 101
ISSN 0070-2153
http://dx.doi.org/10.1016/B978-0-12-394592-1.00001-6

© 2012 Elsevier Inc.
All rights reserved.

Abstract

Drosophila has been the key model system for studies on planar cell polarity (PCP). The rich morphology of the insect exoskeleton contains many structures that display PCP. Among these are the trichomes (cuticular hairs) that cover much of the exoskeleton, sensory bristles, and ommatidia. Many genes have been identified that must function for the development of normal PCP. Among these are the genes that comprise the *frizzled/starry night* (*fz/stan*) and *dachsous/fat* pathways. The mechanisms that underlie the function of the *fz/stan* pathway are best understood. All of the protein products of these genes accumulate asymmetrically in wing cells and there is good evidence that this involves local intercellular signaling between protein complexes on the distal edge of one cell and the juxtaposed proximal edge of its neighbor. It is thought that a feedback system, directed transport, and stabilizing protein–protein interactions mediate the formation of distal and proximal protein complexes. These complexes appear to recruit downstream proteins that function to spatially restrict the activation of the cytoskeleton in wing cells. This leads to the formation of the array of distally pointing hairs found on wings.

1. INTRODUCTION TO DROSOPHILA PLANAR CELL POLARITY

The cuticular surface of insects has a rich morphology that led them to be the first systems where the development of planar cell polarity (PCP) was studied in depth (see, e.g., Lawrence, 1966). For the past 30 years, genetic studies using Drosophila have provided many of the most important insights that resulted in the blossoming of the field (Gubb & Garcia-Bellido, 1982; Held, Duarte, & Derakhshanian, 1986; Wong & Adler, 1993). These included the identification of many of the genes that play key roles in PCP (Adler, Charlton, & Liu, 1998; Chae et al., 1999; Collier & Gubb, 1997; Collier, Lee, Burgess, & Adler, 2005; Feiguin, Hannus, Mlodzik, & Eaton, 2001; Gubb & Garcia-Bellido, 1982; Klingensmith, Nusse, & Perrimon, 1994; Strutt & Warrington, 2008; Taylor, Abramova, Charlton, & Adler, 1998; Theisen et al., 1994; Usui et al., 1999; Vinson, Conover, & Adler, 1989; Wolff & Rubin, 1998; Yan et al., 2008; Yang, Axelrod, & Simon, 2002; Zeidler, Perrimon, & Strutt, 1999), that the protein products of these genes accumulate asymmetrically in cells (Adler, Zhu, & Stone, 2004; Axelrod, 2001; Bastock, Strutt, & Strutt, 2003; Feiguin et al., 2001; Shimada, Usui, Yanagawa, Takeichi, & Uemura, 2001; Strutt, 2001; Strutt & Warrington, 2008; Tree et al., 2002; Usui et al., 1999; Yan et al., 2008), that many of the proteins interact physically (Bastock et al., 2003; Das, Jenny, Klein, Eaton, & Mlodzik, 2004; Jenny, Darken, Wilson, & Mlodzik, 2003; Jenny, Reynolds-Kenneally, Das,

Burnett, & Mlodzik, 2005; Lu, Yan, & Adler, 2010; Wong et al., 2003), and the evidence for feedback systems (Tree et al., 2002) that are a key for the functioning of the system. As has proved to be true for so much in biology, the genes that function in PCP in flies have similar functions in a wide range of animals (see Goodrich & Strutt, 2011; Wang & Nathans, 2007; other articles in this volume).

2. THE CELLULAR BASIS FOR PCP

Three types of adult structures have been the primary focus of experiments on PCP in Drosophila (Adler, 2002). The most important are the cuticular hairs found over much of the flies body. The wing, which is flat and where each cell elaborates a relatively long distally pointing hair, has been the principal tissue for studies on hairs (Fig. 1.1) (Gubb & Garcia-Bellido, 1982; Vinson & Adler, 1987; Wong & Adler, 1993), although they have also been studied in depth on the abdomen (Casal, Lawrence, & Struhl, 2006).

2.1. Epidermal hairs and the fly wing

The *frizzled* (*fz*) pathway (also called the *starry night* (*stan*) pathway, Casal et al., 2006) controls PCP by restricting the site of hair initiation to the distal most part of wing cells (Wong & Adler, 1993) (Fig. 1.2). The hairs grow out away from the distal cell periphery resulting in a distally pointing hair (Wong & Adler, 1993). Mutations in *fz/stan* pathway genes lead to hairs forming at abnormal subcellular locations and having abnormal polarity (Figs. 1.1 and 1.2). In this review, I will focus on the wing with an emphasis on the *fz/stan* pathway. The simple cell biology and flat shape of the wing make it an ideal tissue for studies on PCP, and the principles deduced from studying it have for the most part proved to be conserved in other tissues in the fly and in other organisms.

2.2. Sensory bristles

The cuticular surface of the fly is also decorated with many bristle sense organs. On the dorsal thorax and abdomen, the bristle shaft points posteriorly and on appendages they point distally (Adler, 2002) (Fig. 1.3A and C). In other body regions (e.g., the head), bristles are locally aligned and show a reproducible polarity. The *fz* pathway regulates bristle polarity by controlling the orientation of the spindle during the determinative cell divisions that

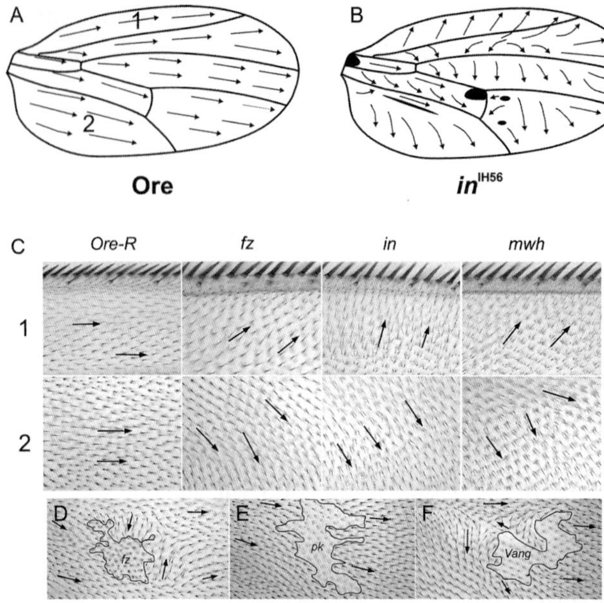

Figure 1.1 Hair polarity on the fly wing. A shows the polarity of hairs on the dorsal surface of a wild-type wing. B shows the stereotypic pattern seen on the dorsal surface of an *inturned* wing. The blackened area is where the polarity was too variable to draw a consensus vector. C shows images of wings from two regions (noted in A) of Oregon R, *fz*, *in*, and *mwh* mutant wings. Note the relative similarity of the polarity patterns. The downstream genes often show a slightly stronger polarity disruption, but with similar directionality. D shows an *fz* clone marked by the cell marker *strb* (causes short deformed and multiplied hairs). E a *pk* clone marked by *sha* (causes a loss of hairs) and F a *Vang* clone marked by *sha*. Arrows show local hair polarity.

give rise to the cells of the sense organ (Gho & Schweisguth, 1998; Lu, Usui, Uemura, Jan, & Jan, 1999) (Fig. 1.3B). On the leg, some bristle sense organs signal to a neighboring epidermal cell to form a bract (Fig. 1.3C). This involves the EGFR pathway, and this system displays planar polarity as the bract is routinely proximal to the bristle (Held, 2002) (Fig. 1.3). Most leg bristles do not show strongly abnormal polarity in *fz/stan* pathway mutants but in those that do the bract is coordinately positioned "upsteam" (Fig. 1.3D).

2.3. The eye

The third developmental unit that has been extensively studied with respect to PCP is the ommatidia of the compound eye (Zheng, Zhang, & Carthew, 1995). Each ommatidia is comprised of about 20 cells including

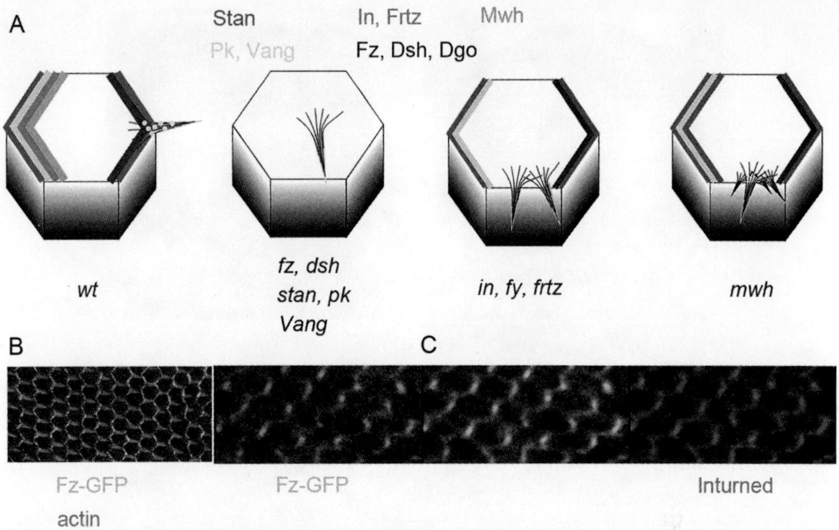

Figure 1.2 Polarized protein accumulation and mutant phenotypes. A shows cartoons of wild-type and mutant cells. In the wild-type cell, all of the *fz/stan* pathway proteins accumulate asymmetrically and hairs are formed at the distal vertex. In cells, mutant for *fz* or other core genes hairs form at a relatively central location and no proteins accumulate asymmetrically. In cells, mutant for *in* or any of the other PPE genes multiple hairs form at an abnormal location on the cell periphery. The core proteins accumulate asymmetrically but the PPE proteins and Mwh do not. In cells, mutant for *mwh* genes multiple hairs form at an abnormal location on the cell periphery. The core and PPE proteins accumulate asymmetrically but Mwh does not. B shows the accumulation of Fz-GFP by direct imaging of GFP (green) (no immunostaining) and actin in red (hairs). C shows the coordinate asymmetric accumulation of Fz (green) and In (red). (See Color Insert.)

8 photoreceptor cells. The arrangement of the photoreceptor cells is asymmetric and chiral giving each ommatidia and the eye as a whole a polarity (Fig. 1.4A). As each ommatidia develops, there is a stepwise recruitment of photoreceptor cells (for a recent review of this, see Kumar, 2012). The R3 and R4 cells are recruited at the same time. The cell located closer to the equator becomes R3 and the other R4 (Fig. 1.4A). The ommatidia undergo a subsequent rotation and this is guided by the position of the R3 and R4 cells (Fig. 1.4B). The *fz* pathway controls PCP by regulating the R3/R4 cell fate decision (Zheng et al., 1995). This is based on the R3 cell having higher Fz activity and R4 higher Vang (Strutt, Johnson, Cooper, & Bray, 2002). This leads to the Delta-dependent activation of the Notch (N) receptor in R4, which controls cells fate (Cooper & Bray, 1999; Fanto & Mlodzik, 1999; Tomlinson & Struhl, 1999). There is an argument in the

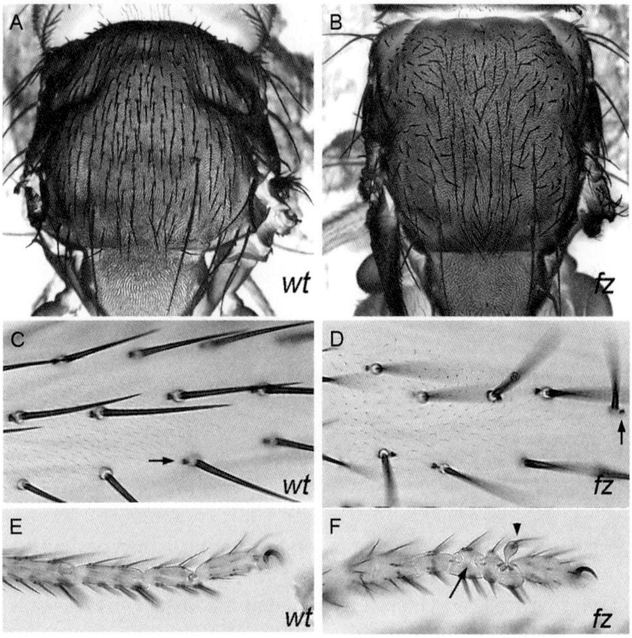

Figure 1.3 Bristles and PCP. A is an SEM of a wild-type thorax and B a *fz* thorax. C is a wild-type femur and D an *fz* femur. Arrows point to bracts. E is a wild-type tarsus and F an *fz* tarsus. The arrow points to a segment with a mirror image duplication and the arrowhead to a bulge.

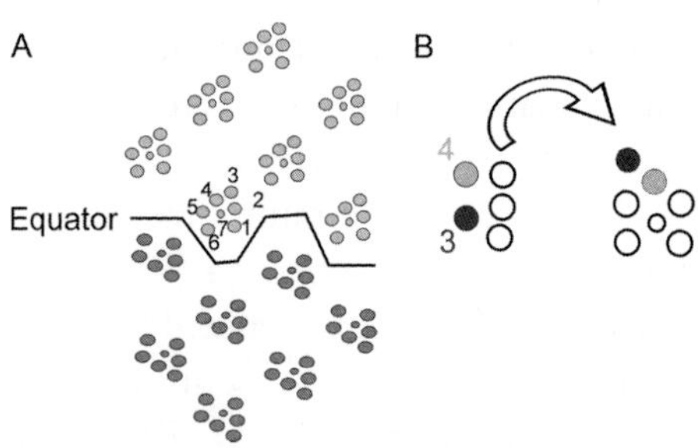

Figure 1.4 The eye and PCP. A is a cartoon of the adult retina with Northern (blue) and Southern (red) hemisphere ommatidia in mirror image across the equator. B shows how the pattern in the adult retina is derived by rotation of the developing retina. The crucial R3 (blue) and R4 (green) cells are marked. (For interpretation of the references to color in this figure legend, the reader is referred to the online version of this chapter.)

literature with regard to the mechanism involved in N activation and how this regulates cell fate (del Alamo & Mlodzik, 2006; Fanto & Mlodzik, 1999; Strutt et al., 2002; Weber, Paricio, & Mlodzik, 2000). Recent data argue that the Ral GTPase also plays a role in R3//R4 cell fate (Cho & Fischer, 2011).

2.4. Other manifestations of PCP

There are several other cell types/structures/tissues where *fz/stan*-based PCP has been studied. Collier and colleagues discovered that the fly wing has parallel cuticular ridges and that their orientation is under the control of the *fz/stan* and *ds/ft* pathways (Fig. 1.5E–G) (Valentine & Collier, 2011). Interestingly, the ridges are oriented differently in the anterior and posterior parts of the wing and this appears to be related to differences between two phases of *fz* signaling that utilize different *pk* isoforms (Doyle, Hogan, Lester, & Collier, 2008; Hogan, Valentine, Cox, Doyle, & Collier, 2011; Valentine & Collier, 2011). Their data also argue that the early phase is dependent on *ds/ft* while the latter is not. The lateral side branches of the arista are another body part where the *fz* pathway has been found to regulate PCP (He & Adler, 2002). The laterals are produced by polyploid cells and share features of both epidermal hairs and bristles. As is the case for epidermal hairs, the *fz* pathway controls the subcellular location for lateral outgrowth although in this cell type the outgrowth is not juxtaposed to the distal most edge of the cell in wild type (He & Adler, 2002).

The denticles that are part of the larval cuticle also display PCP, and they have been intensively used as markers of pattern formation in embryonic segments. In recent years, they have also been used to study PCP (Donoughe & DiNardo, 2011; Price, Roberts, McCartney, Jezuit, & Peifer, 2006; Repiso, Saavedra, Casal, & Lawrence, 2010). Here, as in the adult abdomen, the *fz/stan* and *ds/ft* systems appear to function in parallel.

The *fz/stan* pathway, which plays such a key role in PCP, has also been found to regulate a number of processes that are not always thought of as being related to PCP. The first of these to be identified was the formation of the leg joints (Held et al., 1986). Mutations in genes such as *fz* give rise to defective leg joints that often show mirror image symmetry typical of duplications (Fig. 1.3E and F). One often also sees "blebs." The cellular basis for this has not been established.

Mutations in genes such as *fz* also lead to occasional defects in wing evagination so that the resulting wing points anteriorly in the pupal case (Fig. 1.5A and B) (Lee & Adler, 2002). Such wings end up shorter and deformed,

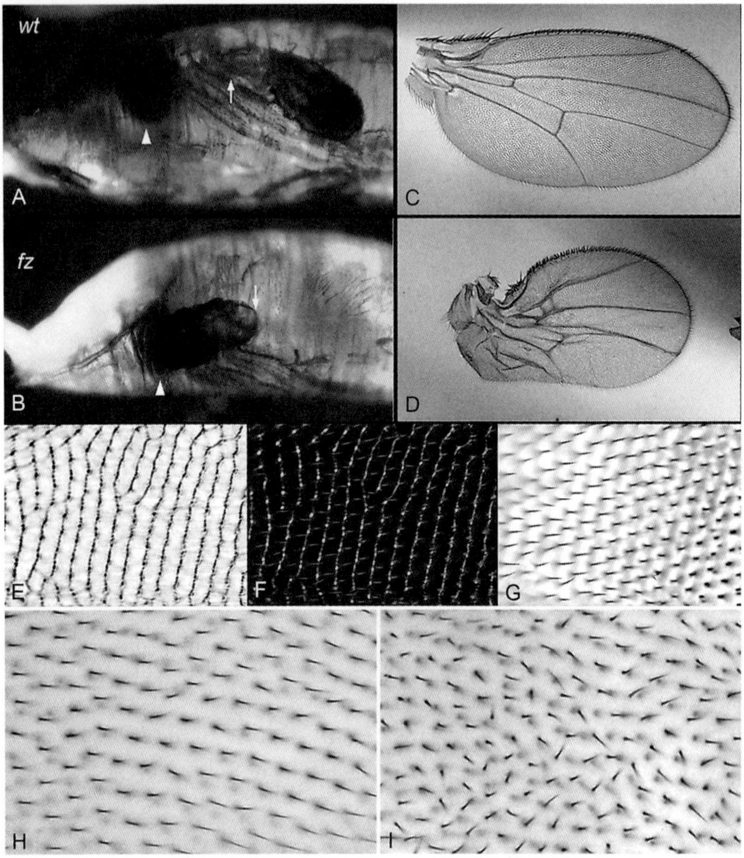

Figure 1.5 Less well-known PCP-related phenotypes. A and B show wild-type and *fz* pupae, respectively. In a modest fraction of *fz* pupae, the wing everts forward instead of posteriorly (as shown in B). Arrowheads point to the eye and arrows to the wing hinge region. C is a wild-type wing and D is a distorted wing that results from the abnormal eversion as shown in B. E–G are wings imaged by cuticle reflection microscopy (E) and normal bright field microscopy (G) and a merged image (F). These panels were generously provided by S. Collier. H is a wild-type wing and I the same region of a wing from a fly homozygous for a mutation in *Gliotactin*. Note the lack of parallel alignment of neighboring hairs in (I). (For color version of this figure, the reader is referred to the online version of this chapter.)

perhaps due to defects in cell rearrangements. This could be related to the effects of the *fz* pathway on the movement of sheets of cells in vertebrates (Goodrich & Strutt, 2011). Mutations in genes such as *fz* also inhibit the repacking of pupal wing cells so that the fraction of hexagonal cells is decreased compared to wild type (Classen, Anderson, Marois, & Eaton, 2005). The connection between the altered hair polarity and the effect on the repacking is unclear (Classen et al., 2005; Ma et al., 2008). An interesting and related observation is that cell rearrangements in the pupal wing lead to changes in the anatomical orientation of *fz/stan* signaling (Aigouy et al., 2011).

Mutations in *fz/stan* pathway genes have also been found to alter the pattern of dendrites and axons in some Drosophila neurons (Matsubara, Horiuchi, Shimono, Usui, & Uemura, 2011; Mrkusich, Flanagan, & Whitington, 2011). In both cases, it remains to be established as to how much of the pathway is conserved in different tissues. A similar very wide range of cell types have been found to utilize *fz* pathway homologs in vertebrates (Goodrich & Strutt, 2011; Wang & Nathans, 2007).

2.5. PCP that is not *fz/stan* pathway dependent

There are also several interesting cases of PCP in Drosophila that are not dependent on the *fz/stan* pathway, the most notable being the polarized cell rearrangements that mediate germ band elongation (Bertet, Sulak, & Lecuit, 2004; Zallen, 2007; Zallen & Wieschaus, 2004) and the polarized actin cytoskeleton seen in the follicle cells of developing egg chambers (Bateman, Reddy, Saito, & Van Vactor, 2001; Frydman & Spradling, 2001). Interestingly, the *fat2* gene is required for actin polarization in follicle cells (Viktorinova, Konig, Schlichting, & Dahmann, 2009).

3. PCP GENES

3.1. Early studies on the *fz/stan* pathway

Several of the genes of the *fz/stan* pathway were discovered long ago and were used as markers in classical genetic experiments (see Table 1.1). These include *fz*, *prickle* (*pk*), *disheveled* (*dsh*), *inturned* (*in*), and *multiple wing hairs* (*mwh*). Other *fz/stan* pathway genes were discovered in screens designed to identify new PCP genes. These include *stan* (also known as *flamingo*) (Chae et al., 1999; Usui et al., 1999), *Van Gogh* (*Vang*) (also known as *strabismus*) (Taylor et al., 1998; Wolff & Rubin, 1998), and *fritz* (*frtz*) (Collier et al., 2005). (Note: in this review, I use the primary FlyBase name for all Drosophila genes.) The first detailed study on the function of these genes in PCP was published about 30 years ago by Gubb and Garcia-Bellido (1982). Several important conclusions came from this work. One was that the mutant hair polarity patterns seen on wings were not random. Rather, the mutant pattern was stereotypic. Thus, cells in the anterior region of an *fz* mutant wing routinely point more anteriorly

Table 1.1 Key Drosophila planar cell polarity genes

Drosophila gene	Synonyms	Vertebrate	Comments	Key Drosophila references
fz		*fz* (1,2,3,6)	Seven transmembrane domains	Strutt (2001), Vinson and Adler (1987), and Vinson et al. (1989)
dsh		*Dvl* (1,2,3)	PDZ, Dix domains	Axelrod (2001), Klingensmith et al. (1994), and Theisen et al. (1994)
dgo		Inversin diversin	Ankyrin repeat	Feiguin et al. (2001)
stan	*fmi*	Celsr	Cadherin domains, seven transmembrane domains	Chae et al. (1999) and Usui et al. (1999)
pk		*pk* (1,2)	Lim domain, PET domain	Gubb et al. (1999) and Tree et al. (2002)
Vang	*stbm*	*Vangl* (1,2)	Four transmembrane domains	Taylor et al. (1998) and Wolff and Rubin (1998)
in		*intu*	PDZ domain	Adler et al. (2004) and Park, Liu, Sharp, and Adler (1996)
fy		*fuz*		Collier and Gubb (1997) and Strutt and Warrington (2008)
frtz		*frtz*	WD40 domain	Collier et al. (2005)
mwh		None	GBD-FH3	Strutt and Warrington (2008) and Yan et al. (2008)
ds		*Dchs1*	Cadherin domains	Adler et al. (1998), Casal et al. (2006), Clark et al. (1995), Ma, Yang, McNeil, Simon, and Axelrod (2003), Matakatsu and Blair (2006), and Simon (2004)

Table 1.1 Key Drosophila planar cell polarity genes—cont'd

Drosophila gene	Synonyms	Vertebrate	Comments	Key Drosophila references
ft		*Fat*	Cadherin domains	Adler et al. (1998), Casal et al. (2006), Ma et al. (2003), Mahoney et al. (1991), Matakatsu and Blair (2006), and Simon (2004)
fj		*Fj*	Golgi kinase	Simon, Xu, Ishikawa, and Irvine (2010), Villano and Katz (1995), and Zeidler et al. (1999)
Rho1		*RhoA*	GTPase	Strutt, Weber, and Mlodzik (1997) and Yan, Lu Fang, and Adler (2009)

than do wild type (Fig. 1.1C). A second important insight was that altering wing shape (by mutation) did not make substantial differences to the stereotypic *fz* mutant polarity pattern. Similarly, the loss of part of the wing margin due to mutation did not affect the mutant polarity pattern.

The insight that a set of genes comprised a regulatory pathway that controlled PCP came from Wong and Adler (1993). They made several key findings. They realized that mutations in all of the *fz/stan* pathway genes gave rise to similar stereotypic polarity patterns (Fig. 1.1A–C) but that genes could be placed into phenotypic groups by the frequency of extra hairs produced (Fig. 1.2A). The three phenotypic groups were also epistasis groups, and it is now clear that they represent different levels in the regulatory hierarchy. The upstream genes (often called core genes) consist of *fz*, *dsh*, *pk*, *stan*, *Vang*, and *dgo*. (Note: *stan* (Chae et al., 1999; Usui et al., 1999), *Vang* (Taylor et al., 1998; Wolff & Rubin, 1998), and *dgo* (Feiguin et al., 2001) had not been discovered at the time of the paper.) Below, the core group are *in*, *fy*, and *frtz*, which are often referred to as the planar polarity effector (PPE) genes (Note: *frtz* had not been discovered at the time of the paper; Collier et al., 2005.) The *mwh* gene was the only member of the third group. They also found that in wild-type pupal wing cells the cytoskeleton was activated in the vicinity of the distal most vertex to promote hair outgrowth and this resulted in a hair that grew distally. Further, mutations in all of the *fz* pathway genes resulted in hairs forming at an alternative subcellular location. An important insight from the analysis of the pupal phenotype was that the *fz/stan* pathway regulated hair initiation.

The observation that *fz* pathway proteins accumulated asymmetrically in wing cells was first observed by Uemura and colleagues studying the *stan/flamingo* gene (Usui et al., 1999) and this breakthrough opened up a new stage in PCP research (Fig. 1.2A and B). Over the next few years, similar results were obtained for all of the *fz/stan* pathway proteins (Adler et al., 2004; Axelrod, 2001; Bastock et al., 2003; Feiguin et al., 2001; Shimada et al., 2001; Strutt, 2001; Strutt & Warrington, 2008; Tree et al., 2002; Usui et al., 1999; Yan et al., 2008). Further biochemical experiments established that several pairs of these proteins could interact directly with one another (Bastock et al., 2003; Das et al., 2004; Jenny et al., 2003, 2005; Lu et al., 2010; Wong et al., 2003). A more detailed discussion of the function of these genes and how the asymmetric accumulation of the proteins is achieved is presented later.

The *fz/stan* pathway functions to limit the region of the cell where the cytoskeleton is activated to form a hair. Neither are all wing cells of the same size, nor are all epithelial cells in other body regions of the same size or shape as wing cells. Further, nutrition and the temperature during development can alter cell size without altering planar polarity. Hence, the pathway has the ability to scale and function in a variety of cell geometries. It is clear, however, that this ability is not infinite. Cells that are much larger than normal can arise from the cells being polyploidy or stretched due to wound healing. Both of these conditions result in the cells often forming multiple hairs that can be of abnormal polarity (Adler, Liu, & Charlton, 2000). The *fz/stan* pathway is still functioning in these large cells, but it is unable to regulate the cytoskeleton well enough so that a single distally pointing hair is produced.

3.2. *ds/ft* pathway

Mutations in several other groups of genes also give rise to PCP phenotypes. The atypical cadherins *dachsous* and *fat* and the Golgi kinase *four jointed* were first discovered to result in PCP phenotypes due to their leg joint phenotypes (Held et al., 1986). Later, it was established that they were also required for wing and eye PCP (Adler et al., 1998; Matakatsu & Blair, 2004; Simon, 2004; Zeidler et al., 1999). A variety of evidence indicates that Ft and Ds act as a receptor ligand pair and that Fj modulates their activity by phosphorylation (Brittle, Repiso, Casal, Lawrence, & Strutt, 2011; Matakatsu & Blair, 2004; Simon, 2004; Simon et al., 2010; Sopko et al., 2009). One observation that does not fit this model is the observation that transgene-supplied Ft that lacks the extracellular domain provides rescue of the *ft* PCP phenotype over most of the wing (Matakatsu & Blair, 2006). How to reconcile these observations is not clear. The Ft and Ds proteins do not appear to accumulate asymmetrically as do *fz/stan* pathway proteins; however, they are able to polarize cells. This is seen both by their effect on hair polarity and by their activity leading to the asymmetric accumulation of the atypical myosin Dachs (Mao et al., 2006, 2011). Early studies found that the *fz/stan* pathway was functional in *ds* and *ft* mutants but that it signaled in an anatomically abnormal way (Adler et al., 1998), and subsequently, it was proposed that the *ds/ft* pathway functioned as a global signal that oriented the activity of the *fz/stan* pathway with respect to the body as a whole in the wing and eye (Ma et al., 2003; Simon, 2004; Yang et al., 2002). However, studies by Casal et al. (2006) provided compelling data that in the abdomen that the *ds/ft* and *fz/stan* pathways functioned in parallel. For example, double mutants of one *ds/ft* and one *fz/stan* pathway gene had a more severe phenotype than any single mutant or double mutant where only one of the pathways was affected. They also showed that *ds/ft* could repolarize cells that were mutant for the *fz/stan* pathway. One interpretation of these observations is that the relationship between the two pathways is different in different tissues. This is not a satisfying explanation; however, it may be correct. More recent experiments suggest that *ds/ft* may alter *fz/stan* signaling in indirect ways by affecting the axis of cell division, cell rearrangements, and the polarization of microtubules that are involved in the transport of Fz-containing vesicles (Harumoto et al., 2010; Mao et al., 2011). There is increasing evidence that *ds/ft* also function in vertebrate PCP (Goodrich & Strutt, 2011).

3.3. Septate junction proteins

A third set of genes essential for the morphogenesis of normal wing PCP are genes such as *Gliotactin*, *Neuroglian*, *Coracle*, and *varicose* (Moyer & Jacobs, 2008; Venema, Zeev-Ben-Mordehai, & Auld, 2004). These genes encode proteins that are associated with the septate junction, and they are required for the alignment of neighboring hairs (Fig. 1.5H and I). The septate junction is the invertebrate equivalent of the vertebrate tight junction. These genes appear to act after hair outgrowth, and unlike the *fz/stan* pathway, mutations in these genes do not result in cells forming multiple hairs. The asymmetric localization of Dsh is not altered in a Gli mutant, and the localization of Coracle is not affected by mutations in *fz/stan* genes (Venema et al., 2004 #71). Further, double mutants of septate junction and *fz/stan* genes show an additive phenotype consistent with their functioning in parallel. How the septate junction genes function in PCP has not been extensively studied. It is also not clear if this set of genes is needed for PCP in other contexts in Drosophila or in other organisms.

3.4. Other genes that impact PCP

In recent years, many additional genes have been identified that are required for PCP in the fly. The literature is somewhat inconsistent as to the properties of a gene required to consider it a PCP gene. Some authors require a hair/bristle orientation phenotype (or ommatidia of the incorrect chiral type), while others have simply required cells form multiple hairs (of normal orientation) or result in misrotated ommatidia of the correct chiral type. As is discussed in more detail below, many aspects of general cell biology are likely involved in polarizing cells; hence, it is not surprising that genes that encode a wide variety of protein types and functions have been found to produce PCP phenotypes. These include some that might be thought of as having rather general functions. Not surprisingly, these genes differ from those of the *fz/stan* pathway components in being much less specific with regard to their mutant phenotypes. The collection of genes includes ones that encode proteins that function in intracellular trafficking such as *Rab23* (Pataki et al., 2010), *Rab5* (Purvanov, Koval, & Katanaev, 2010), and the Rab5 effector *Rabenosyn-5* (Mottola, Classen, Gonzalez-Gaitan, Eaton, & Zerial, 2010). Mutations in genes that encode proteins that function to promote actin filament disassembly *twinstar* (*cofilin*) (Blair et al., 2006) and *flare* (*AIP1*) (Ren, Charlton, & Adler, 2007) have been found to result in PCP phenotypes and disruptions in the normal asymmetric accumulation of *fz/stan* pathway proteins. Also implicated in PCP are genes that encode transcription regulators such as *grainy head* (*grh*) (Lee & Adler, 2004) and *atrophin* (Fanto et al., 2003); kinases such as *misshapened* (Paricio, Feiguin, Boutros, Eaton, & Mlodzik, 1999), *Rho Kinase* (Winter et al., 2001), the Abelson tyrosine kinase (Singh, Yanfeng, Grumolato, Aaronson, & Mlodzik, 2010), *aPKC* (Djiane, Yogev, & Mlodzik, 2005), *casein kinase I* (Strutt, Price, & Strutt, 2006), and the *gilgamesh* kinase (Gault, Olguin, Weber, & Mlodzik, 2012); the *Widerborst* phosphatase (Hannus, Feiguin, Heisenberg, & Eaton,

2002); the *approximated* palmitoyltransferase (Matakatsu & Blair, 2008); and small GTPases such as *Rho1* (Strutt et al., 1997; Yan et al., 2009) and Go (Purvanov et al., 2010). An interesting and unexpected group of genes that are required for normal PCP function in ion transport including the Na^{+}/H^{+} exchanger *Nhe2* (Simons et al., 2009), a component of the vesicle ATPase (Buechling et al., 2010; Hermle, Saltukoglu, Grunewald, Walz, & Simons, 2010), and an organic anion transporter, *oat30B* (Mummery-Widmer et al., 2009). The effect of Nhe2 was suggested to be mediated through an effect on the interaction of Dsh and plasma membrane lipids. Many of the genes noted above produce mutant phenotypes that suggest they could function in concert with or affect the function of either the *fz/stan* or the *ds/ft* pathways. For example, the PCP phenotype that results from a loss of *grh* function in wing cells was shown to be associated with a loss of *stan* expression in the mutant cells (Lee & Adler, 2004). A notable and interesting exception was the identification of *chascon* and *jitterbug*, which affect notum PCP by impacting the tendons that attach the indirect flight muscles to the notum (Olguin, Glavic, & Mlodzik, 2011). In this case, a loss of the ability of the tissue to respond to the mechanical tension produced by the IFMs appears to be the mechanism responsible for the PCP mutant phenotype.

3.5. Genome-wide screen for PCP genes

A genome-wide screen based on transgene-mediated RNAi that was designed to identify genes that played a role in Notch signaling also identified many genes required for PCP (Mummery-Widmer et al., 2009). In this screen, *pannier-Gal4* was used to drive transgene expression. *pannier* is expressed in the central region of the notum; hence, it is an ideal driver to use to look for effects on bristle sense organ determination and differentiation. Effects on PCP in this screen were manifested by the misalignment of notum bristles. Satisfyingly, this screen identified in order of the strength of phenotype *fz*, *Vang*, *frtz*, *pk*, *stan*, *fy*, *in*, *ft*, and *ds*. Thus, most of the *fz/stan* and *ds/ft* pathway genes were identified confirming that the screen was very effective. The screen failed to identify *dgo*, *fj*, and *mwh*. The failure to identify *dgo* and *mwh* is not surprising as null alleles of *dgo* and *mwh* do not show a bristle polarity phenotype (Feiguin et al., 2001; Strutt & Warrington, 2008; Yan et al., 2008). This is not the case for *fj* and the failure to identify *fj* may simply reflect the transgene not knocking down the expression enough to produce a mutant phenotype. It is unlikely that this screen identified all of the genes that play a role in PCP. Some genes would likely be missed due to the knockdown not being effective (as is presumably the case for *fj*). Other genes would likely be missed due to the knockdowns resulting in lethality or a loss of bristles due to effects on the determinative cell divisions. This seems likely to be the case for genes involved in essential basic cellular functions, and indeed genes that are important for bristle morphogenesis such as *Rab11* and exocyst components were missed in the genome-wide screen for this reason (Nagaraj & Adler, 2012). This screen has already served as a valuable source of candidate PCP genes. For example, *VhaM8.9* (Buechling et al., 2010) and *jitterbug* (Olguin et al., 2011) were both identified in this screen and subsequently shown in other studies to produce PCP phenotypes. It is interesting to consider that many of the genes that produced the strongest phenotypes in the screen have not yet been the focus of PCP papers. A knockdown of the *organic anion transporting polypeptide 30B* produced the strongest PCP phenotype (along with *fz*). In the wing, a knockdown of this gene can reverse hair polarity much as the overexpression of the *spiny leg* isoform of the *pk* gene can. It will be interesting to see if these two proteins function in an antagonistic way. Among the other genes where knockdowns produced very strong phenotypes were *d-Cup* (a Ca^{++}-binding protein), *CG17290* (which contains a gyrase motif), protein tyrosine phosphatase 69D, *CG15649* (a novel protein), and *CG18005* (which contains RED domains suggesting a role in RNA splicing). The roles that these genes and their encoded proteins play in PCP can currently only be guessed at. It is notable that even this small set of candidate genes and proteins suggests a wide range of biochemical functions. It seems likely that much remains to be learned about PCP.

4. WHAT IS THE BASIS FOR THE ASYMMETRIC LOCALIZATION OF *fz/stan* PATHWAY PROTEINS?

All of the *fz/stan* pathway proteins accumulate asymmetrically in wing cells, with all except Stan accumulating on either the proximal or the distal cell membrane (Stan is found on both) (Adler et al., 2004; Axelrod, 2001; Bastock et al., 2003; Feiguin et al., 2001; Shimada et al., 2001; Strutt, 2001; Strutt & Warrington, 2008; Tree et al., 2002; Usui et al., 1999; Yan et al., 2008) (Fig. 1.2A). The accumulation is uneven across these membranes. That is, some locations along these membranes show a higher level of protein than others and all of the *fz/stan* pathway proteins appear to be enriched at the same foci on the membrane (Fig. 1.2B). This suggests that the proteins accumulate in a synergistic way. Further, all of the core group proteins must be functional for the normal asymmetric accumulation to occur (Fig. 1.2A).

4.1. Interactions of transmembrane proteins

The distal edge of one cell is juxtaposed to the proximal edge of the neighboring cell distal to it (Fig. 1.6). This provides an interface for asymmetric signaling that is likely to be part of a positive feedback mechanism that helps establish the distinct proximal and distal membrane domains. Several groups have proposed that interactions between the three *fz/stan* pathway proteins that are transmembrane proteins, Fz, Vang, and Stan, are a key to this process (Chen et al., 2008; Strutt & Strutt, 2008; Wu & Mlodzik, 2008). A variety of data support these models. These include biochemical evidence for an interaction and *in situ* protein localization studies. There are, however, differences between the experimental results and the interpretations reported by these groups, so at the current time, there is not a consensus for how these proteins interact and provide for polarized signaling. It will be important for these differences to be resolved; however, the general conclusion that heterotypic protein interactions provide the basis for the intercellular signaling between cells appears to be on solid ground.

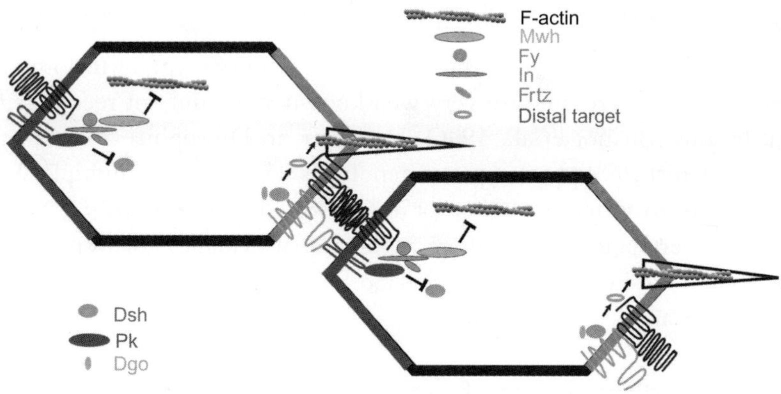

Figure 1.6 A possible model for PCP protein action in wing cells. Two neighboring cells are diagramed. Proteins on the distal membrane of one cell (Fz and Stan) interact with proteins on the proximal membrane of the neighboring cell (Vang and Stan). The core transmembrane proteins recruit the cytoplasmic core proteins (Dsh, Dgo, and Pk). The PPE proteins are recruited to the proximal side where In recruits Mwh. Mwh inhibits the actin cytoskeleton in the distal part of the cell. Dsh activates an activator of the actin cytoskeleton distally. For simplicity, only one of the feedback interactions between cytoplasmic proteins is drawn (Pk inhibiting Dsh on the distal side). (See Color Insert.)

4.2. Intracellular proteins

It is likely that in addition to intercellular signaling an intracellular feedback system plays an important role in the formation of the proximal and distal protein domains (Amonlirdviman et al., 2005; Tree et al., 2002). Several interactions have been detected between proximal and distal proteins. For example, Dsh and Vang, Dsh and Pk, Dgo and Pk, and Dgo and Vang have been found to interact (Bastock et al., 2003; Das et al., 2004; Jenny et al., 2003; Tree et al., 2002). The interaction between proximal and distal proteins appears to be negative. Hence, a negative interaction between Pk and Dsh could prevent Dsh from accumulating on the proximal side and Pk from accumulating on the distal side (Fig. 1.6). Such models suggest the interactions be dependent on cellular location. For example, when Pk and Dsh interact on the proximal side of the cell, Dsh would be lost. A hint as to how this may work comes from experiments on mice where it was found that a Smurf ubiquitin ligase targets Pk for proteasome-mediated degradation (Narimatsu et al., 2009). As of yet, no equivalent experimental evidence has been published that a similar mechanism works in the fly (e.g., Pk interacting with Dsh results in Dsh being degraded in a proteasomal-dependent mechanism), but it is certainly an attractive hypothesis to explain the negative protein interactions.

4.3. Wnts in fly PCP?

The *fz* gene family gets its name from the PCP phenotype of its founding member, but Fz proteins are very well known as a family of receptors for Wnt ligands (Bhanot et al., 1996). However, in Drosophila, studies have failed to detect PCP phenotypes in genotypes where one or multiple Wnts are lost due to mutation (Chen et al., 2008; Lawrence, Casal, & Struhl, 2002). In vertebrate tissues, there is substantial evidence for Wnt ligands functioning in PCP (see Goodrich & Strutt, 2011; Wang & Nathans, 2007; other articles in this volume).

4.4. Mechanisms for the formation of PCP protein complexes

There are several possible mechanisms that could be responsible for the asymmetric accumulation of PCP proteins on the proximal or distal membranes. One would be for the proteins to be preferentially synthesized at those locations (Berleth et al., 1988; Davis & Ish-Horowicz, 1991; Macdonald & Struhl, 1988). A second would be the directed trafficking of the proteins to the correct side of the cell (Shimada, Yonemura, Ohkura, Strutt, & Uemura,

2006; Strutt, Warrington, & Strutt, 2011). A third would be trapping of proteins to a domain by binding to a target (Strutt et al., 2011). A fourth possibility would be for the rate of diffusion of the protein to be regulated spatially across the cell (Griffin, Odde, & Seydoux, 2011).

4.5. mRNA localization?

The data do not support the hypothesis for localized synthesis as *in situ* hybridization experiments have failed to show evidence for mRNA localization. Further, in other systems, the localization of an mRNA is usually mediated by sequences outside the coding region (see e.g., Ferrandon, Elphick, Nusslein-Volhard, & St Johnston, 1994; Macdonald & Struhl, 1988). However, in fly PCP, fully functional transgenes encode mRNAs that lack most of the 5′- and 3′-untranslated regions (see, e.g., Krasnow & Adler, 1994).

4.6. Directional trafficking and trapping

There is evidence supporting the involvement of both directed transport and trapping in the asymmetric accumulation of PCP proteins. Uemura and colleagues have used *in vivo* confocal imaging to determine that Fz and Stan preferentially traffic along the proximal distal axis of wing cells as opposed to along the anterior–posterior axis (Shimada et al., 2006). They also provided evidence that the trafficking is along the web of apical microtubules that show a proximal/distal bias (Turner & Adler, 1998). This bias in the orientation of microtubules appears to be at least partially mediated by the action of the *ds/ft* pathway. This may provide an explanation for the ability of the *ds/ft* pathway to bias *fz/stan* activity (Harumoto et al., 2010). It is not clear, however, if the bias for movement along the proximal distal axis is great enough to explain everything. Further, the *in vivo* observations showed that Fz-containing puncta could fuse with accumulations on the membrane and that this appeared to be reversible (Shimada et al., 2006; Strutt et al., 2011). Recent results of Strutt et al. (2011) obtained by FRAP experiments established that the three transmembrane proteins Fz, Stan, and Vang are all required for the stable asymmetric accumulation of PCP protein complexes and that the cytoplasmic proteins Dsh, Pk, and Dgo are needed for the formation of large puncta. The cytoplasmic proteins could be recruited to the proximal and distal domains by interacting with the complexes of transmembrane domains leading to further stabilization and an increase in the size of the complexes. Consistent with that possibility, Dsh has been shown to bind to Fz (Wong et al., 2003) and Pk has been shown to bind to Vang (Bastock et al., 2003; Jenny et al., 2003).

The negative protein interactions between cytoplasmic proteins noted above could also serve to remove mislocalized proteins and hence to increase the asymmetry.

Taken together with earlier observations, one is lead to a model where directed transport leads to a modest bias in the localization of the three transmembrane proteins. Direct protein interactions across the juxtaposed distal and proximal membrane domains would stabilize proximally localized Vang and distally localized Fz and hence increase the bias in protein localization. The binding of the cytoplasmic proteins would further stabilize and increase the size of the protein complexes. Thus, the formation of the asymmetric protein complexes would involve several layers of mechanisms.

5. DIRECTIONAL CELL NONAUTONOMY

One striking property of *fz* and *Vang* mutant cells is the directional domineering nonautonomy displayed by clones (Taylor et al., 1998; Vinson & Adler, 1987) (Fig. 1.1D and F). This differs from the essentially cell autonomous activity of other *fz* pathway genes (Fig. 1.1E) (these other genes show very limited cell nonautonomy that is rare or limited to neighboring cells) (Chae et al., 1999; Collier & Gubb, 1997; Collier et al., 2005; Gubb & Garcia-Bellido, 1982; Lee & Adler, 2002; Park et al., 1996). The polarity of wild-type cells distal to *fz* clones and proximal to *Vang* clones is affected. This is due to hairs pointing toward cells of lower Fz activity and toward higher Vang activity (Adler, Krasnow, & Liu, 1997; Adler, Taylor, & Charlton, 2000; Casal et al., 2006; Strutt & Strutt, 2007). Complementary effects are seen when cells are manipulated to express higher Fz or Vang levels (Adler et al., 1997; Adler, Taylor, et al., 2000).

5.1. The basis for directional cell nonautonomy

Early evidence suggested that *fz* had both cell autonomous and cell nonautonomous functions (Vinson & Adler, 1987). Several *fz* alleles behaved cell autonomously (Vinson & Adler, 1987) and that the cell nonautonomous function of *fz* was independent of *dsh* and preceded and was temporally separate from the cell autonomous function (Lee & Adler, 2002; Strutt & Strutt, 2002). It is worth noting that the PCP system does not require patterned expression of any of the *fz* pathways genes as even expression of these genes provides complete rescue of null alleles (Axelrod, 2001;

Bastock et al., 2003; Collier & Gubb, 1997; Collier et al., 2005; Feiguin et al., 2001; Krasnow & Adler, 1994; Park et al., 1996; Strutt, 2001; Strutt & Warrington, 2008; Usui et al., 1999; Yan et al., 2008). The basis for the domineering nonautonomy has been studied at length. Early studies established that it was a property of abnormal local signaling between cells (Adler, Taylor, et al., 2000). The current consensus model is that the intercellular signaling between neighboring cells due to the asymmetric accumulation of PCP proteins is propagated on a cell-by-cell basis and this is the basis for the domineering nonautonomy (Amonlirdviman et al., 2005; Goodrich & Strutt, 2011; Maung & Jenny, 2011; Tree et al., 2002; Wu & Mlodzik, 2008). Thus, a cell with no Fz cannot recruit Vang to the juxtaposed membrane (Wu & Mlodzik, 2008). This leads to a high level of Vang on the side of the cell distal to the mutant membrane. This would be propagated for a number of cells and such a system could be at the heart of the intercellular signaling required for PCP. One drawback to such a model is that the function of a gene such as *dsh* that acts cell autonomously is required for the antisymmetric accumulation of other PCP proteins (Axelrod, 2001; Klingensmith et al., 1994; Strutt, 2001; Theisen et al., 1994; Tree et al., 2002; Usui et al., 1999). At first glance, this result would seem to contradict the asymmetry propagation model. However, computer simulations indicate that the model can explain the observations (Amonlirdviman et al., 2005; Schamberg, Houston, Monk, & Owen, 2010) as only modest levels of asymmetry may be required for signaling and this may be difficult to detect in immunostaining experiments. There remains, however, a need for an explanation for the results that indicate the cell nonautonomous function precedes the cell autonomous function (Strutt & Strutt, 2002).

It is worth noting that this mechanism works well to explain both normal PCP development and domineering nonautonomy in a tissue where all cells express the *fz/stan* pathway proteins and show the asymmetric accumulation. In the eye, only the R3 and R4 cells need to express *fz* pathway genes for the development of normal PCP (Zheng et al., 1995). In these cells, *fz/stan* pathway proteins accumulate asymmetrically with proteins that accumulate on the distal side of wing cells accumulating in the R3 cell membrane that is juxtaposed to the R4 cell (Strutt et al., 2002). Similarly, proteins that accumulate on the proximal side of wing cells accumulate on the R4 cell membrane that is juxtaposed to R3 (Strutt et al., 2002). Thus, heterotypic interactions across cell membranes seem likely to be a mechanism used in the generation of the two domains (e.g., see Wu & Mlodzik, 2008).

6. DOWNSTREAM EFFECTORS

6.1. Planar polarity effector genes

The three PPE proteins, In, Fy, and Frtz, are recruited to the proximal side of wing cells in a core group-dependent manner (Fig. 1.2A) (Adler et al., 2004; Strutt & Warrington, 2008). For example, in an *fz* mutant, the asymmetric accumulation of In and Frtz is lost and indeed these proteins are found at much lower levels in such a mutant. It seems likely that at least one of these proteins is recruited by binding (either directly or indirectly) to either Pk or Vang, the two proximal specific proteins. As is the case for the upstream genes, each of the PPE genes needs to be functional for the proteins to preferentially accumulate on the proximal side (Adler et al., 2004; Strutt & Warrington, 2008). In contrast to the situation with the upstream genes, the overexpression of PPE genes has not been reported to produce a gain of function PCP phenotype (Park et al., 1996). This seems somewhat surprising as overexpression often leads to a loss in the spatially restricted accumulation of a protein, which results in altered function. One possibility is that the overexpression does not saturate the system that localizes the proteins to the proximal side. This seems unlikely. A more attractive possibility is that a protein that is not properly localized is not active due to a lack of one or more binding partners.

6.2. *mwh*

The *mwh* gene, which by epistasis tests is the most downstream member of the *fz/stan* pathway, also accumulates on the proximal side of wing cells and this is dependent on both core and PPE proteins and genes (Fig. 1.2A) (Strutt & Warrington, 2008; Yan et al., 2008). Two papers have suggested possible mechanisms. In one, Lu and colleagues found that Mwh and In directly interacted and they suggested that In on the proximal side of wing cells recruited Mwh (Lu et al., 2010). In the other, Strutt and colleagues found that Mwh was phosphorylated in a PPE-dependent manner and suggested this was important in the action of Mwh in ensuring a single distally pointing hair is formed (Strutt & Warrington, 2008). These two mechanisms are, of course, not mutually exclusive. A variety of results argue that Mwh serves as an inhibitor of the actin cytoskeleton (Fig. 1.6). How this is accomplished is unclear, but it is worth noting that the amino terminal half of the large Mwh protein contains and a G protein-binding domain and has sequence similarity to diaphanous family formins. These formins are known to function as dimers, be activated by Rho:GTP, and promote the formation of long actin filaments (Campellone & Welch, 2011; Chesarone, DuPage, & Goode, 2011; Goode & Eck, 2007). Consistent with the sequence similarity between Mwh and Dia, both genetic interactions and coimmunoprecipiation experiments argue that Rho1 activates Mwh (Yan et al., 2009). This raises the possibility that Mwh might inhibit the actin cytoskeleton by acting as a dominant negative formin.

6.3. A distal target of Dsh?

The cytoskeleton is activated to form a hair in the vicinity of the distal most vertex of pupal wing cells (Wong & Adler, 1993). The level of activity is important for insuring a single hair is formed and that it is the proper size and shape. For example, if the actin cytoskeleton is inhibited by drug treatment or by mutations in actin cytoskeleton components such as myosin II, multiple shorter than normal hairs are formed (Franke, Montague, & Kiehart, 2010; Turner & Adler, 1998). Inhibition at the proximal side of wing cells from the recruitment of Mwh seems insufficient to specify the distal vertex. Several lines of data indicate that distal edge proteins can stimulate the cytoskeleton. Although mutations in *in*, *fy*, *frtz*, and *mwh* are epistatic to loss-of-function mutations in distal proteins such as *dsh* and *fz* that is not the case for all gain of functions in those genes. The late overexpression of *dsh* leads to the formation of multiple hair cells and an increase is seen even when cells also lack *in* or *mwh* function (Lee & Adler, 2002). Indeed, the effects of these two genetic changes are additive. Thus, at least when overexpressed Dsh is able to affect cytoskeleton function and it must be doing so by interacting with an unidentified cellular constituent and not known *fz/stan* downstream components such as *in* or *mwh*. Independent evidence supporting the distal edge proteins stimulating the cytoskeleton came from Strutt and colleagues who found that Fz acted to promote hair formation (Strutt & Warrington, 2008). Several experimental results supported this idea including a delay in hair initiation in *fz* mutant cells compared to wild-type neighbors. This was seen even when comparing fz^- and fz^+ cells in a *Vang* mutant background where Fz was not properly localized.

The identity of the putative distal target(s) is currently unknown. In principle, the target could be a cytoskeletal activator that is activated by the distal proteins or, alternatively, a cytoskeletal inhibitor that is inactivated by the distal proteins (Fig. 1.6). If the distal target is an activator, then we can predict that a loss-of-function mutation would result in delayed hair initiation and perhaps small or short hairs. Several such genes have been reported including *shavenoid* which has a strong hair delay/loss phenotype (Ren, He, Stone, Kirakodu, & Adler, 2006). However, the phenotype of null alleles of *sha* is strongly enhanced by cells simultaneously being mutant for *fz/stan* pathway genes. Thus, if Sha is a target it cannot be the only one. Other possible activator targets are proteins that modulate the activity of the actin cytoskeleton such as formins (Campellone & Welch, 2011; Chesarone et al., 2011; Goode & Eck, 2007) or the Arp23 complex (Campellone & Welch, 2011; Zigmond, 2004). No evidence suggesting this is the case has been published. It is also possible that the target is an inhibitor, and if so, we can predict that loss-of-function mutations would result in premature hair initiation. No possible targets with this property have been reported.

ACKNOWLEDGMENTS

The author is supported by a grant from the NIGMS (GM-37136). I thank Simon Collier for generously providing the images of wing ridges.

REFERENCES

Adler, P. N. (2002). Planar signaling and morphogenesis in Drosophila. *Developmental Cell, 2*, 525–535.

Adler, P. N., Charlton, J., & Liu, J. (1998). Mutations in the cadherin superfamily member gene dachsous cause a tissue polarity phenotype by altering frizzled signaling. *Development, 125*, 959–968.

Adler, P. N., Krasnow, R. E., & Liu, J. (1997). Tissue polarity points from cells that have higher Frizzled levels towards cells that have lower Frizzled levels. *Current Biology, 7*, 940–949.

Adler, P. N., Liu, J., & Charlton, J. (2000). Cell size and the morphogenesis of epidermal hairs. *Genesis, 28*, 82–91.

Adler, P., Taylor, J., & Charlton, J. (2000). The domineering non-autonomy of frizzled and van Gogh clones in the Drosophila wing is a consequence of a disruption in local signaling. *Mechanisms of Development, 96*, 197–207.

Adler, P. N., Zhu, C., & Stone, D. (2004). Inturned localizes to the proximal side of wing cells under the instruction of upstream planar polarity proteins. *Current Biology, 14*, 2046–2051.

Aigouy, B., Farhadifar, R., Staple, D. B., Sagner, A., Roper, J. C., Julicher, F., et al. (2011). Cell flow reorients the axis of planar polarity in the wing epithelium of Drosophila. *Cell, 142*, 773–786.

Amonlirdviman, K., Khare, N. A., Tree, D. R., Chen, W. S., Axelrod, J. D., & Tomlin, C. J. (2005). Mathematical modeling of planar cell polarity to understand domineering non-autonomy. *Science, 307*, 423–426.

Axelrod, J. (2001). Unipolar membrane association of Dishevelled mediates Frizzled planar cell polarity signaling. *Genes & Development, 15*, 1182–1187.

Bastock, R., Strutt, H., & Strutt, D. (2003). Strabismus is asymmetrically localised and binds to Prickle and Dishevelled during Drosophila planar polarity patterning. *Development, 130*, 3007–3014.

Bateman, J., Reddy, R. S., Saito, H., & Van Vactor, D. (2001). The receptor tyrosine phosphatase Dlar and integrins organize actin filaments in the Drosophila follicular epithelium. *Current Biology, 11*, 1317–1327.

Berleth, T., Burri, M., Thoma, G., Bopp, D., Richstein, S., Frigerio, G., et al. (1988). The role of localization of bicoid RNA in organizing the anterior pattern of the Drosophila embryo. *The EMBO Journal, 7*, 1749–1756.

Bertet, C., Sulak, L., & Lecuit, T. (2004). Myosin-dependent junction remodelling controls planar cell intercalation and axis elongation. *Nature, 429*, 667–671.

Bhanot, P., Brink, M., Samos, C. H., Hsieh, J. C., Wang, Y., Macke, J. P., et al. (1996). A new member of the frizzled family from Drosophila functions as a Wingless receptor. *Nature, 382*, 225–230.

Blair, A., Tomlinson, A., Pham, H., Gunsalus, K. C., Goldberg, M. L., & Laski, F. A. (2006). Twinstar, the Drosophila homolog of cofilin/ADF, is required for planar cell polarity patterning. *Development, 133*, 1789–1797.

Brittle, A. L., Repiso, A., Casal, J., Lawrence, P. A., & Strutt, D. (2011). Four-jointed modulates growth and planar polarity by reducing the affinity of dachsous for fat. *Current Biology, 20*, 803–810.

Buechling, T., Bartscherer, K., Ohkawara, B., Chaudhary, V., Spirohn, K., Niehrs, C., et al. (2010). Wnt/Frizzled signaling requires dPRR, the Drosophila homolog of the prorenin receptor. *Current Biology, 20*, 1263–1268.

Campellone, K. G., & Welch, M. D. (2011). A nucleator arms race: Cellular control of actin assembly. *Nature Reviews. Molecular Cell Biology, 11*, 237–251.

Casal, J., Lawrence, P. A., & Struhl, G. (2006). Two separate molecular systems, Dachsous/Fat and Starry night/Frizzled, act independently to confer planar cell polarity. *Development, 133*, 4561–4572.

Chae, J., Kim, M. J., Goo, J. H., Collier, S., Gubb, D., Charlton, J., et al. (1999). The Drosophila tissue polarity gene starry night encodes a member of the protocadherin family. *Development, 126*, 5421–5429.

Chen, W. S., Antic, D., Matis, M., Logan, C. Y., Povelones, M., Anderson, G. A., et al. (2008). Asymmetric homotypic interactions of the atypical cadherin flamingo mediate intercellular polarity signaling. *Cell, 133*, 1093–1105.

Chesarone, M. A., DuPage, A. G., & Goode, B. L. (2011). Unleashing formins to remodel the actin and microtubule cytoskeletons. *Nature Reviews. Molecular Cell Biology, 11*, 62–74.

Cho, B., & Fischer, J. A. (2011). Ral GTPase promotes asymmetric Notch activation in the Drosophila eye in response to Frizzled/PCP signaling by repressing ligand-independent receptor activation. *Development, 138*, 1349–1359.

Clark, H. F., Brentrup, D., Schneitz, K., Bieber, A., Goodman, C., & Noll, M. (1995). Dachsous encodes a member of the cadherin superfamily that controls imaginal disc morphogenesis in Drosophila. *Genes & Development, 9*, 1530–1542.

Classen, A. K., Anderson, K. I., Marois, E., & Eaton, S. (2005). Hexagonal packing of Drosophila wing epithelial cells by the planar cell polarity pathway. *Developmental Cell, 9*, 805–817.

Collier, S., & Gubb, D. (1997). Drosophila tissue polarity requires the cell-autonomous activity of the fuzzy gene, which encodes a novel transmembrane protein. *Development, 124*, 4029–4037.

Collier, S., Lee, H., Burgess, R., & Adler, P. (2005). The WD40 repeat protein fritz links cytoskeletal planar polarity to frizzled subcellular localization in the Drosophila epidermis. *Genetics, 169*, 2035–2045.

Cooper, M. T., & Bray, S. J. (1999). Frizzled regulation of Notch signalling polarizes cell fate in the Drosophila eye. *Nature*, *397*, 526–530.

Das, G., Jenny, A., Klein, T. J., Eaton, S., & Mlodzik, M. (2004). Diego interacts with Prickle and Strabismus/Van Gogh to localize planar cell polarity complexes. *Development*, *131*, 4467–4476.

Davis, I., & Ish-Horowicz, D. (1991). Apical localization of pair-rule transcripts requires 3' sequences and limits protein diffusion in the Drosophila blastoderm embryo. *Cell*, *67*, 927–940.

del Alamo, D., & Mlodzik, M. (2006). Frizzled/PCP-dependent asymmetric neuralized expression determines R3/R4 fates in the Drosophila eye. *Developmental Cell*, *11*, 887–894.

Djiane, A., Yogev, S., & Mlodzik, M. (2005). The apical determinants aPKC and dPatj regulate Frizzled-dependent planar cell polarity in the Drosophila eye. *Cell*, *121*, 621–631.

Donoughe, S., & DiNardo, S. (2011). Dachsous and frizzled contribute separately to planar polarity in the Drosophila ventral epidermis. *Development*, *138*, 2751–2759.

Doyle, K., Hogan, J., Lester, M., & Collier, S. (2008). The Frizzled Planar Cell Polarity signaling pathway controls Drosophila wing topography. *Developmental Biology*, *317*, 354–367.

Fanto, M., Clayton, L., Meredith, J., Hardiman, K., Charroux, B., Kerridge, S., et al. (2003). The tumor-suppressor and cell adhesion molecule Fat controls planar polarity via physical interactions with Atrophin, a transcriptional co-repressor. *Development*, *130*, 763–774.

Fanto, M., & Mlodzik, M. (1999). Asymmetric Notch activation specifies photoreceptors R3 and R4 and planar polarity in the Drosophila eye. *Nature*, *397*, 523–526.

Feiguin, F., Hannus, M., Mlodzik, M., & Eaton, S. (2001). The ankyrin repeat protein Diego mediates Frizzled-dependent planar polarization. *Developmental Cell*, *1*, 93–101.

Ferrandon, D., Elphick, L., Nusslein-Volhard, C., & St Johnston, D. (1994). Staufen protein associates with the 3'UTR of bicoid mRNA to form particles that move in a microtubule-dependent manner. *Cell*, *79*, 1221–1232.

Franke, J. D., Montague, R. A., & Kiehart, D. P. (2010). Nonmuscle myosin II is required for cell proliferation, cell sheet adhesion and wing hair morphology during wing morphogenesis. *Developmental Biology*, *345*, 117–132.

Frydman, H. M., & Spradling, A. C. (2001). The receptor-like tyrosine phosphatase lar is required for epithelial planar polarity and for axis determination within drosophila ovarian follicles. *Development*, *128*, 3209–3220.

Gault, W. J., Olguin, P., Weber, U., & Mlodzik, M. (2012). Drosophila CK1-gamma, gilgamesh, controls PCP-mediated morphogenesis through regulation of vesicle trafficking. *The Journal of Cell Biology*, *196*, 605–621.

Gho, M., & Schweisguth, F. (1998). Frizzled signalling controls orientation of asymmetric sense organ precursor cell divisions in Drosophila. *Nature*, *393*, 178–181.

Goode, B. L., & Eck, M. J. (2007). Mechanism and function of formins in the control of actin assembly. *Annual Review of Biochemistry*, *76*, 593–627.

Goodrich, L. V., & Strutt, D. (2011). Principles of planar polarity in animal development. *Development*, *138*, 1877–1892.

Griffin, E. E., Odde, D. J., & Seydoux, G. (2011). Regulation of the MEX-5 gradient by a spatially segregated kinase/phosphatase cycle. *Cell*, *146*, 955–968.

Gubb, D., & Garcia-Bellido, A. (1982). A genetic analysis of the determination of cuticular polarity during development in Drosophila melanogaster. *Journal of Embryology and Experimental Morphology*, *68*, 37–57.

Gubb, D., Green, C., Huen, D., Coulson, D., Johnson, G., Tree, D., et al. (1999). The balance between isoforms of the prickle LIM domain protein is critical for planar polarity in Drosophila imaginal discs. *Genes & Development*, *13*, 2315–2327.

Hannus, M., Feiguin, F., Heisenberg, C. P., & Eaton, S. (2002). Planar cell polarization requires Widerborst, a B' regulatory subunit of protein phosphatase 2A. *Development, 129*, 3493–3503.

Harumoto, T., Ito, M., Shimada, Y., Kobayashi, T. J., Ueda, H. R., Lu, B., et al. (2010). Atypical cadherins Dachsous and Fat control dynamics of noncentrosomal microtubules in planar cell polarity. *Developmental Cell, 19*, 389–401.

He, B., & Adler, P. N. (2002). The frizzled pathway regulates the development of arista laterals. *BMC Developmental Biology, 2*, 7.

Held, L. I., Jr. (2002). Bristles induce bracts via the EGFR pathway on Drosophila legs. *Mechanisms of Development, 117*, 225–234.

Held, L. I., Jr., Duarte, C. M., & Derakhshanian, K. (1986). Extra joints and misoriented bristles on Drosophila legs. *Progress in Clinical and Biological Research, 217A*, 293–296.

Hermle, T., Saltukoglu, D., Grunewald, J., Walz, G., & Simons, M. (2010). Regulation of Frizzled-dependent planar polarity signaling by a V-ATPase subunit. *Current Biology, 20*, 1269–1276.

Hogan, J., Valentine, M., Cox, C., Doyle, K., & Collier, S. (2011). Two frizzled planar cell polarity signals in the Drosophila wing are differentially organized by the Fat/Dachsous pathway. *PLoS Genetics, 7*, e1001305.

Jenny, A., Darken, R. S., Wilson, P. A., & Mlodzik, M. (2003). Prickle and Strabismus form a functional complex to generate a correct axis during planar cell polarity signaling. *The EMBO Journal, 22*, 4409–4420.

Jenny, A., Reynolds-Kenneally, J., Das, G., Burnett, M., & Mlodzik, M. (2005). Diego and Prickle regulate Frizzled planar cell polarity signalling by competing for Dishevelled binding. *Nature Cell Biology, 7*, 691–697.

Klingensmith, J., Nusse, R., & Perrimon, N. (1994). The Drosophila segment polarity gene dishevelled encodes a novel protein required for response to the wingless signal. *Genes & Development, 8*, 118–130.

Krasnow, R. E., & Adler, P. N. (1994). A single frizzled protein has a dual function in tissue polarity. *Development, 120*, 1883–1893.

Kumar, J. P. (2012). Building an ommatidium one cell at a time. *Developmental Dynamics, 241*, 136–149.

Lawrence, P. (1966). Gradients in the insect segment: The orientation of hairs in the milkweed bug Oncopeltus fasciatus. *The Journal of Experimental Biology, 44*, 452–458.

Lawrence, P. A., Casal, J., & Struhl, G. (2002). Towards a model of the organisation of planar polarity and pattern in the Drosophila abdomen. *Development, 129*, 2749–2760.

Lee, H., & Adler, P. N. (2002). The function of the frizzled pathway in the Drosophila wing is dependent on inturned and fuzzy. *Genetics, 160*, 1535–1547.

Lee, H., & Adler, P. N. (2004). The grainy head transcription factor is essential for the function of the frizzled pathway in the Drosophila wing. *Mechanisms of Development, 121*, 37–49.

Lu, B., Usui, T., Uemura, T., Jan, L., & Jan, Y. N. (1999). Flamingo controls the planar polarity of sensory bristles and asymmetric division of sensory organ precursors in drosophila. *Current Biology, 9*, 1247–1250.

Lu, Q., Yan, J., & Adler, P. N. (2010). The Drosophila planar polarity proteins Inturned and Multiple Wing Hairs interact physically and function together. *Genetics, 185*, 549–558.

Ma, D., Amonlirdviman, K., Raffard, R. L., Abate, A., Tomlin, C. J., & Axelrod, J. D. (2008). Cell packing influences planar cell polarity signaling. *Proceedings of the National Academy of Sciences of the United States of America, 105*, 18800–18805.

Ma, D., Yang, C. H., McNeil, H., Simon, M. A., & Axelrod, J. D. (2003). Fidelity in planar cell polarity signalling. *Nature, 421*, 543–547.

Macdonald, P. M., & Struhl, G. (1988). cis-Acting sequences responsible for anterior localization of bicoid mRNA in Drosophila embryos. *Nature, 336*, 595–598.

Mahoney, P. A., Weber, U., Onofrechuk, P., Biessmann, H., Bryant, P. J., & Goodman, C. S. (1991). The fat tumor suppressor gene in Drosophila encodes a novel member of the cadherin gene superfamily. *Cell*, *67*, 853–868.
Mao, Y., Rauskolb, C., Cho, E., Hu, W. L., Hayter, H., Minihan, G., et al. (2006). Dachs: An unconventional myosin that functions downstream of Fat to regulate growth, affinity and gene expression in Drosophila. *Development*, *133*, 2539–2551.
Mao, Y., Tournier, A. L., Bates, P. A., Gale, J. E., Tapon, N., & Thompson, B. J. (2011). Planar polarization of the atypical myosin Dachs orients cell divisions in Drosophila. *Genes & Development*, *25*, 131–136.
Matakatsu, H., & Blair, S. S. (2004). Interactions between Fat and Dachsous and the regulation of planar cell polarity in the Drosophila wing. *Development*, *131*, 3785–3794.
Matakatsu, H., & Blair, S. S. (2006). Separating the adhesive and signaling functions of the Fat and Dachsous protocadherins. *Development*, *133*, 2315–2324.
Matakatsu, H., & Blair, S. S. (2008). The DHHC palmitoyltransferase approximated regulates Fat signaling and Dachs localization and activity. *Current Biology: CB*, *18*, 1390–1395.
Matsubara, D., Horiuchi, S. Y., Shimono, K., Usui, T., & Uemura, T. (2011). The seven-pass transmembrane cadherin Flamingo controls dendritic self-avoidance via its binding to a LIM domain protein, Espinas, in Drosophila sensory neurons. *Genes & Development*, *25*, 1982–1996.
Maung, S. M., & Jenny, A. (2011). Planar cell polarity in Drosophila. *Organogenesis*, 7, 165–179.
Mottola, G., Classen, A. K., Gonzalez-Gaitan, M., Eaton, S., & Zerial, M. (2010). A novel function for the Rab5 effector Rabenosyn-5 in planar cell polarity. *Development*, *137*, 2353–2364.
Moyer, K. E., & Jacobs, J. R. (2008). Varicose: A MAGUK required for the maturation and function of Drosophila septate junctions. *BMC Developmental Biology*, *8*, 99.
Mrkusich, E. M., Flanagan, D. J., & Whitington, P. M. (2011). The core planar cell polarity gene prickle interacts with flamingo to promote sensory axon advance in the Drosophila embryo. *Developmental Biology*, *358*, 224–230.
Mummery-Widmer, J. L., Yamazaki, M., Stoeger, T., Novatchkova, M., Bhalerao, S., Chen, D., et al. (2009). Genome-wide analysis of Notch signalling in Drosophila by transgenic RNAi. *Nature*, *458*, 987–992.
Nagaraj, R., & Adler, P. N. (2012). Dusky-like functions as a Rab11 effector for the deposition of cuticle during Drosophila bristle development. *Development*, *139*, 906–916.
Narimatsu, M., Bose, R., Pye, M., Zhang, L., Miller, B., Ching, P., et al. (2009). Regulation of planar cell polarity by Smurf ubiquitin ligases. *Cell*, *137*, 295–307.
Olguin, P., Glavic, A., & Mlodzik, M. (2011). Intertissue mechanical stress affects Frizzled-mediated planar cell polarity in the Drosophila notum epidermis. *Current Biology*, *21*, 236–242.
Paricio, N., Feiguin, F., Boutros, M., Eaton, S., & Mlodzik, M. (1999). The Drosophila STE20-like kinase misshapen is required downstream of the Frizzled receptor in planar polarity signaling. *The EMBO Journal*, *18*, 4669–4678.
Park, W. J., Liu, J., Sharp, E. J., & Adler, P. N. (1996). The Drosophila tissue polarity gene inturned acts cell autonomously and encodes a novel protein. *Development*, *122*, 961–969.
Pataki, C., Matusek, T., Kurucz, E., Ando, I., Jenny, A., & Mihaly, J. (2010). Drosophila Rab23 is involved in the regulation of the number and planar polarization of the adult cuticular hairs. *Genetics*, *184*, 1051–1065.
Price, M. H., Roberts, D. M., McCartney, B. M., Jezuit, E., & Peifer, M. (2006). Cytoskeletal dynamics and cell signaling during planar polarity establishment in the Drosophila embryonic denticle. *Journal of Cell Science*, *119*, 403–415.

Purvanov, V., Koval, A., & Katanaev, V. L. (2010). A direct and functional interaction between Go and Rab5 during G protein-coupled receptor signaling. *Science Signaling*, *3*, ra65.

Ren, N., Charlton, J., & Adler, P. N. (2007). The flare gene, which encodes the AIP1 protein of Drosophila, functions to regulate F-actin disassembly in pupal epidermal cells. *Genetics*, *176*, 2223–2234.

Ren, N., He, B., Stone, D., Kirakodu, S., & Adler, P. N. (2006). The shavenoid gene of Drosophila encodes a novel actin cytoskeleton interacting protein that promotes wing hair morphogenesis. *Genetics*, *172*, 1643–1653.

Repiso, A., Saavedra, P., Casal, J., & Lawrence, P. A. (2010). Planar cell polarity: The orientation of larval denticles in Drosophila appears to depend on gradients of Dachsous and Fat. *Development*, *137*, 3411–3415.

Schamberg, S., Houston, P., Monk, N. A., & Owen, M. R. (2010). Modelling and analysis of planar cell polarity. *Bulletin of Mathematical Biology*, *72*, 645–680.

Shimada, Y., Usui, T., Yanagawa, S., Takeichi, M., & Uemura, T. (2001). Asymmetric colocalization of Flamingo, a seven-pass transmembrane cadherin, and Dishevelled in planar cell polarization. *Current Biology*, *11*, 859–863.

Shimada, Y., Yonemura, S., Ohkura, H., Strutt, D., & Uemura, T. (2006). Polarized transport of Frizzled along the planar microtubule arrays in Drosophila wing epithelium. *Developmental Cell*, *10*, 209–222.

Simon, M. A. (2004). Planar cell polarity in the Drosophila eye is directed by graded Four-jointed and Dachsous expression. *Development*, *131*, 6175–6184.

Simon, M. A., Xu, A., Ishikawa, H. O., & Irvine, K. D. (2010). Modulation of FAt:Dachsous binding by the cadherin domain kinase four-jointed. *Current Biology*, *20*, 811–817.

Simons, M., Gault, W. J., Gotthardt, D., Rohatgi, R., Klein, T. J., Shao, Y., et al. (2009). Electrochemical cues regulate assembly of the Frizzled/Dishevelled complex at the plasma membrane during planar epithelial polarization. *Nature Cell Biology*, *11*, 286–294.

Singh, J., Yanfeng, W. A., Grumolato, L., Aaronson, S. A., & Mlodzik, M. (2010). Abelson family kinases regulate Frizzled planar cell polarity signaling via Dsh phosphorylation. *Genes & Development*, *24*, 2157–2168.

Sopko, R., Silva, E., Clayton, L., Gardano, L., Barrios-Rodiles, M., Wrana, J., et al. (2009). Phosphorylation of the tumor suppressor fat is regulated by its ligand Dachsous and the kinase discs overgrown. *Current Biology*, *19*, 1112–1117.

Strutt, D. (2001). Asymmetric localization of frizzled and the establishment of cell polarity in the Drosophila wing. *Molecular Cell*, *7*, 367–375.

Strutt, D., Johnson, R., Cooper, K., & Bray, S. (2002). Asymmetric localization of frizzled and the determination of notch-dependent cell fate in the Drosophila eye. *Current Biology*, *12*, 813–824.

Strutt, H., Price, M. A., & Strutt, D. (2006). Planar polarity is positively regulated by casein kinase Iepsilon in Drosophila. *Current Biology*, *16*, 1329–1336.

Strutt, H., & Strutt, D. (2002). Nonautonomous planar polarity patterning in Drosophila: Dishevelled-independent functions of frizzled. *Developmental Cell*, *3*, 851–863.

Strutt, D., & Strutt, H. (2007). Differential activities of the core planar polarity proteins during Drosophila wing patterning. *Developmental Biology*, *302*, 181–194.

Strutt, H., & Strutt, D. (2008). Differential stability of flamingo protein complexes underlies the establishment of planar polarity. *Current Biology*, *18*, 1555–1564.

Strutt, D., & Warrington, S. J. (2008). Planar polarity genes in the Drosophila wing regulate the localisation of the FH3-domain protein Multiple Wing Hairs to control the site of hair production. *Development*, *135*, 3103–3111.

Strutt, H., Warrington, S. J., & Strutt, D. (2011). Dynamics of core planar polarity protein turnover and stable assembly into discrete membrane subdomains. *Developmental Cell*, *20*, 511–525.

Strutt, D. I., Weber, U., & Mlodzik, M. (1997). The role of RhoA in tissue polarity and Frizzled signalling. *Nature*, *387*, 292–295.

Taylor, J., Abramova, N., Charlton, J., & Adler, P. N. (1998). Van Gogh: A new Drosophila tissue polarity gene. *Genetics*, *150*, 199–210.

Theisen, H., Purcell, J., Bennett, M., Kansagara, D., Syed, A., & Marsh, J. L. (1994). Dishevelled is required during wingless signaling to establish both cell polarity and cell identity. *Development*, *120*, 347–360.

Tomlinson, A., & Struhl, G. (1999). Decoding vectorial information from a gradient: Sequential roles of the receptors Frizzled and Notch in establishing planar polarity in the Drosophila eye. *Development*, *126*, 5725–5738.

Tree, D. R. P., Shulman, J. M., Rousset, R., Scott, M. P., Gubb, D., & Axelrod, J. D. (2002). Prickle mediates feedback amplification to generate asymmetric planar cell polarity signaling. *Cell*, *109*, 1–11.

Turner, C. M., & Adler, P. N. (1998). Distinct roles for the actin and microtubule cytoskeletons in the morphogenesis of epidermal hairs during wing development in Drosophila. *Mechanisms of Development*, *70*, 181–192.

Usui, T., Shima, Y., Shimada, Y., Hirano, S., Burgess, R. W., Schwarz, T. L., et al. (1999). Flamingo, a seven-pass transmembrane cadherin, regulates planar cell polarity under the control of Frizzled. *Cell*, *98*, 585–595.

Valentine, M., & Collier, S. (2011). Planar cell polarity and tissue design: Shaping the Drosophila wing membrane. *Fly*, *5*, 316–321.

Venema, D. R., Zeev-Ben-Mordehai, T., & Auld, V. J. (2004). Transient apical polarization of Gliotactin and Coracle is required for parallel alignment of wing hairs in Drosophila. *Developmental Biology*, *275*, 301–314.

Viktorinova, I., Konig, T., Schlichting, K., & Dahmann, C. (2009). The cadherin Fat2 is required for planar cell polarity in the Drosophila ovary. *Development*, *136*, 4123–4132.

Villano, J. L., & Katz, F. N. (1995). Four-jointed is required for intermediate growth in the proximal-distal axis in Drosophila. *Development*, *121*, 2767–2777.

Vinson, C. R., & Adler, P. N. (1987). Directional non-cell autonomy and the transmission of polarity information by the frizzled gene of Drosophila. *Nature*, *329*, 549–551.

Vinson, C. R., Conover, S., & Adler, P. N. (1989). A Drosophila tissue polarity locus encodes a protein containing seven potential transmembrane domains. *Nature*, *338*, 263–264.

Wang, Y., & Nathans, J. (2007). Tissue/planar cell polarity in vertebrates: New insights and new questions. *Development*, *134*, 647–658.

Weber, U., Paricio, N., & Mlodzik, M. (2000). Jun mediates Frizzled-induced R3/R4 cell fate distinction and planar polarity determination in the Drosophila eye. *Development*, *127*, 3619–3629.

Winter, C. G., Wang, B., Ballew, A., Royou, A., Karess, R., Axelrod, J. D., et al. (2001). Drosophila Rho-associated kinase (Drok) links frizzled-mediated planar cell polarity signaling to the actin cytoskeleton. *Cell*, *105*, 81–91.

Wolff, T., & Rubin, G. (1998). Strabismus, a novel gene that regulates tissue polarity and cell fate decisions in Drosophila. *Development*, *125*, 1149–1159.

Wong, L. L., & Adler, P. N. (1993). Tissue polarity genes of Drosophila regulate the subcellular location for prehair initiation in pupal wing cells. *The Journal of Cell Biology*, *123*, 209–221.

Wong, H. C., Bourdelas, A., Krauss, A., Lee, H. J., Shao, Y., Wu, D., et al. (2003). Direct binding of the PDZ domain of Dishevelled to a conserved internal sequence in the C-terminal region of Frizzled. *Molecular Cell*, *12*, 1251–1260.

Wu, J., & Mlodzik, M. (2008). The frizzled extracellular domain is a ligand for Van Gogh/Stbm during nonautonomous planar cell polarity signaling. *Developmental Cell*, *15*, 462–469.

Yan, J., Huen, D., Morely, T., Johnson, G., Gubb, D., Roote, J., et al. (2008). The multiple-wing-hairs gene encodes a novel GBD-FH3 domain-containing protein that functions both prior to and after wing hair initiation. *Genetics, 180*, 219–228.

Yan, J., Lu, Q., Fang, X., & Adler, P. N. (2009). Rho1 has multiple functions in Drosophila wing planar polarity. *Developmental Biology, 333*, 186–199.

Yang, C. H., Axelrod, J. D., & Simon, M. A. (2002). Regulation of Frizzled by fat-like cadherins during planar polarity signaling in the Drosophila compound eye. *Cell, 108*, 675–688.

Zallen, J. A. (2007). Planar polarity and tissue morphogenesis. *Cell, 129*, 1051–1063.

Zallen, J. A., & Wieschaus, E. (2004). Patterned gene expression directs bipolar planar polarity in Drosophila. *Developmental Cell, 6*, 343–355.

Zeidler, M. P., Perrimon, N., & Strutt, D. I. (1999). The four-jointed gene is required in the Drosophila eye for ommatidial polarity specification. *Current Biology, 9*, 1363–1372.

Zheng, L., Zhang, J., & Carthew, R. W. (1995). Frizzled regulates mirror-symmetric pattern formation in the Drosophila eye. *Development, 121*, 3045–3055.

Zigmond, S. H. (2004). Beginning and ending an actin filament: Control at the barbed end. *Current Topics in Developmental Biology, 63*, 145–188.

NOTE ADDED IN PROOF

Several quite important papers on fly PCP were published recently and interested readers should take a look at them (see below).

Ambegaonkar, A., Pan, G., Mani, M., Feng, Y., & Irvine, K. D. (2012). Propagation of Dachsous-fat Planar Cell Polarity. *Current Biology, 22*, 1302–1308.

Brittle, A., Thomas, C., & Strutt, D. (2012). Planar Polarity Specification through Asymmetric Localization of Fat and Dachsous. *Current Biology, 22*, 907–914.

Peng, Y., Han, C., & Axelrod, J. D. (2012). Planar Polarized Protrusions Break the Symmetry of EGFR Signaling during *Drosophila* Bract Cell Fate Induction. *Developmental Cell, 23*, 1–12.

Sagner, A., Merkel, M., Aigouy, B., Gaebel, J., Brankatschk, M., Julicher, F., & Eaton, S. (2012). Establishment of Global Patterns of Planar Polarity during Growth of the *Drosophila* Wing Epithelium. *Current Biology* 22, 1296–1301.

Struhl, G., Casal, J., & Lawrence, P. A. (2012). Dissecting the molecular bridges that mediate the function of Frizzled in planar cell polarity. *Development, 139*, 3665–3674.

CHAPTER TWO

Asymmetric Protein Localization in Planar Cell Polarity: Mechanisms, Puzzles, and Challenges

Ying Peng*, Jeffrey D. Axelrod[†,1]
*Department of Biochemistry and Molecular Biology, Mayo Clinic, Rochester, Minnesota, USA
[†]Department of Pathology, Stanford University School of Medicine, Stanford, California, USA
[1]Corresponding author: e-mail address: jaxelrod@stanford.edu

Contents

Abstract

The polarization of epithelial cells along an axis orthogonal to their apical–basal axis is increasingly recognized for roles in a variety of developmental events and physiological functions. While now studied in many model organisms, mechanistic understanding is rooted in intensive investigations of planar cell polarity (PCP) in *Drosophila*. Consensus has emerged that two molecular modules, referred to here as the global and core modules, operate upstream of effector proteins to produce morphological PCP. Proteins of the core module develop subcellular asymmetry, accumulating in two groups on opposite sides of cells, consistent with proposed functions in producing cell polarity and in communicating that polarity between neighboring cells. Less clear are the molecular and cell biological mechanisms underlying core module function in the generation and communication of subcellular asymmetry and the relationship between the global and the core modules. In this review, we discuss these two unresolved questions, highlighting important studies and potentially enlightening avenues for further

Current Topics in Developmental Biology, Volume 101
ISSN 0070-2153
http://dx.doi.org/10.1016/B978-0-12-394592-1.00002-8
© 2012 Elsevier Inc.
All rights reserved.

investigation. It is likely that results from *Drosophila* will continue to inform our views of the growing list of examples of PCP in vertebrate systems.

1. INTRODUCTION

It is well appreciated that most cells assemble highly polarized structures that are essential for their specialized functions. In epithelial cells, the most obvious polarized feature is the universal apical–basal polarity that distinguishes the cell surface facing the external environment or lumen from that adjacent to the basal lamina. Extensive studies have revealed essential roles of apical–basal polarity in carrying out epithelial function and maintaining tissue homeostasis. At the same time, it has also been appreciated that epithelial cells can be polarized along the tissue surface, on an axis perpendicular to the apical–basal axis. This polarity, called planar cell polarity (PCP), is apparent in many epithelia in multicellular organisms. Understanding of the physiological significance of PCP, though often less apparent, has been steadily growing with the recent intensification of molecular genetic studies in various model organisms (Goodrich & Strutt, 2011). These efforts have shown that regulation of cellular function by PCP is important for processes including tissue morphogenesis (Keller, 2002), directional cell migration (Wada & Okamoto, 2009), and directional mechanosensing (Kelly & Chen, 2007) and will be discussed in other reviews in this volume.

While control of PCP is largely distinct from that of apical–basal polarity, the core families of PCP proteins localize and appear to act apically at the adherens junctions (Goodrich & Strutt, 2011). As first discovered during acquisition of planar polarity in the *Drosophila* wing epithelium (Axelrod, 2001; Strutt, 2001), those proteins become asymmetrically localized in a highly stereotypical manner, such that a distal subset localizes at the distal cell cortex and interacts with a proximal subset in the neighboring cell, and vice versa, resulting in the polarized localization of both proximal and distal components within each cell (Vladar, Antic, & Axelrod, 2009). In this review, we focus on our current understanding of the mechanisms that give rise to this asymmetric protein localization. We suggest that asymmetric protein localization is a characteristic and essential feature of planar-polarized epithelia, based on a growing list of examples from both invertebrate and vertebrate systems. The majority of this review discusses possible cell-autonomous and non-cell-autonomous mechanisms through which asymmetric protein localization arises. Our current understanding is based largely on experimental studies with *Drosophila* wing epithelium, in combination with mathematical simulations that examine the properties of proposed models. While studies with vertebrate models have to date yielded less mechanistic insight, numerous observations suggest substantial mechanistic conservation (Mitchell et al., 2009; Sienknecht, Anderson, Parodi, Fantetti, & Fekete, 2011).

We begin with a brief discussion of the three modules of planar polarity genes and propose a hierarchical structure, a model first developed almost a decade ago (Tree, Ma, & Axelrod, 2002). Despite the elapsed time, the mechanisms underlying this organization have not been revealed. We believe that the model has proved to be a valid general framework for understanding PCP, despite some recent challenges, and that clarifying mechanisms will soon emerge. Given its controversial nature, the model deserves a quick revisit here.

2. REVISITING THE THREE-TIERED HIERARCHY OF PCP

2.1. Original three-tiered hierarchy model

The existence of planar-polarized features of many types of epithelial structures has enabled extensive genetic studies of the genes and molecular pathways that control PCP. On the basis of phenotype, as well as genetic interaction, cell biological, and biochemical studies, these components can be classified as belonging to one of three distinct functional modules. We have argued previously that these three modules interact hierarchically (Tree, Ma, et al., 2002).

A highly conserved core module including the proteins Frizzled (Fz), Disheveled (Dsh), Van Gogh (Vang, aka Strabismus), Prickle (Pk), Flamingo (Fmi), and Diego (Dgo) produces molecular asymmetry within and between cells. Distinct proximal (Vang, Pk, and Fmi) and distal (Fz, Dsh, Dgo, and Fmi) complexes segregate to opposite sides of the cell, where they interact with the opposite complex in the neighboring cell at or near the adherens junctions. Feedback mechanisms between components of this key module ensure exclusive asymmetric protein localization during planar polarization by exclusion of oppositely oriented complexes from adjacent regions of the cell cortex and by recruitment of the opposite complex in the neighboring cell (Goodrich & Strutt, 2011) The result is an amplification of localization asymmetry of these core PCP proteins, producing steep intracellular gradients from any initial biasing input. This polarity amplification is necessarily coupled with local alignment of polarity between neighboring cells (Amonlirdviman et al., 2005). The model predicts the observed interdependence of the asymmetric localization of each core PCP proteins (Bastock, Strutt, & Strutt, 2003; Tree, Shulman, et al., 2002), although depending on specific molecular mechanisms involved allows for different degrees of residual function and polarization in the individual mutants (Axelrod, 2009). The majority of this review focuses on this core signal amplification module.

As originally proposed, a global module acts at the top of the hierarchy to provide directional information to orient polarization with respect to the tissue axes. In many vertebrate systems, PCP signaling relies on secreted Wnts, leading to the proposal that a global Wnt concentration gradient might directly provide such a directional cue (Goodrich & Strutt, 2011; Vladar et al., 2009; Wansleeben & Meijlink, 2011). The strength of the data for these assertions varies but in at least some cases makes a reasonably strong argument. In contrast, in *Drosophila*, strong evidence argues against a direct contribution of Wnts to planar polarity (Casal, Struhl, & Lawrence, 2002; Chen et al., 2008; Lawrence, Casal, & Struhl, 2002). Instead, global directional cues are provided by a system involving oppositely oriented gradients of differential gene expression across the tissue axes. This module comprises the proteins Fat and Dachsous (Ft and Ds; both atypical cadherins) and Four-jointed (Fj; a golgi ectokinase) (Strutt, 2009). Ft and Ds form heterodimers that can orient in either direction at a given cell–cell interface. Fj acts on both Ft and Ds, making Ft a stronger ligand for Ds and Ds a weaker ligand for Ft (Brittle, Repiso, Casal, Lawrence, & Strutt, 2010; Simon, Xu, Ishikawa, & Irvine, 2010). Ds and Fj are expressed in opposing gradients in each of the well-studied polarizing tissues in *Drosophila* (Casal et al., 2002; Ma, Yang, McNeill, Simon, & Axelrod, 2003; Yang, Axelrod, & Simon, 2002) and are proposed to result in a biased orientation of Ft–Ds heterodimers at intercellular boundaries reflecting the direction of the Fj and Ds expression gradients (Simon, 2004; Strutt, 2009). This mechanism produces a subtle gradient of Ft activity within each cell, and epistasis studies in the eye suggest that Ft provides the critical output signal from this module (Yang et al., 2002). In other words, the mechanism converts tissue-wide expression gradients into subcellular gradients of Ft activity.

A distinguishing characteristic of phenotypes displayed by global module mutants is the presence of significant defects in planar polarity at the level of orientation, but with essentially all cells achieving full molecular and morphological asymmetry, thus distinguishing these mutants from mutants in the core- and tissue-specific modules, in which molecular and morphological asymmetry are typically reduced or abolished. Perhaps, the most illustrative example is the phenotype seen in large *ft* clones in the wing (Ma et al., 2003). Within these clones, the prehairs form complex swirling patterns, whereas wild-type hairs form parallel arrays. Furthermore, molecular polarization at the level of the core proteins remains intact, although misoriented, indicating that the mutant cells polarize but no longer recognize the tissue axes (Ma et al., 2003). Additional evidence that the core module continues to polarize cells when the global module is disrupted comes from the observation that small *ft* mutant clones on *Drosophila* wings display normal polarity (Ma et al., 2003); the local alignment property of the core module aligns the mutant cells with each other (to form swirls) and with the polarity of the nearby wild-type cells outside the clone (to align with the tissue axis if the clone is small). Thus, the global module provides directionality to PCP but is not required for cell polarization *per se*, provided the core module is intact.

It is important to point out that, although the Ft/Ds/Fj global module as described contributes to PCP in all *Drosophila* tissues so far studied, several observations indicate that the current model is missing important pieces. When the Ds and Fj gradients are experimentally flattened, polarity of the ommatidia in the fly eye is severely disrupted as predicted (Simon, 2004). In contrast, planar polarity of the wing epithelium is more modestly affected, with much of the wing retaining a predominantly distal polarity. What accounts for the remaining distal directionality? While Ft is clearly important for global polarization, as evidenced by the disrupted polarity in loss-of-function *ft* clones, an extracellularly truncated fragment of Ft that cannot interact with Ds, and should therefore not be responsive to the Ds or Fj gradients, can rescue both the viability of *ft* mutant animals and produce distal polarity in a substantial, predominantly distal, portion of the wing (Matakatsu & Blair, 2006; our unpublished observations). From these observations, it is reasonable to conclude that there is an additional directional signal distinct from the gradients of Ds and Fj. Two more recent studies also provide support for the notion that additional polarizing signals are present, being organized by unknown cues at the dorsal–ventral boundary and perhaps also the anterior–posterior boundary (Brittle, Thomas, & Strutt, 2012; Sagner et al., 2012). The nature of such a signal(s) is unknown. Resolution of these mysteries awaits further studies, as does determining whether any of these signals contribute to global PCP signaling in vertebrates, and if so, how their role relates to that of Wnt proteins.

Potential mechanisms by which the global module, as well as additional unidentified global cues, might transmit directional information to the core module remain poorly understood. One plausible postulate derives from the observation of a polarized apical microtubule network that is seen to have a slight excess of plus-ends on one side of the cell (Eaton, Wepf, & Simons, 1996; Shimada, Yonemura, Ohkura, Strutt, & Uemura, 2006). Vesicles containing Fz have been observed to traverse the cell in a microtubule-dependent fashion, moving in a plus-end-biased direction (Shimada et al., 2006). Preliminary evidence suggests that the subcellular Ft gradient produced by the action of graded Ds and Fj expression (and recently visualized where the gradients are steep; Brittle et al., 2012) may produce the bias of this microtubule network. Perturbation of Ds, both by mutation and overexpression, produces alteration of microtubule orientation, providing at least some evidence that the global module orients the core module by organizing microtubules (Harumoto et al., 2010). Although appealing, this potential mechanism requires substantial additional validation. Whether additional unidentified global cues might work through regulating apical microtubules remains purely speculative.

At the bottom of the hierarchy reside the tissue-specific effectors. Distinct sets of components are responsible for translating the molecular asymmetry of the core PCP proteins to the specific polarized outputs required in each tissue. These range from asymmetric cytoskeletal organization to asymmetric cell fate determination. Mutating these components therefore causes planar-polarization defects of certain structures. Because these effectors are at the bottom of the hierarchy, their malfunction, in general, affects the polarized readouts without compromising the protein asymmetry, and thus the function, of the core PCP genes (Adler, Zhu, & Stone, 2004). Recently, some core PCP proteins have been suggested to be directly involved in polarizing cellular structures. For example, in vertebrate multiciliated cells, Dvl2 (a Dsh homologue) is associated with basal bodies of apical motile cilia, though direct evidence that it contributes to polarization of those cilia is thus far lacking (Park, Mitchell, Abitua, Kintner, & Wallingford, 2008).

2.2. Reasons to reconsider?

A three-tiered hierarchy was initially proposed based on a variety of observations. The core module was proposed to regulate the tissue-specific components in part based on their tissue specificity (Adler, Taylor, & Charlton, 2000; Lee & Adler, 2002; Strutt, Johnson, Cooper, & Bray, 2002) and on epistasis suggestive of this architecture (Lee & Adler, 2002; Yang et al., 2002). Upon its description, the global module was placed upstream of the core module by epistasis experiments in the eye (Yang et al., 2002) and by the simple yet powerful observation that core PCP proteins are incorrectly aligned within global mutant wing clones (Ma et al., 2003).

This three-tiered hierarchy model suggests a linear relationship between the global-, core-, and tissue-specific modules, in which the global module translates relatively shallow transcriptional gradients into subtle subcellular gradients, the core simultaneously amplifies subcellular asymmetry and locally aligns polarity, and the tissue-specific modules read out polarity cues to produce morphological or cell fate asymmetry. Though the linear relationship of the modules can be inferred from the above observations, the nature of the molecular interactions between the three tiers remains largely unknown, and the lack of detailed mechanistic knowledge of the information flow between modules leaves open the possibility of other architectures for the relationship between modules.

The most direct challenges to a strictly linear three-tiered hierarchy model come from genetic studies of denticle polarity in the *Drosophila* adult abdomen and larval epidermis (Casal, Lawrence, & Struhl, 2006; Donoughe & DiNardo, 2011; Repiso, Saavedra, Casal, & Lawrence, 2010). Two main observations have been proposed to be incompatible with the linear model. First, in both larval denticles and adult abdomen, double mutants constructed between components of the upstream global module and the core amplification module display stronger polarity defects than single mutants of each module alone (Casal et al., 2006; Donoughe & DiNardo, 2011; Repiso et al., 2010). Such enhancement of mutant phenotype has been argued to suggest that the upstream module and the core module can affect the downstream effectors (controlling denticle polarity) in parallel. Second, overexpression of upstream module components has been shown to alter denticle polarity in the abdomen even in the absence of an intact core signal amplification module (Casal et al., 2006). This was similarly interpreted to suggest the existence of a direct link from global directional cue to the tissue-specific polarity readout. While it is plausible that the linear three-tiered model is indeed an incorrect universal description of planar polarity signaling, as these interpretations suggest, we argue that there is also an important interpretive flaw that renders the conclusions of both experiments ambiguous. A parallel relationship between the modules was inferred from the absence of an epistasis relationship between the global and the core components;

however, the mutants of the core module selected for these experiments might not fully abolish core module function, and therefore, the observed results could be compatible with either a linear or nonlinear (parallel) architecture between the modules. Given the available evidence, we think that the linear model remains most valid as a blueprint of the relationship between modules, but it is likely that only the emergence of molecular-level understanding of signal transmission between modules will solidify (or eliminate) the linear three-tiered model. A more detailed discussion of this problem has been published elsewhere (Axelrod, 2009). A summary of the revised hierarchical model as described earlier can be found in Fig. 2.1.

3. ASYMMETRIC PROTEIN LOCALIZATION: A HALLMARK OF PCP

Because regular planar-polarized arrays of asymmetrically constructed cellular structures on the surface of epithelial cells have been appreciated for some time, the asymmetrically localized distribution of core PCP proteins suggested a striking feature of the PCP signaling mechanism that might underlie the molecular polarization of these cells. Indeed, we now believe that the segregation of these proteins to opposite sides of the cell is intimately linked to the mechanism that amplifies an initial input bias and locally aligns polarity between cells. Since its initial discovery in the *Drosophila* pupal wing epithelium (Axelrod, 2001; Strutt, 2001), similar asymmetrically localized distributions of core PCP proteins have been found in a substantial number of planar-polarized epithelial types in both invertebrates and vertebrates.

In the developing fly eye disk, PCP proteins are found differentially enriched at the adherens junction between the two cells that will differentiate asymmetrically to become the R3 and R4 photoreceptor cells in each developing ommatidium (Das, Reynolds-Kenneally, & Mlodzik, 2002; Rawls & Wolff, 2003; Strutt et al., 2002). The enrichment of the Fz/Dsh complex on the prospective R3 cell side is thought to impose an initial bias that inhibits Notch signaling in this cell, biasing the asymmetric, lateral inhibition-mediated R3/R4 fate decision between the two equipotent precursors (Strutt et al., 2002). Asymmetric PCP protein localization was also found in the sensory organ precursor (SOP) cells on the developing thorax (Bellaiche, Beaudoin-Massiani, Stuttem, & Schweisguth, 2004; Bellaiche, Gho, Kaltschmidt, Brand, & Schweisguth, 2001). Vang

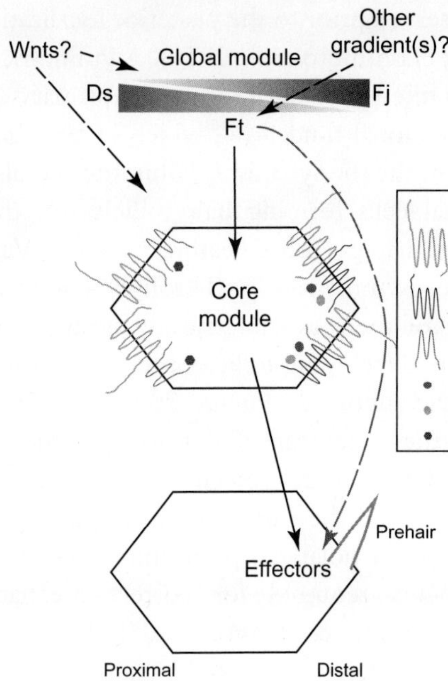

Figure 2.1 Hierarchical model of the PCP signaling pathway. The pathway consists of three modules, the global-, core-, and tissue-specific effector modules. According to the series model, the global module provides directional input to the core module (blue arrow) that establishes and amplifies subcellular asymmetry. This subcellular asymmetry is used to direct tissue-specific effector module function within the cell. According to the parallel model, the global module communicates directly with the tissue-specific effector module (green dashed arrow), without signaling to the core module. Directional information for the global module comes from tissue-level expression gradients of Ds and Fj, but it is likely that other gradients are also important, at least in the wing (black dashed arrow). While Wnt proteins seem not to play a direct role in PCP signaling in *Drosophila*, they appear to do so in vertebrates (gray dashed arrows). Precisely, how Wnts affect vertebrate PCP is unclear. Asymmetrically segregated core PCP proteins are shown. Various effector modules produce different tissue-specific responses. Here, effectors establish the distal location for growth of a prehair, as on the wing and abdomen. (See Color Insert.)

is enriched on the anterior cortex, while Fz is predominantly localized on the posterior side of the SOP cell. Their asymmetric localization predefines the anterior/posterior membrane domains, which subsequently determine the axis of SOP division and the asymmetric distribution of determinants of the daughter cell fates (Bellaiche et al., 2004, 2001).

The generality of such polarized protein localization was supported by recent investigations on various vertebrate epithelia. In the ventral node of early mouse embryos, restricted localization of Vangl1, Vangl2, Pk2,

Dvl2, and Dvl3 was observed prior to the posterior localization of the basal body (Antic et al., 2010; Hashimoto et al., 2010). Asymmetric localization of the basal body results in tilted growth of the motile primary cilium, thereby generating leftward directional fluid flow, which in turn gives rise to the left/right asymmetry of the body axis (Hashimoto et al., 2010; Song et al., 2010). Epidermal cells forming hair follicles on the mammalian skin also demonstrate planar-polarized features: Celsr1, Vangl2, and Fz6 were found to localize asymmetrically. Through a mechanism not yet understood, the action of these asymmetrically localized proteins results in asymmetry within the developing follicle, producing a planar-polarized tilt to hair growth (Devenport & Fuchs, 2008). In the auditory and vestibular epithelia in the inner ear of developing chicken and mouse embryos, asymmetric PCP protein crescents were observed both in the mechanosensing hair cells and the surrounding supporting cells, which is thought to be responsible for polarized positioning and orientation of the kinocilium and stereocilia required for correct mechanotransduction (Davies, Formstone, Mason, & Lewis, 2005; Deans et al., 2007; Montcouquiol et al., 2006; Wang, Guo, & Nathans, 2006).

Similar protein asymmetry was also evident in several epithelial types with multiple, planar-polarized, motile cilia. Ependymal cells on the ventricular lining of the brain, for example, demonstrate asymmetric Vangl2 localization, which is dependent on the Fmi orthologues Celsr2 and Celsr3 (Guirao et al., 2010; Tissir et al., 2010). Another well-characterized example is the multiciliated tracheal epithelial cells, where almost all known PCP proteins show asymmetric localization to opposite apical proximal and distal crescents, closely mimicking what we have learned from developing fly wing epithelium (Eszter Vladar et al., unpublished). It should be noted, however, that numerous other developmental events in vertebrates have been described to be under control of the PCP genes, or a set thereof, but for which robust asymmetric localization has not been observed (Vladar et al., 2009). These are typically nonepithelial, and how mechanistically similar they are to the epithelial PCP described here remains to be determined.

4. THE WAYS AND MEANS TO PLANAR POLARIZE A CELL: MECHANISMS OF ACHIEVING ASYMMETRY

The process of segregating the proximal and distal PCP proteins to opposite regions of the adherens junction creates distinct domains at the cell cortex. Achieving this segregation requires an energy investment to overcome entropy. In this section, we discuss the active mechanisms through which such asymmetry is achieved and maintained.

4.1. Required components as revealed by genetics

As discovered in the *Drosophila* wing epithelium, a well-recognized and apparently conserved feature of the asymmetric protein localization mechanism is interdependence among the core PCP proteins for their asymmetric localizations (Goodrich & Strutt, 2011). The fully manifest asymmetric localization of each core PCP protein depends on the intact function of each of the other core proteins, suggesting a tight feedback-based mechanism.

Six proteins have been identified as important for establishing PCP and whose localization satisfies these criteria. The atypical cadherin Flamingo (Fmi) was the first protein observed in a PCP-specific asymmetric localization, enriched on both the proximal and the distal cortices of every cell during planar polarization of the wing (Shimada, Usui, Yanagawa, Takeichi, & Uemura, 2001; Usui et al., 1999). Unipolar asymmetry was first independently seen for the seven-pass transmembrane protein Frizzled (Fz) (Strutt, 2001) and the cytosolic protein Disheveled (Dsh) (Axelrod, 2001), both of which are enriched on the distal cortex of each wing epithelial cell. This distally localized PCP complex was later found to include Diego, another cytosolic protein with ankyrin repeats (Das, Jenny, Klein, Eaton, & Mlodzik, 2004). On the opposite cell cortex, the four-pass transmembrane protein Vang (also known as Stbm) and the LIM-domain cytosolic protein Prickle (Pk) are enriched proximally (Bastock et al., 2003; Tree, Shulman, et al., 2002). Importantly, correct apical localization of all six of these core PCP proteins at the adherens junction depends on the presence and function of the others. Protein localization asymmetry builds up slowly during fly wing development, beginning in the third instar and showing the most prominent asymmetry during the hours just prior to the outgrowth of trichomes. After the planar-polarized outgrowth of wing hairs, asymmetry is quickly lost.

4.2. Domineering nonautonomy: How to talk to your neighbor

Well prior to the characterization of PCP mutants in flies and the availability of modern genetic tools to make mosaic clones, transplantation experiments in larger insects showed that planar polarities in neighboring cells can influence each other in a non-cell-autonomous manner (Lawrence, 1975).

Nonautonomy is now most readily observed using genetic mosaics to examine polarity in and around a clone of cells missing or overexpressing a PCP gene of interest. Of the six core proteins, loss-of-function clones of *fz* and *vang* strongly influence the polarity of neighboring nonmutant tissue, referred to as domineering nonautonomy (Adler et al., 2000; Casal et al., 2002; Taylor, Abramova, Charlton, & Adler, 1998; Vinson & Adler, 1987). These observations clearly indicate that polarity disruption at a clone boundary can propagate between cells and the influence of these effects can reach as far as tens of cells away from the clone boundary.

Associated with and underlying the nonautonomy seen in hair polarity patterns, Fz and Vang have been experimentally shown to recruit each other to adjacent sides of intercellular junctions in neighboring cells (Fig. 2.2A;

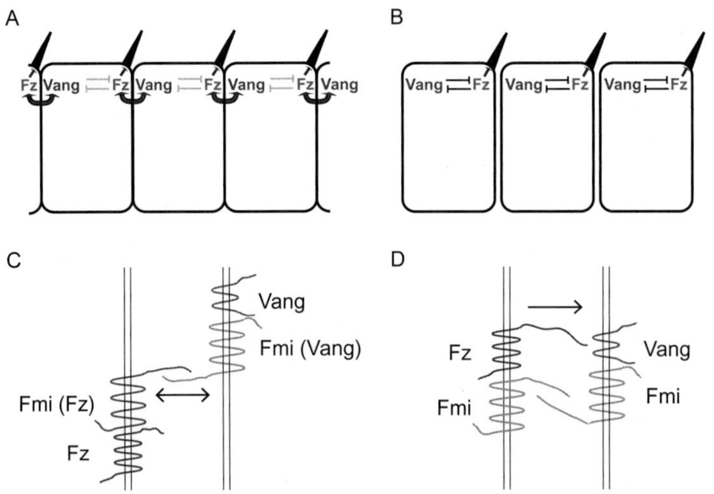

Figure 2.2 Generation and amplification of asymmetry by the core module. (A) Intercellular signaling. Mutual intercellular recruitment between Fz in one cell and Vang in the neighboring cell (red arrows) is required for core module function. Initially, Fz and Vang are distributed uniformly around the adherens junctions but become segregated to proximal and distal sides, as shown. (B) Intracellular signaling. Segregation of Fz and Vang requires mutual repulsion that requires activity of the cytoplasmic core proteins. The mechanism of this repulsion is not well defined. (C and D) Two models for the mutual intercellular recruitment between Fz and Vang. Chen et al. (2008) (C) propose that recruitment depends on asymmetric signaling through Flamingo homodimers in which each monomer adopts different functional states. The state associated with Vang (orange) is the basal state, while the state associated with Fz (red) is induced by interaction with Fz. No contact between Fz and Vang is required. In contrast, Wu et al. (2008) (D) propose unidirectional signaling that requires physical interaction between Fz and Vang, while symmetric Fmi homodimers scaffold the interaction. (For interpretation of the references to color in this figure legend, the reader is referred to the online version of this chapter.)

Amonlirdviman et al., 2005; Bastock et al., 2003; Chen et al., 2008; Strutt & Strutt, 2009). In wild-type polarized fly wing tissue, Fz crescents, on the distal edge of the cell, lie juxtaposed to the Vang-enriched crescents on the proximal edge of the neighboring cells. This relationship between Fz and Vang holds true for each case of clonal manipulation examined in fly wings: Fz or Vang is recruited to the clonal boundary when its counterpart is overexpressed within the clone and is absent on the clonal boundary if its counterpart is missing within the clone. A great deal of attention has been paid to this highly specific interaction between neighboring cells, which provides an elegant model to explain how a planar polarity signal propagates from cell to cell, serving to align polarity within a field and produce a coherent and error-free response (Amonlirdviman et al., 2005; Ma et al., 2003). Recent observations on both morphological and molecular polarity propagation in vertebrate systems suggest that this mechanism of domineering nonautonomy is conserved beyond the insect kingdom (Mitchell et al., 2009; Sienknecht et al., 2011).

Because of the central importance of this unique feature of PCP signaling, several groups independently investigated how specific interactions might mediate recruitment between the proximal and distal complexes. Through a variety of genetic experiments including manipulating PCP genes at clone boundaries, it was concluded that only three core PCP components are necessary for the mutual recruitment between Fz and Vang. Besides Fz and Vang themselves, the atypical cadherin Fmi is needed on both sides of the adherens junction for this interaction to occur (Chen et al., 2008; Strutt, 2001; Strutt & Strutt, 2007, 2008). As mutant forms of Fz and Vang lacking the majority of their extracellular domains can still recruit each other in the presence of Fmi and as Fmi in a cell lacking either Fz or Vang recruits Fz from the neighboring cell, Chen et al. (2008) favored the proposal that the signal for the mutual recruitment between distal and proximal complexes is transmitted via Fmi dimers formed at the adherens junction but not likely via direct interactions between Fz and Vang (Fig. 2.2C). This hypothesis raises the puzzling problem of how Fmi homodimers might transduce information asymmetrically such that a dimer associated with Fz on one side of the adherens junction selectively recruits Vang in the neighboring cell, and vice versa. Functional evidence for this asymmetry comes from the observation that the behavior of Fmi expressed in one cell to recruit Fz in the neighboring cell can be shifted by the presence of Fz to favor recruiting Vang in the neighboring cell (Chen et al., 2008). Though the physical basis for this asymmetry is not yet known, asymmetric conformations, posttranslational modifications, or unequal stoichiometries are possible explanations. While this asymmetric Fmi homodimer model remains our favored blueprint for further exploration of PCP signaling across the adherens junction, several groups have suggested alternative mechanisms. While it is universally agreed that the mutual recruitment between Fz and Vang complexes depends on the involvement of Fmi homodimers, there is disagreement regarding whether Fmi dimers asymmetrically transmit information between cells or act solely as a structural bridge to stabilize the Fz–Vang interaction, through which such information might flow. Wu et al. have presented evidence for a direct interaction between the extracellular CRD domain of Fz and Vang *in vitro* (Lawrence, Casal, & Struhl, 2004; Wu & Mlodzik, 2008). Physiological relevance of this interaction was argued based on the loss of PCP signaling observed when the CRD of Fz was replaced with that of the PCP-irrelevant *Drosophila* Fz2 gene. This negative result is discrepant with Chen et al.'s observation that an Fz derivative lacking the CRD can provide some PCP signaling activity. These dramatic differences in rescuing efficiency might reflect differences in protein expression or folding efficiency and remain an important puzzle to be resolved.

Another clear distinction between the two models just described involves the directionality of polarity information transmission. According to the asymmetric Fmi homodimer model, polarity information flows in both directions such that cells on either side respond to the other (Fig. 2.2C). The direct Fz–Vang interaction model as articulated by several groups argues for a unidirectional information flow in which Vang does not send information across the adherens junction but instead acts only as signaling receptor for the CRD domain of Fz (Fig. 2.2D) (Lawrence et al., 2004; Wu & Mlodzik, 2008). This unidirectional model, however, does not readily explain how (at borders of *fz–vang* twin clones) cells expressing Vang but not Fz cause repolarization of adjacent cells expressing Fz but not Vang (Strutt & Strutt, 2007).

A third, independent, investigation, by Strutt et al., contributed observations to this discussion that, while not providing resolution, must be accommodated by a correct model (Strutt & Strutt, 2008). Consistent with the asymmetric Fmi homodimer model, Strutt et al. presented evidence suggesting that Fmi exists in two alternative forms, preferentially interacting either with Fz or with Vang, with the bias involving differential binding to the C-terminal cytoplasmic tail of Fmi. Furthermore, they found that Fmi preferentially binds Fz, rather than Vang, when present in excess. This differential affinity is in accordance with the observation that when produced in excess, Fmi molecules behave as though associated with Vang and recruit Fmi–Fz complexes on the other side of the adherens junction (Chen et al., 2008). Furthermore, Strutt and Strutt observed that stable intercellular Fmi complexes form when neither cell expresses Vang and one cell expresses Fz, but not when neither cell expresses Fz and one expresses Vang. Because one side of the complex must have Fz, this result is consistent with the model of Chen et al. in which Fz induces a form of Fmi that selectively interacts with Fmi that is or is not associated with Vang.

The debate about how Fz and Vang recruit each other across intercellular boundaries is likely to continue as additional studies explore the interactions between Fz, Vang, and Fmi at the adherens junctions. We expect that elucidating the biochemical characteristics and conformational nature of Fmi dimers will be a key milestone pushing our understanding of planar polarity propagation to a new mechanistic level.

4.3. Autonomous choices: Focusing within a single cell

While much has been learned about the mutual recruitment between the distal Fz complexes and proximal Vang complexes (Fig. 2.2A), it has yet to be rigorously determined whether these cross-junctional interactions can be solely responsible for achieving asymmetric protein localization in a self-organizing way. Mathematical modeling can simulate the acquisition of PCP protein asymmetry based on the mutual recruitment between Fz- and Vang-containing complexes in neighboring cells. To do so, the model must include repulsive interactions between the proximal and distal complexes within the same cell, and indeed, data implicating such interactions in part motivated the initial proposal of the feedback loop (Tree, Shulman, et al., 2002). Additional genetic analyses have begun to define the component requirement and mechanism of this cell-autonomous repulsion (Fig. 2.2B). Three cytoplasmic core PCP genes, Dsh, Dgo, and Pk, which are not required for the assembly of asymmetric Fz-, Vang-, and Fmi-containing intercellular complexes, are almost certainly playing essential roles in the cell-autonomous repulsion between proximal and distal complexes required for amplifying asymmetry in a self-organizing fashion. When any of these three genes' function is compromised, the mutant cell loses PCP protein asymmetry in a largely cell-autonomous way.

These three cytoplasmic PCP genes also differ from Fz/Vang/Fmi in terms of the temporal requirement for their function (Strutt & Strutt, 2007). Fz/Vang/Fmi function is required from ~6 h APF in fly pupal development for proper planar polarization of the wing tissue. Although they may act earlier, proper polarization can be achieved when Pk and Dsh function are present only after 16 h APF.

Multiple lines of evidence indicate that the cytoplasmic core proteins stabilize or enhance accumulation of intercellular complexes, and because they act within the cell, the simplest explanation is that they act cell autonomously. The first comes from genetic experiments in which any one of these three PCP proteins is overexpressed within a clone of wing cells (Bastock et al., 2003; Das et al., 2004; Tree, Shulman, et al., 2002). A substantial enhancement of protein localization at adherens junctions within the clone is observed, for most, if not all, other core PCP proteins. This suggests that these cytoplasmic factors can reinforce stable intercellular complex formation, either directly or indirectly. Among these, it is especially interesting to point out that overexpressing Dgo autonomously enhances clustering of Fmi molecules at the adherens junction (Jenny, Reynolds-Kenneally, Das, Burnett, & Mlodzik, 2005). Notably, accumulation of PCP complexes is similarly nonuniform at wild-type junctions. These clusters likely resemble the recently reported discrete membrane domains undergoing unique protein turnover dynamics (Strutt, Warrington, & Strutt, 2011).

These observations of enhancement of cortical PCP complex localization do not offer any direct hints toward how the cytoplasmic PCP proteins promote the amplification of asymmetry during polarization. One hint comes from the finding that at least some of these factors demonstrate mutually exclusive biochemical interactions. For example, both Pk and Dgo bind to Dsh *in vitro*. However, the Dsh–Pk interaction is strongly inhibited by the presence of Dgo, likely through competition for the same Dsh binding site (Jenny et al., 2005). We are just at the beginning of understanding what is likely a complex core PCP protein interaction network. Furthermore, it will be a considerable challenge to correlate the *in vitro* findings to the physiological roles of any specific interaction *in vivo*, where the network structure includes feedback mechanisms making predictions of pattern outcomes nonintuitive. We postulate that at least some of these protein interactions involving the cytoplasmic core PCP factors mediate the cell-autonomous repulsive interactions between the Fz-containing distal complex and the Vang-containing proximal complex.

Additional cell-autonomous mechanisms are likely required by each polarizing cell to interpret and transduce directional signals from the global module, or from other sources of directional information, to downstream modules. Some recent results suggest that a polarized apical microtubule network in fly wing cells may be oriented by signals from the global module and in turn influence directional trafficking of vesicles containing core PCP proteins (Harumoto et al., 2010). Whether and how this aspect of cell-autonomous regulation contributes to the establishment of core PCP protein asymmetry awaits more detailed exploration.

4.4. Building a unifying mechanism

Given the robust establishment of asymmetric core PCP protein localization in numerous polarized cell types in the animal kingdom, we postulate that the set of cell-autonomous and nonautonomous mechanisms described in this review are conserved, at least in their essential mechanistic logic. During the decade, as PCP protein localization asymmetry was initially discovered, the contribution of this process to the establishment of asymmetry has been intensively studied. Mechanistic molecular details of many of these processes have begun to emerge. Yet, despite the considerable progress, a relatively large amount is still to be learned about this fascinating process.

Genetic experiments, relying heavily on clonal analyses, have led to a model of the core PCP mechanism in which asymmetric complexes assembled between neighboring cells transmit polarity information between cells, serving both to amplify asymmetry once it is initiated and to produce local polarity alignment between neighbors. Based on this basic biological principle, mathematical modeling, in various forms, has shown by mimicking the characteristic patterns of PCP protein localization in wild-type as well as mutant genetic mosaic wings that this kind of mechanism is a plausible description of how the core PCP module functions (Amonlirdviman et al., 2005; Burak & Shraiman, 2009; Le Garrec, Lopez, & Kerszberg, 2006; Schamberg, Houston, Monk, & Owen, 2010; Webb & Owen, 2004; Zhu, 2009). By continuing to combine modeling with biological experimentation, it should be possible to derive a much deeper understanding of the specific molecular interactions that underlie the feedback mechanism.

The nature of the global directional cue remains controversial. While we prefer a series model in which overall directional information comes from the Ft/Ds/Fj module and acts to bias the directionality of the core module, others argue for a parallel architecture in which each of these modules acts directly on the readout modules. In the series model, directionality comes from expression gradients, and while this could explain a variety of observations, it has shortcomings that must be addressed. In contrast, the parallel model requires that the core module acquire directionality via another, as yet undescribed, mechanism.

In the upcoming decade, we expect to see an even more integrative and interactive approach between experimental approaches including genetic, cell biological, and biochemical methods, and increasingly sophisticated mathematical modeling techniques. Increasingly, precise understanding will allow us to devise sophisticated genetic manipulations *in vivo*, which will aim to isolate and test a specific process that contributes to the polarization mechanism. More powerful experimental methods will begin to yield a much more detailed understanding of each molecular pathway and specific protein interaction, and ever more sophisticated modeling methods will contextualize their contributions to the PCP protein localization process and the eventual asymmetric outcome.

ACKNOWLEDGMENTS

Work in the Axelrod lab is supported by grants from NIH/NIGMS. We thank Dr. Yi Guo for her artistic input and assistance preparing the figures.

REFERENCES

Adler, P. N., Taylor, J., & Charlton, J. (2000). The domineering non-autonomy of frizzled and van Gogh clones in the Drosophila wing is a consequence of a disruption in local signaling. *Mechanisms of Development*, *96*, 197–207.

Adler, P. N., Zhu, C., & Stone, D. (2004). Inturned localizes to the proximal side of wing cells under the instruction of upstream planar polarity proteins. *Current Biology*, *14*, 2046–2051.

Amonlirdviman, K., Khare, N. A., Tree, D. R., Chen, W. S., Axelrod, J. D., & Tomlin, C. J. (2005). Mathematical modeling of planar cell polarity to understand domineering non-autonomy. *Science*, *307*, 423–426.

Antic, D., Stubbs, J. L., Suyama, K., Kintner, C., Scott, M. P., & Axelrod, J. D. (2010). Planar cell polarity enables posterior localization of nodal cilia and left–right axis determination during mouse and Xenopus embryogenesis. *PLoS One*, *5*, e8999.

Axelrod, J. D. (2001). Unipolar membrane association of Dishevelled mediates Frizzled planar cell polarity signaling. *Genes & Development*, *15*, 1182–1187.

Axelrod, J. D. (2009). Progress and challenges in understanding planar cell polarity signaling. *Seminars in Cell & Developmental Biology*, *20*, 964–971.

Bastock, R., Strutt, H., & Strutt, D. (2003). Strabismus is asymmetrically localised and binds to Prickle and Dishevelled during Drosophila planar polarity patterning. *Development*, *130*, 3007–3014.

Bellaiche, Y., Beaudoin-Massiani, O., Stuttem, I., & Schweisguth, F. (2004). The planar cell polarity protein Strabismus promotes Pins anterior localization during asymmetric division of sensory organ precursor cells in Drosophila. *Development*, *131*, 469–478.

Bellaiche, Y., Gho, M., Kaltschmidt, J. A., Brand, A. H., & Schweisguth, F. (2001). Frizzled regulates localization of cell-fate determinants and mitotic spindle rotation during asymmetric cell division. *Nature Cell Biology*, *3*, 50–57.

Brittle, A. L., Repiso, A., Casal, J., Lawrence, P. A., & Strutt, D. (2010). Four-jointed modulates growth and planar polarity by reducing the affinity of dachsous for fat. *Current Biology*, *20*, 803–810.

Brittle, A., Thomas, C., & Strutt, D. (2012). Planar polarity specification through asymmetric subcellular localization of Fat and Dachsous. *Current Biology*, *22*, 907–914.

Burak, Y., & Shraiman, B. I. (2009). Order and stochastic dynamics in Drosophila planar cell polarity. *PLoS Computational Biology*, *5*, e1000628.

Casal, J., Lawrence, P. A., & Struhl, G. (2006). Two separate molecular systems, Dachsous/Fat and Starry night/Frizzled, act independently to confer planar cell polarity. *Development*, *133*, 4561–4572.

Casal, J., Struhl, G., & Lawrence, P. A. (2002). Developmental compartments and planar polarity in Drosophila. *Current Biology*, *12*, 1189–1198.

Chen, W. S., Antic, D., Matis, M., Logan, C. Y., Povelones, M., Anderson, G. A., et al. (2008). Asymmetric homotypic interactions of the atypical cadherin flamingo mediate intercellular polarity signaling. *Cell*, *133*, 1093–1105.

Das, G., Jenny, A., Klein, T. J., Eaton, S., & Mlodzik, M. (2004). Diego interacts with Prickle and Strabismus/Van Gogh to localize planar cell polarity complexes. *Development*, *131*, 4467–4476.

Das, G., Reynolds-Kenneally, J., & Mlodzik, M. (2002). The atypical cadherin Flamingo links Frizzled and Notch signaling in planar polarity establishment in the Drosophila eye. *Developmental Cell*, *2*, 655–666.

Davies, A., Formstone, C., Mason, I., & Lewis, J. (2005). Planar polarity of hair cells in the chick inner ear is correlated with polarized distribution of c-flamingo-1 protein. *Developmental Dynamics*, *233*, 998–1005.

Deans, M. R., Antic, D., Suyama, K., Scott, M. P., Axelrod, J. D., & Goodrich, L. V. (2007). Asymmetric distribution of prickle-like 2 reveals an early underlying polarization of vestibular sensory epithelia in the inner ear. *The Journal of Neuroscience*, *27*, 3139–3147.

Devenport, D., & Fuchs, E. (2008). Planar polarization in embryonic epidermis orchestrates global asymmetric morphogenesis of hair follicles. *Nature Cell Biology*, *10*, 1257–1268.

Donoughe, S., & DiNardo, S. (2011). Dachsous and frizzled contribute separately to planar polarity in the Drosophila ventral epidermis. *Development*, *138*, 2751–2759.

Eaton, S., Wepf, R., & Simons, K. (1996). Roles for Rac1 and Cdc42 in planar polarization and hair outgrowth in the wing of Drosophila. *The Journal of Cell Biology*, *135*, 1277–1289.

Goodrich, L. V., & Strutt, D. (2011). Principles of planar polarity in animal development. *Development*, *138*, 1877–1892.

Guirao, B., Meunier, A., Mortaud, S., Aguilar, A., Corsi, J. M., Strehl, L., et al. (2010). Coupling between hydrodynamic forces and planar cell polarity orients mammalian motile cilia. *Nature Cell Biology*, *12*, 341–350.

Harumoto, T., Ito, M., Shimada, Y., Kobayashi, T. J., Ueda, H. R., Lu, B., et al. (2010). Atypical cadherins Dachsous and Fat control dynamics of noncentrosomal microtubules in planar cell polarity. *Developmental Cell*, *19*, 389–401.

Hashimoto, M., Shinohara, K., Wang, J., Ikeuchi, S., Yoshiba, S., Meno, C., et al. (2010). Planar polarization of node cells determines the rotational axis of node cilia. *Nature Cell Biology*, *12*, 170–176.

Jenny, A., Reynolds-Kenneally, J., Das, G., Burnett, M., & Mlodzik, M. (2005). Diego and Prickle regulate Frizzled planar cell polarity signalling by competing for Dishevelled binding. *Nature Cell Biology*, 7, 691–697.

Keller, R. (2002). Shaping the vertebrate body plan by polarized embryonic cell movements. *Science, 298*, 1950–1954.

Kelly, M., & Chen, P. (2007). Shaping the mammalian auditory sensory organ by the planar cell polarity pathway. *The International Journal of Developmental Biology, 51*, 535–547.

Lawrence, P. A. (1975). The structure and properties of a compartment border: The intersegmental boundary in Oncopeltus. *Ciba Foundation Symposium, 29*, 3–23.

Lawrence, P. A., Casal, J., & Struhl, G. (2002). Towards a model of the organisation of planar polarity and pattern in the Drosophila abdomen. *Development, 129*, 2749–2760.

Lawrence, P. A., Casal, J., & Struhl, G. (2004). Cell interactions and planar polarity in the abdominal epidermis of Drosophila. *Development, 131*, 4651–4664.

Le Garrec, J. F., Lopez, P., & Kerszberg, M. (2006). Establishment and maintenance of planar epithelial cell polarity by asymmetric cadherin bridges: A computer model. *Developmental Dynamics, 235*, 235–246.

Lee, H., & Adler, P. N. (2002). The function of the frizzled pathway in the Drosophila wing is dependent on inturned and fuzzy. *Genetics, 160*, 1535–1547.

Ma, D., Yang, C. H., McNeill, H., Simon, M. A., & Axelrod, J. D. (2003). Fidelity in planar cell polarity signalling. *Nature, 421*, 543–547.

Matakatsu, H., & Blair, S. S. (2006). Separating the adhesive and signaling functions of the Fat and Dachsous protocadherins. *Development, 133*, 2315–2324.

Mitchell, B., Stubbs, J. L., Huisman, F., Taborek, P., Yu, C., & Kintner, C. (2009). The PCP pathway instructs the planar orientation of ciliated cells in the Xenopus larval skin. *Current Biology, 19*, 924–929.

Montcouquiol, M., Sans, N., Huss, D., Kach, J., Dickman, J. D., Forge, A., et al. (2006). Asymmetric localization of Vangl2 and Fz3 indicate novel mechanisms for planar cell polarity in mammals. *The Journal of Neuroscience, 26*, 5265–5275.

Park, T. J., Mitchell, B. J., Abitua, P. B., Kintner, C., & Wallingford, J. B. (2008). Dishevelled controls apical docking and planar polarization of basal bodies in ciliated epithelial cells. *Nature Genetics, 40*, 871–879.

Rawls, A. S., & Wolff, T. (2003). Strabismus requires Flamingo and Prickle function to regulate tissue polarity in the Drosophila eye. *Development, 130*, 1877–1887.

Repiso, A., Saavedra, P., Casal, J., & Lawrence, P. A. (2010). Planar cell polarity: The orientation of larval denticles in Drosophila appears to depend on gradients of Dachsous and Fat. *Development, 137*, 3411–3415.

Sagner, A., Merkel, M., Aigouy, B., Gaebel, J., Brankatschk, M., Julicher, F., et al. (2012). Establishment of Global Patterns of Planar Polarity during Growth of the Drosophila Wing Epithelium. *Current Biology, 22*, 1296–1301.

Schamberg, S., Houston, P., Monk, N. A., & Owen, M. R. (2010). Modelling and analysis of planar cell polarity. *Bulletin of Mathematical Biology, 72*, 645–680.

Shimada, Y., Usui, T., Yanagawa, S., Takeichi, M., & Uemura, T. (2001). Asymmetric colocalization of Flamingo, a seven-pass transmembrane cadherin, and Dishevelled in planar cell polarization. *Current Biology, 11*, 859–863.

Shimada, Y., Yonemura, S., Ohkura, H., Strutt, D., & Uemura, T. (2006). Polarized transport of Frizzled along the planar microtubule arrays in Drosophila wing epithelium. *Developmental Cell, 10*, 209–222.

Sienknecht, U. J., Anderson, B. K., Parodi, R. M., Fantetti, K. N., & Fekete, D. M. (2011). Non-cell-autonomous planar cell polarity propagation in the auditory sensory epithelium of vertebrates. *Developmental Biology, 352*, 27–39.

Simon, M. A. (2004). Planar cell polarity in the Drosophila eye is directed by graded Four-jointed and Dachsous expression. *Development, 131*, 6175–6184.

Simon, M. A., Xu, A., Ishikawa, H. O., & Irvine, K. D. (2010). Modulation of fat: Dachsous binding by the cadherin domain kinase four-jointed. *Current Biology, 20*, 811–817.

Song, H., Hu, J., Chen, W., Elliott, G., Andre, P., Gao, B., et al. (2010). Planar cell polarity breaks bilateral symmetry by controlling ciliary positioning. *Nature, 466*, 378–382.

Strutt, D. I. (2001). Asymmetric localization of frizzled and the establishment of cell polarity in the Drosophila wing. *Molecular Cell, 7*, 367–375.

Strutt, D. (2009). Gradients and the specification of planar polarity in the insect cuticle. *Cold Spring Harbor Perspectives in Biology, 1*, a000489.

Strutt, D., Johnson, R., Cooper, K., & Bray, S. (2002). Asymmetric localization of frizzled and the determination of notch-dependent cell fate in the Drosophila eye. *Current Biology, 12*, 813–824.

Strutt, D., & Strutt, H. (2007). Differential activities of the core planar polarity proteins during Drosophila wing patterning. *Developmental Biology, 302*, 181–194.

Strutt, H., & Strutt, D. (2008). Differential stability of flamingo protein complexes underlies the establishment of planar polarity. *Current Biology, 18*, 1555–1564.

Strutt, H., & Strutt, D. (2009). Asymmetric localisation of planar polarity proteins: Mechanisms and consequences. *Seminars in Cell & Developmental Biology, 20*, 957–963.

Strutt, H., Warrington, S. J., & Strutt, D. (2011). Dynamics of core planar polarity protein turnover and stable assembly into discrete membrane subdomains. *Developmental Cell, 20*, 511–525.

Taylor, J., Abramova, N., Charlton, J., & Adler, P. N. (1998). Van Gogh: A new Drosophila tissue polarity gene. *Genetics, 150*, 199–210.

Tissir, F., Qu, Y., Montcouquiol, M., Zhou, L., Komatsu, K., Shi, D., et al. (2010). Lack of cadherins Celsr2 and Celsr3 impairs ependymal ciliogenesis, leading to fatal hydrocephalus. *Nature Neuroscience, 13*, 700–707.

Tree, D. R., Ma, D., & Axelrod, J. D. (2002). A three-tiered mechanism for regulation of planar cell polarity. *Seminars in Cell & Developmental Biology, 13*, 217–224.

Tree, D. R., Shulman, J. M., Rousset, R., Scott, M. P., Gubb, D., & Axelrod, J. D. (2002). Prickle mediates feedback amplification to generate asymmetric planar cell polarity signaling. *Cell, 109*, 371–381.

Usui, T., Shima, Y., Shimada, Y., Hirano, S., Burgess, R. W., Schwarz, T. L., et al. (1999). Flamingo, a seven-pass transmembrane cadherin, regulates planar cell polarity under the control of Frizzled. *Cell, 98*, 585–595.

Vinson, C. R., & Adler, P. N. (1987). Directional non-cell autonomy and the transmission of polarity information by the frizzled gene of Drosophila. *Nature, 329*, 549–551.

Vladar, E. K., Antic, D., & Axelrod, J. D. (2009). Planar cell polarity signaling: The developing cell's compass. *Cold Spring Harbor Perspectives in Biology, 1*, a002964.

Wada, H., & Okamoto, H. (2009). Roles of noncanonical Wnt/PCP pathway genes in neuronal migration and neurulation in zebrafish. *Zebrafish, 6*, 3–8.

Wang, Y., Guo, N., & Nathans, J. (2006). The role of Frizzled3 and Frizzled6 in neural tube closure and in the planar polarity of inner-ear sensory hair cells. *The Journal of Neuroscience, 26*, 2147–2156.

Wansleeben, C., & Meijlink, F. (2011). The planar cell polarity pathway in vertebrate development. *Developmental Dynamics, 240*, 616–626.

Webb, S. D., & Owen, M. R. (2004). Intra-membrane ligand diffusion and cell shape modulate juxtacrine patterning. *Journal of Theoretical Biology, 230*, 99–117.

Wu, J., & Mlodzik, M. (2008). The frizzled extracellular domain is a ligand for Van Gogh/Stbm during nonautonomous planar cell polarity signaling. *Developmental Cell, 15*, 462–469.

Yang, C. H., Axelrod, J. D., & Simon, M. A. (2002). Regulation of Frizzled by fat-like cadherins during planar polarity signaling in the Drosophila compound eye. *Cell, 108*, 675–688.

Zhu, H. (2009). Is anisotropic propagation of polarized molecular distribution the common mechanism of swirling patterns of planar cell polarization? *Journal of Theoretical Biology, 256*, 315–325.

CHAPTER THREE

Control of Cell Polarity and Asymmetric Division in *C. elegans*

Hitoshi Sawa[1]

Multicellular Organization Laboratory, National Institute of Genetics, 1111 Yata, Mishima, Japan
[1]Corresponding author: e-mail address: hisawa@nig.ac.jp

Contents

Abstract

During development of *Caenorhabditis elegans*, most somatic cells divide asymmetrically to produce daughter cells with distinct fates. A Wnt signaling pathway called Wnt/β-catenin asymmetry pathway controls both polarity of mother cells and distinct fates of daughter cells. Unlike the PCP pathway that regulates cell polarity in other organisms, this Wnt pathway in *C. elegans* requires β-catenin. However, similar to the PCP pathway, signaling components including Dishevelled proteins are asymmetrically localized to the cell cortex. I will review current knowledge about the mechanism of this regulation and how the orientation of cell polarity is controlled by Wnt proteins.

Current Topics in Developmental Biology, Volume 101
ISSN 0070-2153
http://dx.doi.org/10.1016/B978-0-12-394592-1.00003-X

© 2012 Elsevier Inc.
All rights reserved.

1. CORTICAL ASYMMETRY OF Wnt SIGNALING COMPONENTS

The mechanism of Wnt/β-catenin asymmetry pathway has been mostly studied in some specific cells: embryonic EMS blastomere, larval T cell, and somatic gonadal precursors (SGPs) (Mizumoto & Sawa, 2007b). These cells as well as most other mitotic cells in *Caenorhabditis elegans* divide along the anterior–posterior axis. Asymmetric division of EMS produces anterior MS and posterior E daughters that generate mesoderm and endoderm, respectively, while the postembryonic T cell divides asymmetrically to produce anterior hypodermal and posterior neural precursors. Among them, the T cell as well as seam cells (see below) are hypodermal cells and divide along planer rather than apical–basal orientation. Therefore, their polarity can be considered as planer polarity. In fact, similar to the PCP pathway, Frizzled receptors LIN-17 and MOM-5 are localized to the posterior cell cortex in the T cell and embryonic cells, respectively (Goldstein, Takeshita, Mizumoto, & Sawa, 2006; Park, Tenlen, & Priess, 2004) (Fig. 3.1). Similarly, Dishevelled proteins DSH-2 and MIG-5 (collectively called DSHs) are localized to the posterior cortex in the T, seam, and EMS cells (Mizumoto & Sawa, 2007a). However, involvements of the other core PCP components, such as homologues of Van Gogh, Prickle, and Flamingo in the asymmetric divisions have not been clearly demonstrated. Further, components of the canonical Wnt pathway, WRM-1/β-catenin, APR-1/APC, and PRY-1/Axin are involved in the pathway and localized to the anterior cell cortex, opposite sides of the DSH localization (Fig. 3.1B and C) (Mizumoto & Sawa, 2007a; Nakamura et al., 2005; Sugioka, Mizumoto, & Sawa, 2011; Takeshita & Sawa, 2005). Therefore, in terms of signaling components, the Wnt/β-catenin asymmetry pathway is similar to the canonical Wnt pathway, even though the asymmetric localization reminds us of the PCP regulation. These asymmetric localizations are regulated by Wnt and Frizzled proteins. However, detailed molecular mechanisms remain elusive. Unlike β-catenin of other species, WRM-1 does not have GSK3β phosphorylation sites and does not bind to HMR-1/cadherin (Korswagen, Herman, & Clevers, 2000; Natarajan, Witwer, & Eisenmann, 2001). Therefore, it is still a mystery how WRM-1 localizes to the cortex and how it is regulated by Wnts.

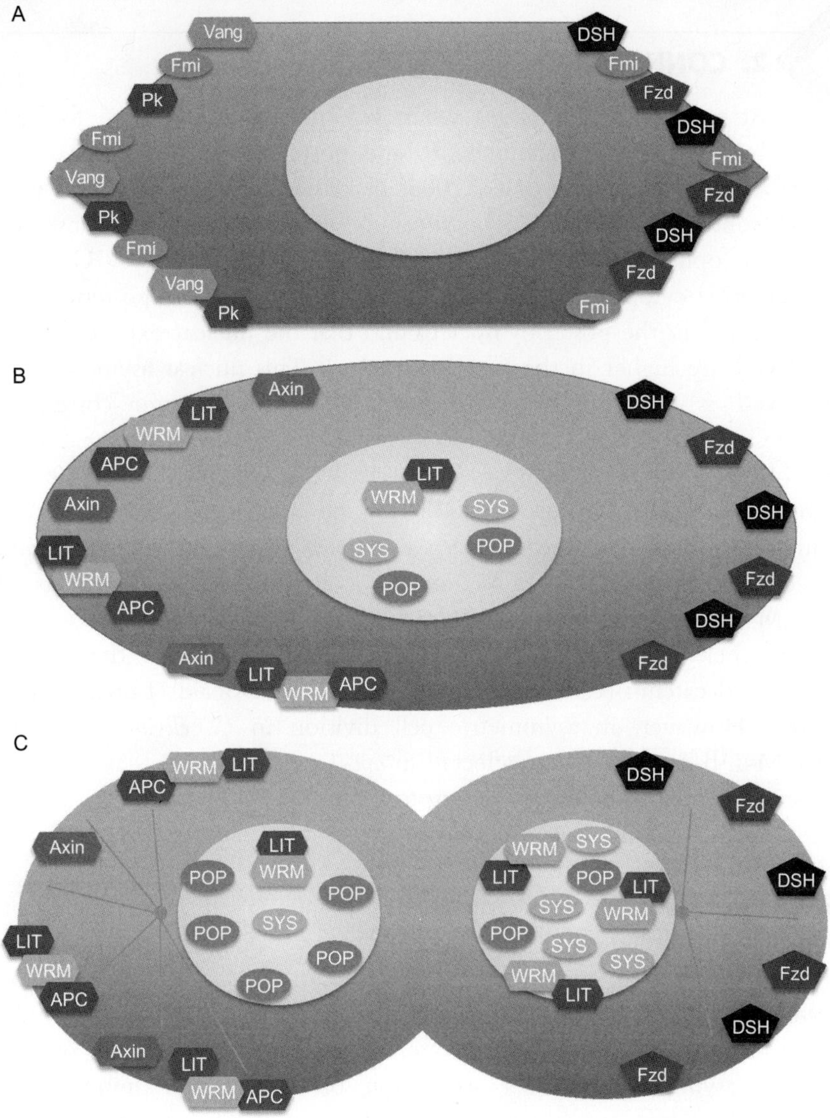

Figure 3.1 Asymmetric localizations of signaling components. (A) The PCP pathway in *Drosophila* wing. (B) and (C) the Wnt/β-catenin asymmetry pathway in *C. elegans* before cell division (B) and at telophase (C). Fzd, Frizzled; DSH, Dishevelled; Vang, Van Gogh; Pk, Prickle; Fmi, Flamingo; WRM, WRM-1; LIT, LIT-1; SYS, SYS-1; POP, POP-1. Green lines represent spindle microtubules. (For color version of this figure, the reader is referred to the online version of this chapter.)

2. CONNECTING CELL CORTEX TO NUCLEI

At telophase of divisions when two daughter nuclei are formed, WRM-1 localized preferentially to the posterior than anterior nuclei (Fig. 3.1C) (Takeshita & Sawa, 2005; Nakamura et al., 2005). This is in good contrast to its anterior cortical localization that is still observed during telophase. Photobleaching experiments revealed that WRM-1 in the anterior cytoplasm and nucleus as well as that in the posterior side accumulates in the posterior nucleus and that the nuclear export rates of WRM-1 are higher in the anterior nucleus. This nuclear asymmetry of WRM-1 is regulated by WRM-1 itself on the anterior cortex, as expression of WRM-1::CAAX that uniformly localized to the cortex inhibits WRM-1 localization in both nuclei (Mizumoto & Sawa, 2007a). Cortical WRM-1 recruits APR-1 to the anterior cortex. In *apr-1* mutants, WRM-1 nuclear export is inhibited, resulting in its localization in both nuclei. Thus, APR-1 on the cortex mediates the effects of WRM-1 in the inhibition of WRM-1 nuclear localization.

In other organisms, it is well known that APC functions in the degradation of β-catenin (Cadigan & Peifer, 2009; MacDonald, Tamai, & He, 2009). However, in asymmetric cell division in *C. elegans*, levels of WRM-1/β-catenin are not affected in *apr-1* mutants. APC is also known to stabilize microtubules by binding to their plus ends in mammalian cells (Dikovskaya, Zumbrunn, Penman, & Näthke, 2001). Although this function of APC has not been shown to regulate β-catenin, we have recently showed that APC regulates β-catenin nuclear localization through microtubules in the EMS blastomere (Sugioka et al., 2011). APR-1 on the anterior cortex stabilizes astral microtubules, creating asymmetry of spindle (more astral microtubules from the anterior spindle pole than from the posterior one) (Fig. 3.1C). Disruption of this spindle asymmetry by laser irradiation of the anterior spindle pole disrupted nuclear asymmetry of WRM-1, while the enhancement of the spindle asymmetry by the posterior irradiation caused concomitant increase of WRM-1 nuclear asymmetry. Further, the posterior irradiation in *mom-2*/Wnt mutants in which asymmetry of spindle and nuclear WRM-1 is disrupted rescued asymmetric POP-1/TCF localization (see below) regulated by nuclear WRM-1. These results showed that spindle microtubules stabilized by APR-1 enhance export of WRM-1 from the anterior nucleus, creating its nuclear asymmetry. How spindle regulates

WRM-1 nuclear export is not known. The requirements of kinesin for WRM-1 localization raised the model that microtubule-dependent transport of WRM-1 toward the cell cortex removes it from the perinuclear region, enhancing its nuclear export (Sugioka et al., 2011).

3. ASYMMETRIC REGULATION OF TRANSCRIPTION IN NUCLEI

In other organisms, β-catenin functions as a coactivator of transcription factors of the TCF/LEF family. However, WRM-1 in *C. elegans* has a weak, if any, activity to enhance transcription by POP-1/TCF (Korswagen et al., 2000; Natarajan et al., 2001). Instead, WRM-1 in the nucleus regulates nuclear export of POP-1 through LIT-1 kinase that is homologous to Nemo-like kinase in mammals (Lo, Gay, Odom, Shi, & Lin, 2004) (Fig. 3.2). WRM-1 can bind to both LIT-1 and POP-1 (Rocheleau et al., 1999; Yang et al., 2011). Upon WRM-1 binding, LIT-1 is activated to phosphorylate POP-1. Phosphorylated POP-1 binds to PAR-5/14-3-3 and is exported out of nucleus (Lo et al., 2004). Similar to WRM-1, LIT-1 is also enriched to the posterior nucleus after asymmetric divisions, probably through its tight binding to WRM-1 (Lo et al., 2004; Takeshita & Sawa, 2005). Therefore, the phosphorylation and nuclear export of POP-1 occurs asymmetrically in the posterior nucleus, creating high/low POP-1 asymmetry (higher in the anterior nucleus) that is opposite to the nuclear low/high WRM-1 asymmetry (higher in the posterior nucleus) (Fig. 3.1C) (Lin, Thompson, & Priess, 1995).

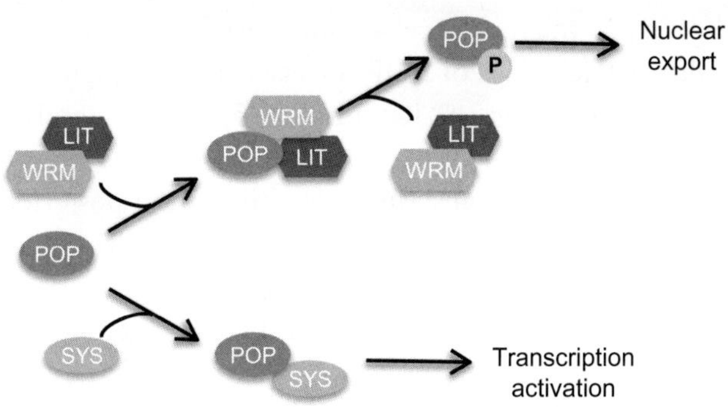

Figure 3.2 Regulation of transcription by POP-1. Free POP-1 binds either to SYS-1 to activate transcription or to the WRM-1/LIT-1 complex to be exported out of nuclei. (For color version of this figure, the reader is referred to the online version of this chapter.)

Including WRM-1, *C. elegans* has four β-catenins, one of them, SYS-1, functions as a coactivator of POP-1 in the Wnt/β-catenin asymmetry pathway (Huang, Shetty, Robertson, & Lin, 2007; Phillips, Kidd, King, Hardin, & Kimble, 2007). Similar to WRM-1 and LIT-1, SYS-1 is asymmetrically enriched in the posterior than in the anterior nuclei (Fig. 3.1). This SYS-1 asymmetry requires Wnt and Dishevelled proteins and suggested to be regulated through asymmetric protein degradation, raising the possibility that the regulation of SYS-1 is similar to that of the canonical Wnt pathway, even though SYS-1 does not have phosphorylation sites of GSK3β that is required for degradation of β-catenin in other species. As the SYS-1 asymmetry does not require WRM-1 or LIT-1, the Wnt/β-catenin asymmetry pathway independently regulates nuclear asymmetry of SYS-1 and POP-1. This results in reciprocal asymmetry of POP-1 and its coactivator SYS-1.

The reciprocal asymmetry of POP-1 and SYS-1 as well as WRM-1 appears to be counterintuitive to the observation that POP-1 binds to both SYS-1 and WRM-1. It was recently shown that the reciprocal asymmetry is achieved in the following way (Fig. 3.2) (Yang et al., 2011). First, binding of SYS-1 and WRM-1 is mutual inhibitory so that only free POP-1 unbound to SYS-1 but not the POP-1–SYS-1 complex binds to WRM-1 and is exported out of nuclei. Second, phosphorylated POP-1 has a weaker affinity to WRM-1. Therefore, upon phosphorylation, POP-1 is released from the WRM-1–LIT-1 complex, resulting in its nuclear export without coexport of WRM-1 and SYS-1. This mechanism allows posterior nucleus to have mostly the POP-1–SYS-1 complex but not free POP-1.

It was shown that the reciprocal asymmetry of nuclear POP-1 and SYS-1 produces distinct activities of POP-1 between the daughter cells (Kidd, Miskowski, Siegfried, Sawa, & Kimble, 2005). In the anterior daughter with high POP-1 and low SYS-1, most POP-1 is free from SYS-1 and functions as a repressor of transcription, while in the posterior daughter, most POP-1 binds to SYS-1 and functions as a transcriptional activator. Such difference of POP-1 functions produces distinct fates of the daughter cells. For example, after the EMS division, POP-1 represses transcription of the *end-1* gene that promotes endoderm formation in the anterior MS daughter that produces mesoderm, while activating it in the posterior E daughter that becomes endoderm (Maduro, Lin, & Rothman, 2002; Shetty, Lo, Robertson, & Lin, 2005).

4. SPINDLE ORIENTATION AND MTOC POSITIONING

In other organisms, the PCP pathway controls spindle orientation. In *C. elegans*, some components of the Wnt/β-catenin asymmetry pathway regulate spindle orientation in at least some of the asymmetrically dividing cells. In the normal EMS cell, spindle is initially aligned along the left–right axis and rotates to adopt the anterior–posterior orientation until anaphase. In mutants of *mom-2*/Wnt and *mom-5*/Fzd, spindle fails to rotate (Schlesinger, Shelton, Maloof, Meneghini, & Bowerman, 1999). In addition, two of the three Dishevelled homologues (DSH-2 and MIG-5) redundantly control spindle orientation in the EMS cell (Walston et al., 2006). In contrast, spindle is normal in *wrm-1*/β-catenin and *apr-1*/APC mutants. Therefore, the pathway appears to be split into three at the DSH proteins: regulation of spindle orientation, WRM-1 localization, and SYS-1 localization. Although *gsk-3*/GSK3β was previously reported to be required for EMS spindle orientation (Schlesinger et al., 1999), *gsk-3(RNAi)* embryos have earlier defects in the transition from meiosis to mitosis in fertilized eggs (Nishi & Lin, 2005; Shirayama et al., 2006). Therefore, its function in the EMS division needs to be reexamined.

The EMS spindle orientation is also controlled by another cell–cell interaction between the EMS and P2 blastomeres that involves a transmembrane protein MES-1 and SRC family tyrosine kinase SRC-1 (Bei et al., 2002). MES-1 is required in both EMS and P2, while SRC-1 is required in EMS. Interestingly, both MES-1 and SRC-1 regulate polarity and spindle orientation in P2 in response to MES-1 itself in EMS (Arata, Lee, Goldstein, & Sawa, 2010; Berkowitz & Strome, 2000). It has been proposed that a homophilic interaction between MES-1 at the boundary of EMS and P2 aligns spindle of EMS and P2 in the same orientation, which resembles the PCP regulation (Arata et al., 2010). However, the mechanism of spindle regulation appears to be distinct between these cells. It has been recently reported that GPR (G protein regulator; homologues of Pins in *Drosophila*) proteins that control spindle pulling forces preferentially localize to the EMS–P2 boundary in P2 but not in EMS (Werts, Roh-Johnson, & Goldstein, 2011). Therefore, in P2, spindle alignment is likely to be regulated by cortical pulling forces produced by GPR proteins on the P2–EMS boundary. However, the mechanism in EMS is still obscure.

During telophase of the EMS division, the posterior but not the anterior centrosome–nuclear complex moves toward the posterior cell cortex to anchor the nucleus to the cortex in the posterior E daughter cell (Sugioka & Sawa, 2010). This process is also regulated by Wnt and MES-1/SRC-1 signaling. Although this nuclear anchoring is not essential in Wnt signaling, it has significant effects on POP-1/TCF nuclear localization in the *mom-2*/Wnt mutant background.

The effects of Wnt signaling in spindle orientation have recently been reported also in seam cells that are considered to be epithelial stem cells because of their self-renewing asymmetric divisions (Wildwater, Sander, de Vreede, & van den Heuvel, 2011). The anterior–posterior orientations of seam cell divisions are not affected even in quintuple Wnt mutants that have mutations in all the five Wnt genes (Yamamoto, Takeshita, & Sawa, 2011). However, the orientations become abnormal in situations which perturb both Wnt signaling and shape of seam cells that is normally elongated along the anterior–posterior axis, for example, in double mutants between *mig-14*/Wntless required for Wnt secretion and *dpy-10* whose mutations cause shorter body and rounded shape of seam cells. The results indicate that orientations of seam cell divisions are redundantly regulated by Wnt signaling and their elongated shape.

5. Wnts INSTRUCTIVELY ORIENT POLARITY OF THE EMS AND T CELLS

Important feature of the PCP regulation is the coordination of cell polarity in a group of cells or tissues (Gray, Roszko, & Solnica-Krezel, 2011). During *C. elegans* development, most cell divisions occur along the anterior–posterior axis with asymmetric localization of Wnt signaling components described above. In particular, POP-1 asymmetry that is higher in the anterior daughter nuclei has been observed in most embryonic and postembryonic cells. Therefore, most *C. elegans* cells are polarized in the same orientation along the anterior–posterior axis at least when they divide.

One exciting possibility for polarity coordination is that Wnt signals function as a global cue for orienting cell polarity. Wnts were first shown to orient polarity in the embryonic EMS and postembryonic T cells (Goldstein et al., 2006). Blastomere isolation and recombination experiments have shown that EMS polarity is determined by the position of the P2 blastomere, as recombining EMS and P2 results in one of the EMS daughters attaching to P2 to produce endoderm (Goldstein, 1993). Although P2–EMS signaling involves two proteins MES-1 and MOM-2/Wnt, it was shown that MOM-2 is the orienting signal by recombining two P2 (one lacks the *mom-2* function and the other lacks the *mes-1* function) to the opposite sides of EMS (Fig. 3.3A). This results in the endoderm

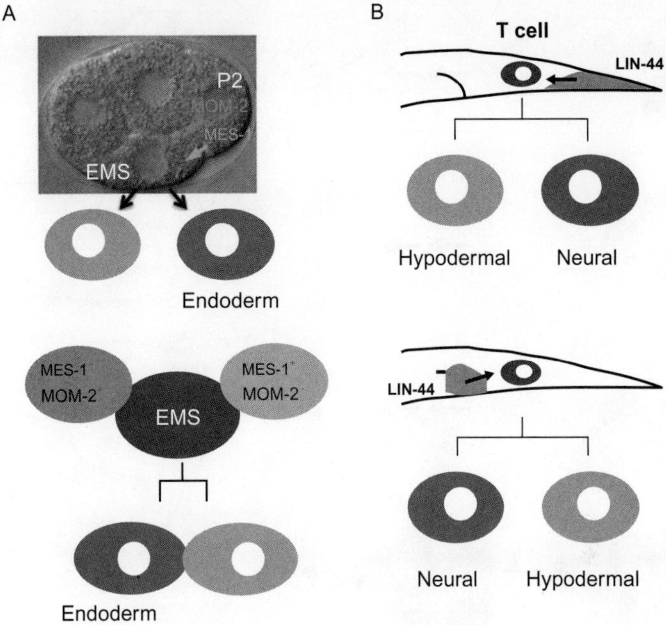

Figure 3.3 Wnts function as polarity cues. (A) The EMS polarity is controlled by the MES-1 and MOM-2/Wnt signals from the P2 cell. When isolated wild-type EMS is combined with two P2 cell mutants of *mes-1* or *mom-2*, endoderm produced from an EMS daughter produced on the side with *mes-1* P2 was attached. (B) The T-cell polarity is controlled by LIN-44 from its posterior neighbor cell. When LIN-44 is expressed on the anterior side of the T cell in *lin-44* mutants, the T-cell polarity is nearly completely reversed. (See Color Insert.)

production from the EMS daughter contacting with *mom-2*-positive P2 lacking *mes-1*, showing that *mom-2* rather than *mes-1* functions as a polarity cue. Postembryonic T-cell polarity is controlled by *lin-44*/Wnt that is expressed in hypodermal cells just posterior to the T cell (Fig. 3.3B). Ectopic expression of *lin-44* anterior to the T cell in the absence of endogenous *lin-44* function causes strong reversal of the T-cell polarity. These observations show that Wnt signals determine the orientation of cell polarity at least when Wnts are produced close to the polarizing cells.

6. COORDINATION OF SPINDLE ORIENTATION BY MOM-2/Wnt

In terms of global cell polarization, at least spindle orientation in embryonic cells was shown to be coordinated by the Wnt signal from the P2 blastomere or its descendants that positions at the posterior end of embryos (Bischoff & Schnabel, 2006). The progeny of isolated AB cell daughters

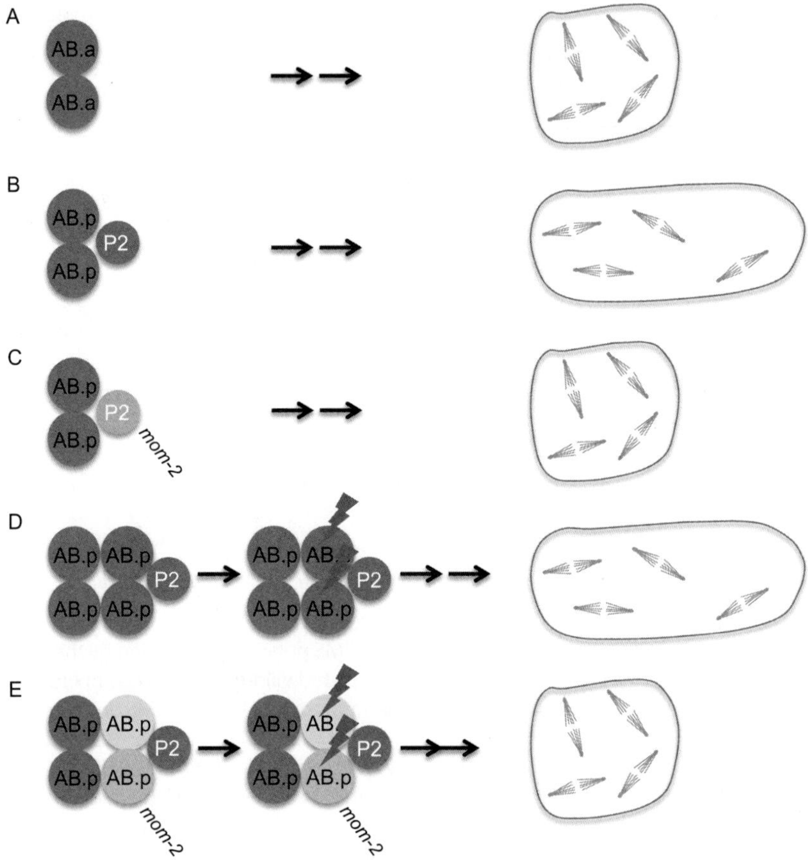

Figure 3.4 Coordination of spindle orientation by MOM-2/Wnt. (A) Isolated AB blastomere cultured alone produce embryonic fragments without elongation and with random spindle orientation (without the *glp-1* signal from P2, AB daughters adopt the AB.a fate). Embryonic fragments elongated with biased spindle orientation toward P2 when AB is combined with wild-type P2 (B) but not with *mom-2* P2 (C). The elongation occurs when two ABs were combined followed by laser ablation of the central AB daughters (D) but not when central AB is from a *mom-2* embryo (E). (For color version of this figure, the reader is referred to the online version of this chapter.)

without P2 cell divides in random orientation (Fig. 3.4A). However, combining the P2 blastomere with AB causes division orientations of AB descendants as well as those of P2 to be biased toward P2, resulting in elongation of the embryonic fragment (Fig. 3.4B). This signaling requires MOM-2/Wnt in P2 and MOM-5/Frizzled in AB daughters (Fig. 3.4C). When P2 is combined with two pairs of AB daughters followed by laser ablation of an AB daughter pair proximal to P2, progeny of the AB daughters that have not directly contacted with P2 undergoes oriented cell divisions, suggesting that the P2 signal works from a distance (Fig. 3.4D). Interestingly, this distant effect of P2 signal was not observed when the AB daughters that contact with P2 do not have *mom-2*/Wnt function (Fig. 3.4E), suggesting that the Wnt signal is transmitted by a relay mechanism so that cells that received the MOM-2 signal send the MOM-2 signal to their neighbors to align spindle orientation toward the signaling cells. As *mom-2* homozygous embryos from heterozygous mothers have normal embryogenesis, the relay mechanism is unlikely to depend on transcription of the *mom-2* gene in the signal-receiving cells which instead should utilize preexisting MOM-2 protein or mRNA maternally supplied by the mother to relay the signal to their neighbors. As recent single molecule mRNA FISH analyses showed that *mom-2* transcripts are nearly restricted to the posterior P1-derived cells (Harterink et al., 2011), AB progeny may secrete preexisting MOM-2 protein only after the reception of the MOM-2 signal.

7. COORDINATION OF ASYMMETRIC CELL DIVISIONS DURING EMBRYOGENESIS

In terms of asymmetric cell fate determination represented by POP-1 asymmetry in each pair of daughter cells, *mom-2* has a minor role in the third divisions of AB that are the first divisions with POP-1 asymmetry in this lineage (Fig. 3.5A), even though, in the blastomere recombination experiments, their mother cells (AB granddaughters: AB^4) can properly respond to *mom-2* activity from the P2 or C blastomeres to create POP-1 asymmetry (Park & Priess, 2003). It was suggested that this is due to *mom-2*-independent activity (possibly dependent on Wnts other than *mom-2*) of the MS blastomere that also can induce POP-1 asymmetry in the AB lineage. After the next divisions of AB^8 cells, POP-1 asymmetry is also independent of *mom-2* and can be observed in the progeny of isolated AB that has no prior exposure to signals from P1 descendants (Fig. 3.5B). However, orientation of the asymmetry can be abnormal in *mom-2* mutants or in the progeny of isolated AB blastomere. In particular, when cells divide perpendicular to the cell cluster, the daughter proximal to the cell cluster tends to have higher POP-1 signal than the distal one that does not contact with the cluster (Park & Priess, 2003; Park et al., 2004) (Fig. 3.5B). MOM-5/Frizzled::GFP that localizes to the posterior cortex of dividing cell in normal

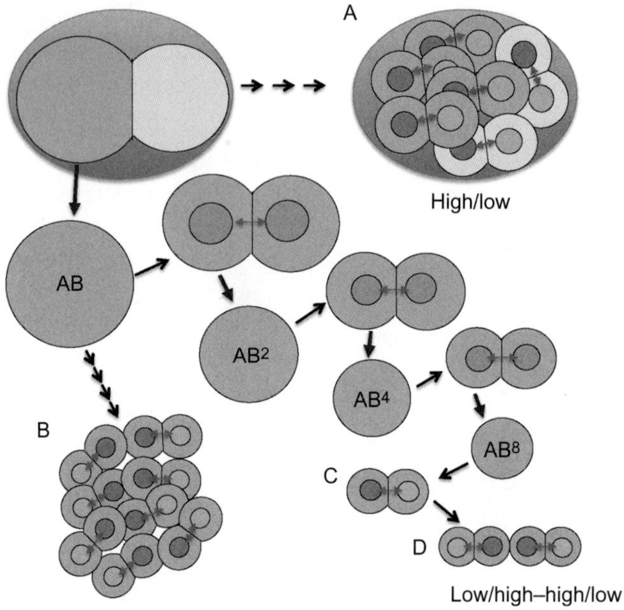

Figure 3.5 Regulation of POP-1 asymmetry in the AB lineage. In normal development, (A) POP-1 asymmetry is first observed after the division of AB granddaughters (AB^4). (B) When P1 is isolated and cultured alone, POP-1 asymmetry is observed after the division of AB^4, but the daughter proximal to the cell cluster tends to have higher POP-1 signal than the distal one that does not contact with the cluster. (C) The sequentially isolated AB^8 cell shows POP-1 asymmetry. (D) In the next division, AB^8 daughters show low/high–high/low POP-1 asymmetry. Green cells, pink double arrows, blue arrows, and black arrows represent the AB lineage, daughter pairs, cell isolation, and cell divisions, respectively. (See Color Insert.)

embryo, tends to be enriched on the distal sides of cells, consistent with the orientation of POP-1 asymmetry.

The POP-1 asymmetry was observed even in daughters of sequentially isolated AB^4 cell but not in those of AB or AB^2, indicating that cells can be autonomously polarized without extrinsic Wnt signals after the AB^8 stage (Fig. 3.5C). After the next division of sequentially isolated AB^4 cells, POP-1 asymmetry in daughter pairs becomes opposite to each other with low/high–high/low POP-1 localization which is not observed in normal development (Fig. 3.5D). This observation together with higher POP-1 in proximal daughters in isolated AB (Fig. 3.5B) suggests that polarity orientation is influenced by signals from or contacts with neighbor cells and that this mechanism should be suppressed to orient polarity of all cells in the same high/low POP-1 asymmetry during normal embryonic development.

8. COORDINATION OF ASYMMETRIC CELL DIVISIONS DURING POSTEMBRYONIC DEVELOPMENT

Similar to embryonic cell polarity, polarity orientations are coordinated with high/low POP-1 asymmetry after most postembryonic cell divisions through the functions of multiple Wnt genes (Yamamoto et al., 2011). Six seam cells (V1–V6) on the lateral side of animals undergo self-renewing asymmetric division. Even though these cells are aligned along the anterior–posterior axis and contact with each other, their polarity does not appear to depend on communication between neighboring cells as in the PCP regulation in *Drosophila* wing, as reversal of polarity in one of the seam cells does not affect that of its neighbors (Whangbo, Harris, & Kenyon, 2000; Yamamoto et al., 2011). Instead, the analyses of Wnt compound mutants showed that their polarity orientation is independently controlled by redundant functions of multiple Wnt genes (Fig. 3.6). Polarity of the V1–V4 seam cells is affected in triple Wnt mutants (*cwn-1, cwn-2 egl-20*) but not in the double mutants among them (minor defects were observed

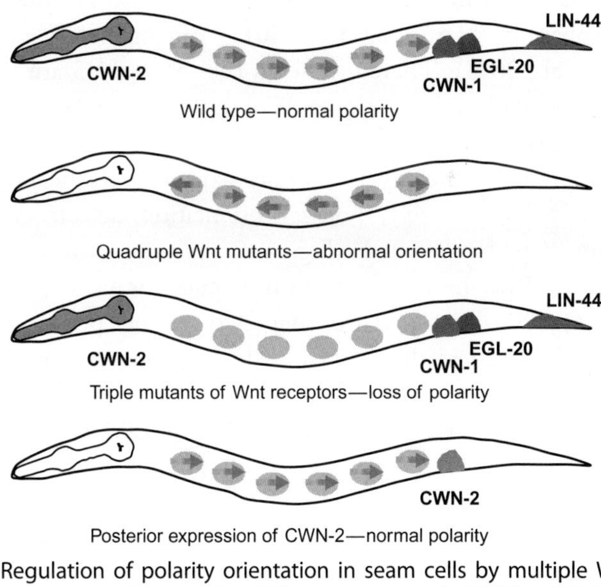

Figure 3.6 Regulation of polarity orientation in seam cells by multiple Wnts. In wild type, Wnts are expressed anteriorly (CWN-2) or posteriorly (CWN-1, EGL-20, LIN-44) to seam cells. In quadruple Wnt mutants as in quintuple ones, polarity orientations were randomized. In contrast, after knock down of three Wnt receptors, cells were not polarized. Posterior expression of CWN-2 rescued polarity defects in a compound Wnt mutant. (See Color Insert.)

for V4 in *cwn-1 egl-20*). Similarly, V6 polarity is disrupted in quadruple Wnt mutants (*lin-44, cwn-1, cwn-2 egl-20*), but not in any triple combinations. Therefore, these cells are properly polarized in the presence of one Wnt (*cwn-1, cwn-2,* or *egl-20* for V1–V4, and *lin-44, cwn-1, cwn-2, or egl-20* for V6). In contrast, V5 polarity is affected in *egl-20* single mutant (Whangbo et al., 2000), although the phenotype is modified by further mutations of *cwn-1* and *cwn-2.* In any combinations of Wnt mutations including quintuple Wnt mutants that have mutations in all five Wnt genes present in *C. elegans* genome, seam cells are mostly polarized in normal or reverse orientation, while loss of polarity is less frequent. The observation is consistent with apparently Wnt-independent POP-1 asymmetry in sequentially isolated AB^4 daughters (Fig. 3.5C). In contrast, in triple Wnt receptor mutants (*lin-17*/Fzd *mom-5*/Fzd, *cam-1*/Ror), all seam cells lose their polarity, resulting in symmetric division and loss of seam cells (Fig. 3.6). Therefore, Wnts are required for proper orientation of polarity, while their receptors function in the polarization process (in addition to orientation) even in the absence of Wnts.

Among these four Wnts, three (*lin-44*, *cwn*-1, and *egl-20*) are expressed in the posterior side of seam cells, while *cwn-2* is expressed anteriorly strongest around the pharynx (Harterink et al., 2011; Whangbo & Kenyon, 1999; Yamamoto et al., 2011). Therefore, seam cells except V5 are polarized in the normal orientation by a single Wnt gene that is expressed either anteriorly or posteriorly. When anteriorly (CWN-2) and posteriorly (CWN-1) expressed Wnts are ectopically expressed in the posterior and anterior regions, respectively, polarity orientation defects of the triple Wnt mutant can be rescued (Fig. 3.6) (Yamamoto et al., 2011), suggesting that Wnts may not function as cues for polarity orientation. Similarly, EGL-20 functions are suggested to be permissive, as the expression of EGL-20 in the pharynx using the *myo-2* promoter rescues polarity reversal of the V5 cell in *egl-20* mutants, although it was suggested that the *myo-2* promoter may drive EGL-20 expression in posterior regions close or posterior to V5, arguing against the validity of this experiment.

In contrast to these experiments that appear to suggest permissive functions of Wnts, LIN-44 instructively orients polarity of the T cell (a posterior seam cell), as described above. Similarly, it was suggested that EGL-20 instructs polarity orientation of the P7.p cell that positions anterior to the EGL-20-expressing cell, as its ectopic expression in the anteriorly positioned anchor cell causes polarity reversal in some genetic backgrounds (see below)

(Green, Inoue, & Sternberg, 2008). Even for seam cells, as described above, Wnts are required for proper polarity orientation of seam cells rather than for their polarization itself. To solve these contradictory observations, it has been proposed that functions of Wnts in polarity orientation may be determined by Wnt-expressing cells (Yamamoto et al., 2011). In this model, when CWN-1 is expressed anteriorly in the pharynx, it may function like CWN-2 rather than normal CWN-1 by the activities of Wnt-modifying enzymes or Wnt cofactors that are specifically expressed in the anterior but not the posterior Wnt-expressing cells. This model or other possibilities should be further explored to understand how the polarity orientation is coordinated in a number of cells during development.

9. POLARITY REVERSAL IN NORMAL DEVELOPMENT

Although most dividing cells show the same polarity orientation (high/low POP-1 asymmetry), two cell divisions show reversed (low/high) polarity. Both of them are involved in the production of mirror-symmetric structures of organs along the anterior–posterior axis.

The mirror-symmetric structure of the vulva is associated with mirror-symmetric lineages of its precursor cells, P5.p and P7.p (Sternberg, 2005). While daughters of P5.p show high/low POP-1 asymmetry as in other daughter pairs, P7.p daughters show reversed low/high POP-1 asymmetry (Deshpande, Inoue, Priess, & Hill, 2005). The polarity reversal is caused by functions of MOM-2/Wnt and LIN-44/Wnt that is expressed in the gonadal anchor cell anteriorly to P7.p but posteriorly to the P5.p cell (Green et al., 2008). These Wnts antagonize the instructive functions of EGL-20/Wnt that promotes high/low POP-1 asymmetry. The P7.p cell appears to discriminate EGL-20 from the posterior region and LIN-44 and MOM-2 from the anterior cell using distinct Wnt receptors, CAM-1/Ror for EGL-20 and LIN-17/Fzd+LIN-18/Derailed for LIN-44+MOM-2. This is in good contrast with seam cells in which functions of three receptors (LIN-17/Fzd, MOM-5/Fzd, and CAM-1/Ror) appear to be redundant.

Another mirror-symmetric organ is gonad that consists of germ cells and somatic gonad. Similar to the vulva, lineage of SGPs Z1 and Z4 are mirror-symmetric with reversal of POP-1 asymmetry in Z1 (low/high) compared with most other cells including Z4 (high/low) (Siegfried, Kidd, Chesney, & Kimble, 2004). The polarity of Z1 and Z4 does not appear to require Wnts, as it is not affected in quintuple Wnt mutants or mutants of *mom-1*/Porcupine that is required for secretion of Wnts (Siegfried et al., 2004; Yamamoto et al., 2011). Instead, it was suggested that the polarity is regulated, at least partly, by signals from the germ cells (Z2 and Z3) that position between Z1 and Z4, as germ cell-less mutants show abnormal polarity of Z1 and Z4 (Yamamoto et al., 2011). Although this germ cell signal has not been identified, it was reported recently that mRNA of *mom-2*/Wnt is present specifically in the germ cells of L1 larvae (Harterink et al., 2011). Considering that a hypomorphic *mom-2* mutation was used in quintuple Wnt mutants which have normal polarity of Z1 and Z4, it may still be possible that *mom-2* in the germ cells causes polarity reversal of Z1 as it does so in P7.p in the vulva. However, in this case, *mom-2* from germ cells should induce opposite polarity (POP-1 is higher in the daughter proximal to the *mom-2*-expressing germ cells) compared to that induced by *mom-2* in EMS or P7.p (POP-1 is higher in the daughter distal to the *mom-2*-expressing cells). Therefore, it is more plausible to imagine that non-Wnt signals from the germ cells orient polarity of Z1 and Z4.

10. FUNCTIONS OF PCP COMPONENTS IN *C. ELEGANS*

Although the coordination of spindle orientation and polarity in *C. elegans* is conceptually similar to the PCP regulation in *Drosophila* and vertebrates, knowledge about functions of PCP-specific components, such as Van Gogh and Prickle, in the processes described above has been limited so far. RNAi of *prkl-1*/Prickle weakly affects polarity of male B cell division that also involves POP-1 asymmetry (Wu & Herman, 2006). A *vang-1*/Van Gogh mutation can suppress abnormal polarity of P7.p caused by *lin-17*/Fzd mutations, although the *vang-1*/Van Gogh mutation by itself does not affect P5.p or P7.p polarity (Green et al., 2008). As deletion mutants of *vang-1*, *prkl-1*, and *fmi-1*/Flamingo are viable, it is unlikely that these genes are globally required for polarity regulation.

vang-1 is also involved in morphogenesis of intestine (Hoffmann, Segbert, Helbig, & Bossinger, 2010). In *vang-1* mutants, some parts of the intestinal lumen, which is normally surrounded by two cells, are instead surrounded by three cells (Fig. 3.7A). As intestine morphogenesis involves cell intercalation, it was suggested that *vang-1* might regulate cell intercalation as the case for its vertebrate homologues. VANG-1 is recruited to the apical surface of intestinal cells by direct interaction with junctional protein DLG-1/Discs large. The VANG-1 localization also requires DSH-2/

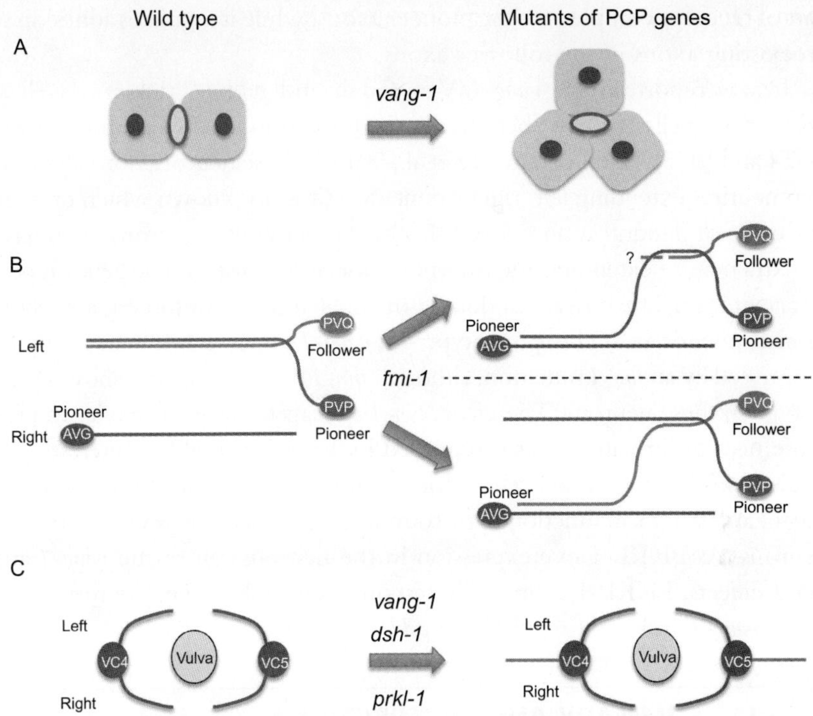

Figure 3.7 Functions of core PCP components in *C. elegans*. (A) Intestinal lumen that is normally surrounded by two cells is instead surrounded by three cells in *vang-1* mutants. (B) Pioneer and follower axons of PVP and PVQ neurons, respectively, that are normally on the left side crossed the ventral midline and joined the right axon track. (C) VC4 and VC5 neurons that are normally bipolar have extra neurites in *vang-1*, *dsh-1*, or *prkl-1* mutants. (For color version of this figure, the reader is referred to the online version of this chapter.)

Dishevelled, suggesting that VANG-1 functions in a PCP-like pathway in intestinal morphogenesis.

The core PCP components are also involved in the formation of axons or neurites of some neurons. *fmi-1* functions in the navigation of pioneer axons and axons that follow preexisting axons with distinct genetic interaction with *lin-17*/Fzd (Steimel et al., 2010) (Fig. 3.7B). Although *lin-17*/Fzd mutants similarly affect both the pioneer and the follower axons, *lin-17* enhances *fmi-1* phenotype in the pioneer axons, while it does not in the follower axons, suggesting that *lin-17* acts in parallel to *fmi-1* in the pioneer axons, while they act together in the follower axons. Intracellular domain of FMI-1 is essential for the guidance of pioneer axon but dispensable for that of follower axons. It was proposed that FMI-1 functions as a receptor or a part of receptor complex for the pioneer axons, while it regulates adhesion to preexisting axons in the follower axons.

It was reported that *vang-1*/Van Gogh and *prkl-1*/Prickle as well as *dsh-1*/Dishevelled inhibit aberrant neurite formation in some motor neurons (VC4 and VC5) (Sanchez-Alvarez et al., 2011). These neurons normally have two neurites extending left–right orientation (it is not known which of them are axons or dendrites) (Fig. 3.7C). In these mutants, the neurons often have an extra neurite extending anterior–posterior orientation. These genes appear to act in a single pathway, as double mutants among the three genes show similar phenotype. The phenotype in *vang-1* or *prkl-1* mutants can be suppressed by an *fmi-1* mutation, although *fmi-1* by itself does not show a phenotype in these neurons. Therefore, *fmi-1* appears to have an activity to promote neurite formation that is repressed by *vang-1* and *prkl-1*. Interestingly, while *prkl-1* function is cell autonomously required in these neurons, *vang-1* and *dsh-1* can function in surrounding epithelial cells as well as in these neurons. As PRKL-1 overexpression in the neurons can rescue *vang-1* and *dsh-1* defects, PRKL-1 is primarily responsible to inhibit neurite formation, while *vang-1* and *dsh-1* regulate the *prkl-1* activity.

11. SUMMARY AND PERSPECTIVE

In terms of the molecular components, regulation of neurite formation by *prkl-1*, *vang-1*, *dsh-1*, and *fmi-1* most resembles to the PCP pathway so far in *C. elegans*. However, how this regulation is related to cell polarity is not clear yet. In contrast, global regulation of spindle orientation and polarity of asymmetrically dividing cells by the Wnt/β-catenin asymmetry pathway are conceptually quite similar to the PCP regulation, even though the involvements of core PCP components appear to be minor and their asymmetric localizations have not been reported. At least, Frizzled and Dishevelled are asymmetrically localized as in the PCP regulation in other organisms. In *Drosophila* wing, the PCP regulation involves asymmetric cortical localizations of Van Gogh and Prickle on the opposite side to those of Frizzled and Dishevelled (Fig. 3.1A). Although asymmetric localizations of their homologues have not been reported, other components of Wnt/β-catenin asymmetry pathway such as WRM-1/β-catenin and APR-1/APC localize to the opposite side of the Frizzled and Dishevelled localizations (Fig. 3.1B). Therefore, one attractive possibility may be that the Wnt/β-catenin asymmetry pathway is the origin of the PCP regulation. During evolution, functions of WRM-1 and APR-1 might be taken over by

Van Gogh and Prickle so that β-catenin mainly regulates transcription in nuclei in the canonical Wnt signaling. (Note that the Wnt/β-catenin asymmetry pathway controls both cell polarity and transcription.) The acquisition of the asymmetric localization of Van Gogh and Prickle might enable cells to communicate with their neighbor to coordinate their polarity, while, at least in *C. elegans* seam cells, polarity of individual cells is independently controlled by Wnts. Such functions of Wnt might be still retained in other organisms to coordinate cell polarity regulated by the PCP pathway. Although functions of Wnt in the PCP regulation are often considered as permissive signals, it was recently suggested that Wnt functions as global cue of cell polarity in chicken limb bud (Gao et al., 2011). In *Drosophila* wing in which polarity is synchronized primarily by cooperation between the FAT-Dachsous system and the PCP signaling, the presence of additional extracellular signals that regulate polarity orientation has been predicted (Strutt, 2008; Wu & Mlodzik, 2009). Although the PCP regulation in *Drosophila* is thought to be Wnt independent, based on the lack of the PCP phenotype in *wingless* mutants, it may be possible that multiple Wnts redundantly control polarity as is the case in *C. elegans* seam cells. Even in *C. elegans*, it is not yet clear whether Wnts function as global cues for orientating polarity. Therefore, it will be important to clarify roles of Wnts in the control of polarity orientation in *C. elegans*.

ACKNOWLEDGMENTS

I would like to thank Ralf Schnabel and members of Sawa lab. for comments on the chapter. This work was supported by Grants-in-Aid for Scientific Research from the Ministry of Education, Culture, Sports, Science, and Technology of Japan.

REFERENCES

Arata, Y., Lee, J. -Y., Goldstein, B., & Sawa, H. (2010). Extracellular control of PAR protein localization during asymmetric cell division in the *C. elegans* embryo. *Development*, *137*, 3337–3345.

Bei, Y., Hogan, J., Berkowitz, L. A., Soto, M. C., Rocheleau, C. E., Pang, K. M., et al. (2002). SRC-1 and Wnt signaling act together to specify endoderm and to control cleavage orientation in early *C. elegans* embryos. *Developmental Cell*, *3*, 113–125.

Berkowitz, L. A., & Strome, S. (2000). MES-1, a protein required for unequal divisions of the germline in early *C. elegans* embryos, resembles receptor tyrosine kinases and is localized to the boundary between the germline and gut cells. *Development*, *127*, 4419–4431.

Bischoff, M., & Schnabel, R. (2006). A posterior centre establishes and maintains polarity of the *Caenorhabditis elegans* embryo by a Wnt-dependent relay mechanism. *PLoS Biology*, *4*, e396.

Cadigan, K. M., & Peifer, M. (2009). Wnt signaling from development to disease: Insights from model systems. *Cold Spring Harbor Perspectives in Biology*, *1*, a002881.

Deshpande, R., Inoue, T., Priess, J. R., & Hill, R. J. (2005). *lin-17*/Frizzled and *lin-18* regulate POP-1/TCF-1 localization and cell type specification during *C. elegans* vulval development. *Developmental Biology*, *278*, 118–129.

Dikovskaya, D., Zumbrunn, J., Penman, G. A., & Näthke, I. S. (2001). The adenomatous polyposis coli protein: In the limelight out at the edge. *Trends in Cell Biology*, *11*, 378–384.

Gao, B., Song, H., Bishop, K., Elliot, G., Garrett, L., English, M. A., et al. (2011). Wnt signaling gradients establish planar cell polarity by inducing Vangl2 phosphorylation through Ror2. *Developmental Cell*, *20*, 163–176.

Goldstein, B. (1993). Establishment of gut fate in the E lineage of *C. elegans*: The roles of lineage-dependent mechanisms and cell interactions. *Development*, *118*, 1267–1277.

Goldstein, B., Takeshita, H., Mizumoto, K., & Sawa, H. (2006). Wnt signals can function as positional cues in establishing cell polarity. *Developmental Cell*, *10*, 391–396.

Gray, R. S., Roszko, I., & Solnica-Krezel, L. (2011). Planar cell polarity: Coordinating morphogenetic cell behaviors with embryonic polarity. *Developmental Cell*, *21*, 120–133.

Green, J. L., Inoue, T., & Sternberg, P. W. (2008). Opposing Wnt pathways orient cell polarity during organogenesis. *Cell*, *134*, 646–656.

Harterink, M., Kim, D. H., Middelkoop, T. C., Doan, T. D., van Oudenaarden, A., & Korswagen, H. C. (2011). Neuroblast migration along the anteroposterior axis of *C. elegans* is controlled by opposing gradients of Wnts and a secreted Frizzled-related protein. *Development*, *138*, 2915–2924.

Hoffmann, M., Segbert, C., Helbig, G., & Bossinger, O. (2010). Intestinal tube formation in *Caenorhabditis elegans* requires *vang-1* and *egl-15* signaling. *Developmental Biology*, *339*, 268–279.

Huang, S., Shetty, P., Robertson, S. M., & Lin, R. (2007). Binary cell fate specification during *C. elegans* embryogenesis driven by reiterated reciprocal asymmetry of TCF POP-1 and its coactivator beta-catenin SYS-1. *Development*, *134*, 2685–2695.

Kidd, A. R., Miskowski, J. A., Siegfried, K. R., Sawa, H., & Kimble, J. (2005). A beta-catenin identified by functional rather than sequence criteria and its role in Wnt/MAPK signaling. *Cell*, *121*, 761–772.

Korswagen, H. C., Herman, M. A., & Clevers, H. C. (2000). Distinct beta-catenins mediate adhesion and signalling functions in *C. elegans*. *Nature*, *406*, 527–532.

Lin, R., Thompson, S., & Priess, J. R. (1995). pop-1 encodes an HMG box protein required for the specification of a mesoderm precursor in early *C. elegans* embryos. *Cell*, *83*, 599–609.

Lo, M. -C., Gay, F., Odom, R., Shi, Y., & Lin, R. (2004). Phosphorylation by the beta-catenin/MAPK complex promotes 14-3-3-mediated nuclear export of TCF/POP-1 in signal-responsive cells in *C. elegans*. *Cell*, *117*, 95–106.

MacDonald, B. T., Tamai, K., & He, X. (2009). Wnt/beta-catenin signaling: Components, mechanisms, and diseases. *Developmental Cell*, *17*, 9–26.

Maduro, M. F., Lin, R., & Rothman, J. H. (2002). Dynamics of a developmental switch: Recursive intracellular and intranuclear redistribution of *Caenorhabditis elegans* POP-1 parallels Wnt-inhibited transcriptional repression. *Developmental Biology*, *248*, 128–142.

Mizumoto, K., & Sawa, H. (2007a). Cortical beta-catenin and APC regulate asymmetric nuclear beta-catenin localization during asymmetric cell division in *C. elegans*. *Developmental Cell*, *12*, 287–299.

Mizumoto, K., & Sawa, H. (2007b). Two betas or not two betas: Regulation of asymmetric division by beta-catenin. *Trends in Cell Biology*, *17*, 465–473.

Nakamura, K., Kim, S., Ishidate, T., Bei, Y., Pang, K., Shirayama, M., et al. (2005). Wnt signaling drives WRM-1/beta-catenin asymmetries in early *C. elegans* embryos. *Genes & Development*, *19*, 1749–1754.

Natarajan, L., Witwer, N. E., & Eisenmann, D. M. (2001). The divergent *Caenorhabditis elegans* beta-catenin proteins BAR-1, WRM-1 and HMP-2 make distinct protein interactions but retain functional redundancy in vivo. *Genetics*, *159*, 159–172.

Nishi, Y., & Lin, R. (2005). DYRK2 and GSK-3 phosphorylate and promote the timely degradation of OMA-1, a key regulator of the oocyte-to-embryo transition in C. elegans. *Developmental Biology*, *288*, 139–149.

Park, F. D., & Priess, J. R. (2003). Establishment of POP-1 asymmetry in early *C. elegans* embryos. *Development*, *130*, 3547–3556.

Park, F. D., Tenlen, J. R., & Priess, J. R. (2004). *C. elegans* MOM-5/frizzled functions in MOM-2/Wnt-independent cell polarity and is localized asymmetrically prior to cell division. *Current Biology*, *14*, 2252–2258.

Phillips, B. T., Kidd, A. R., King, R. S., Hardin, J., & Kimble, J. (2007). Reciprocal asymmetry of SYS-1/beta-catenin and POP-1/TCF controls asymmetric divisions in *Caenorhabditis elegans*. *Proceedings of the National Academy of Sciences of the United States of America*, *104*, 3231–3236.

Rocheleau, C. E., Yasuda, J., Shin, T. H., Lin, R., Sawa, H., Okano, H., et al. (1999). WRM-1 activates the LIT-1 protein kinase to transduce anterior/posterior polarity signals in *C. elegans*. *Cell*, *97*, 717–726.

Sanchez-Alvarez, L., Visanuvimol, J., McEwan, A., Su, A., Imai, J. H., & Colavita, A. (2011). VANG-1 and PRKL-1 cooperate to negatively regulate neurite formation in *Caenorhabditis elegans*. *PLoS Genetics*, 7, e1002257.

Schlesinger, A., Shelton, C. A., Maloof, J. N., Meneghini, M. D., & Bowerman, B. (1999). Wnt pathway components orient a mitotic spindle in the early *Caenorhabditis elegans* embryo without requiring gene transcription in the responding cell. *Genes & Development*, *13*, 2028–2038.

Shetty, P., Lo, M. -C., Robertson, S. M., & Lin, R. (2005). *C. elegans* TCF protein, POP-1, converts from repressor to activator as a result of Wnt-induced lowering of nuclear levels. *Developmental Biology*, *285*, 584–592.

Shirayama, M., Soto, M. C., Ishidate, T., Kim, S., Nakamura, K., Bei, Y., et al. (2006). The conserved kinases CDK-1, GSK-3, KIN-19, and MBK-2 promote OMA-1 destruction to regulate the oocyte-to-embryo transition in *C. elegans*. *Current Biology*, *16*, 47–55.

Siegfried, K. R., Kidd, A. R., Chesney, M. A., & Kimble, J. (2004). The sys-1 and sys-3 genes cooperate with Wnt signaling to establish the proximal-distal axis of the *Caenorhabditis elegans* gonad. *Genetics*, *166*, 171–186.

Steimel, A., Wong, L., Huarcaya Najarro, E., Ackley, B. D., Garriga, G., & Hutter, H. (2010). The Flamingo ortholog FMI-1 controls pioneer-dependent navigation of follower axons in *C. elegans*. *Development*, *137*, 3663–3673.

Sternberg, P.W. (2005). Vulval development. *WormBook: the Online Review of C Elegans Biology*. 1–28.

Strutt, D. (2008). The planar polarity pathway. *Current Biology*, *18*, R898–R902.

Sugioka, K., Mizumoto, K., & Sawa, H. (2011). Wnt regulates spindle asymmetry to generate asymmetric nuclear β-catenin in *C. elegans*. *Cell*, *146*, 942–954.

Sugioka, K., & Sawa, H. (2010). Regulation of asymmetric positioning of nuclei by Wnt and Src signaling and its roles in POP-1/TCF nuclear asymmetry in *Caenorhabditis elegans*. *Genes to Cells*, *15*, 397–407.

Takeshita, H., & Sawa, H. (2005). Asymmetric cortical and nuclear localizations of WRM-1/beta-catenin during asymmetric cell division in *C. elegans*. *Genes & Development*, *19*, 1743–1748.

Walston, T. D., Guo, C., Proenca, R., Wu, M., Herman, M. A., Hardin, J., et al. (2006). *mig-5*/Dsh controls cell fate determination and cell migration in *C. elegans*. *Developmental Biology*, *298*, 485–497.

Werts, A. D., Roh-Johnson, M., & Goldstein, B. (2011). Dynamic localization of *C. elegans* TPR-GoLoco proteins mediates mitotic spindle orientation by extrinsic signaling. *Development*, *138*, 4411–4422.

Whangbo, J., Harris, J. M., & Kenyon, C. (2000). Multiple levels of regulation specify the polarity of an asymmetric cell division in *C. elegans*. *Development*, *127*, 4587–4598.

Whangbo, J., & Kenyon, C. (1999). A Wnt signaling system that specifies two patterns of cell migration in *C. elegans*. *Molecular Cell*, *4*, 851–858.

Wildwater, M., Sander, N., de Vreede, G., & van den Heuvel, S. (2011). Cell shape and Wnt signaling redundantly control the division axis of *C. elegans* epithelial stem cells. *Development*, *138*, 4375–4385.

Wu, M., & Herman, M. A. (2006). A novel noncanonical Wnt pathway is involved in the regulation of the asymmetric B cell division in *C. elegans*. *Developmental Biology*, *293*, 316–329.

Wu, J., & Mlodzik, M. (2009). A quest for the mechanism regulating global planar cell polarity of tissues. *Trends in Cell Biology*, 7, 295–305.

Yamamoto, Y., Takeshita, H., & Sawa, H. (2011). Multiple Wnts redundantly control polarity orientation in *Caenorhabditis elegans* epithelial stem cells. *PLoS Genetics*, 7, e1002308.

Yang, X. -D., Huang, S., Lo, M. -C., Mizumoto, K., Sawa, H., Xu, W., et al. (2011). Distinct and mutually inhibitory binding by two divergent β-catenins coordinates TCF levels and activity in *C. elegans*. *Development*, *138*, 4255–4265.

CHAPTER FOUR

Planar Cell Polarity in Coordinated and Directed Movements

Masazumi Tada*,[1], Masatake Kai[†]

*Department of Cell and Developmental Biology, University College London, London, United Kingdom
[†]Department of Anatomy and Cell Biology, Graduate School of Medicine, Osaka City University, Abeno-ku, Osaka, Japan
[1]Corresponding author: e-mail address: m.tada@ucl.ac.uk

Contents

Current Topics in Developmental Biology, Volume 101
ISSN 0070-2153
http://dx.doi.org/10.1016/B978-0-12-394592-1.00004-1

© 2012 Elsevier Inc.
All rights reserved.

Abstract

Planar cell polarity is a fundamental concept to understanding the coordination of cell movements in the plane of a tissue. Since the planar cell polarity pathway was discovered in mesenchymal tissues involving cell interaction during vertebrate gastrulation, there is an emerging evidence that a variety of mesenchymal and epithelial cells utilize this genetic pathway to mediate the coordination of cells in directed movements. In this review, we focus on how the planar cell polarity pathway is mediated by migrating cells to communicate with one another in different developmental processes.

1. INTRODUCTION

A variety of modes of collective cell migration shape the body axis in animal development in that both epithelial and mesenchymal cells have the ability to coordinate morphogenetic movements but achieve in different ways (reviewed in Friedl & Gilmour, 2009). During amphibian and teleost gastrulation, polarized mesenchymal cells undergo directional cell intercalations in a coordinated fashion, a morphogenetic process called convergent extension (CE), contributing to the elongation of the presumptive notochord (reviewed in Keller, 2002). Another mesenchymal cell population, called prechordal plate progenitors, utilizes to direct the cells as a coherent cluster in a mode different to notochord progenitors. In contrast, simple epithelial cells of the anterior visceral endoderm (AVE) in mice undergo directed movement as a cluster but use different strategies that involve coordinated cell rearrangement. Despite the fact that the regulation of both cell polarity and cell adhesion is fundamental to the orientation and alignment of the cells during tissue elongation in all the cases, there are significant mechanistic divergences.

The genetic pathway that mediates such coordinated cell behaviors is planar cell polarity (PCP). As its identification of this pathway in a plan of epithelial tissue in *Drosophila*, the PCP pathway has been implicated in the regulation of CE in mesenchymal cells of the vertebrate gastrula. Further, there is increasing evidence that the PCP pathway is utilized in a variety of different biological processes, in which the coordination and orientation of cells are required within both epithelial and mesenchymal tissues. In this review, we highlight new insights into fundamental roles for PCP in regulating coordinated and directed cell movements in different developmental processes. Because of space constraints, we refer for details on the roles for the PCP pathway in other contexts to recent excellent reviews (Gray et al., 2011; Simons & Mlodzik, 2008; Wallingford & Mitchell, 2011).

2. COORDINATED AND DIRECTED MOVEMENTS DURING VERTEBRATE GASTRULATION

2.1. Different types of cell movements—Collective cell migration and cell intercalation

During vertebrate gastrulation, coordinated and directed movements of both epithelial and mesenchymal cells shape the body axis of the embryo. Basically, there are two different types of movements: collective cell migration and cell intercalation. In collective migration, cells move either individually (Fig. 4.1A; e.g., neural crest cells) or as a cohesive clump (Fig. 4.1B; e.g., prechordal plate progenitors) without neighbor exchanges. During cell intercalation, cells exchange neighbors in a directed manner, thereby allowing directed movement of epithelial cells (Fig. 4.1C; e.g., the *Drosophila* germband, which is mediated by multicellular rosettes, and similarly the mouse AVE) or elongation of a mesenchymal tissue (Fig. 4.1D; e.g., notochord progenitors, which is mediated by mediolateral intercalation). However, different species utilize these two modes to different degrees in both epithelial and mesenchymal tissues.

2.2. Convergent extension in *Xenopus*

At the onset of gastrulation, soon after involution, chordamesoderm cells (the presumptive notochord and somites) are highly polarized, which elongate and orient in their mediolateral axis with bipolar lamellipodia being stabilized at both ends, and undergo mediolateral cell intercalation (Shih & Keller, 1992). These polarized cells intercalate between one another to redistribute their positions along the anteroposterior (AP) axis, thereby contributing to the extension of the forming body axis. Convergence and extension movements simultaneously occur in *Xenopus*, and therefore, this process is called CE. Isolated chordamesoderm tissues, called Keller explants, can undergo CE movements in the absence of external substrates, and thus CE is a cell-autonomous force-generating process (Keller & Danilchik, 1988). Likewise, CE behaviors associated with the elongation of tissues can be observed in isolated naïve ecotodermal tissues, called animal cap explants, when treated with the mesoderm inducer Activin (e.g., Tada & Smith, 2000) (Fig. 4.2).

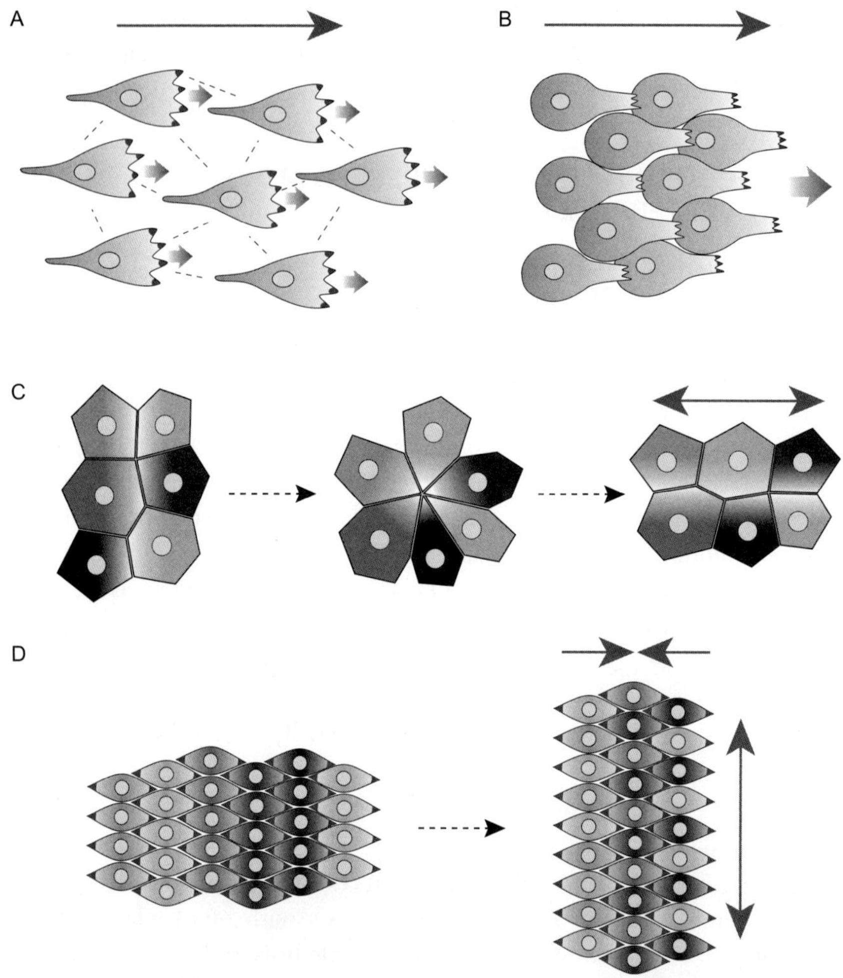

Figure 4.1 Different types of cell movements. (A) Collective cell migration in which cells move individually (e.g., neural crest cells). Cells are loosely associated with each other and migrate in a uniform direction. (B) Collective cell migration as a cohesive clump (e.g., prechordal plate progenitors). Cells are tightly packed and migrate in one direction without changing neighbors. Cells at the leading edge (blue) form actin-based protrusions (shown in gray). (C) Cell intercalation mediated by formation and resolving of rosette (e.g., *Drosophila* germband). Five or more cells meet at one point ("rosette" formation) and resolve to make a new cell–cell boundary in relation to the direction of their movement. (D) Cell intercalation in convergent extension (e.g., notochord progenitors). Cells intercalate mediolaterally to simultaneously narrow and elongate the tissue along the anteroposterior axis. (See Color Insert.)

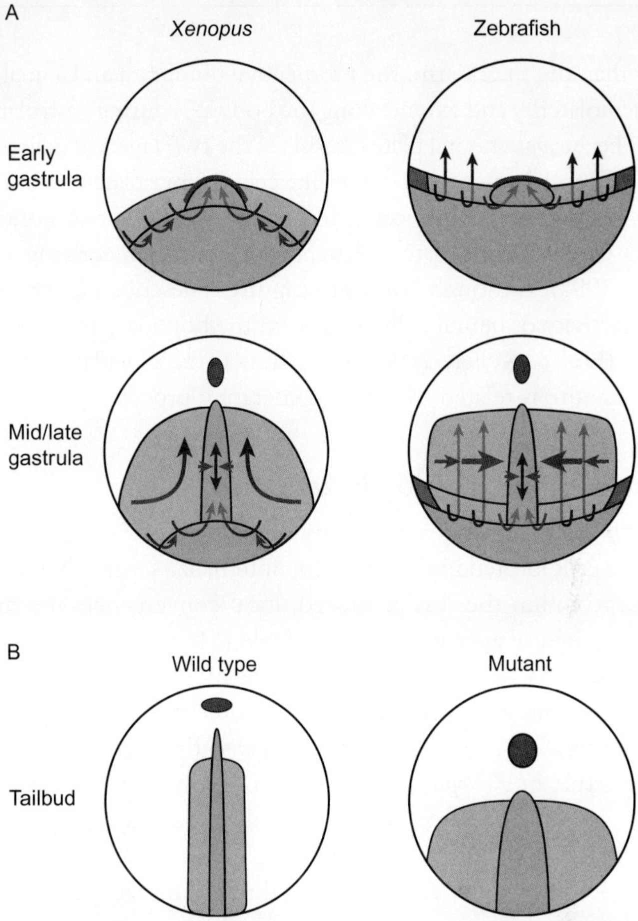

Figure 4.2 Schematic illustrations of cell movements during gastrulation in *Xenopus* and zebrafish. (A) At the early gastrula stage in *Xenopus*, notochord (shown in orange) and somite (green) progenitors involute at the marginal zone and simultaneously converge (blue to red arrows) toward the dorsal midline by cell intercalation. At the mid/late gastrula stage, extensive mediolateral intercalation drives convergent extension (red and blue arrows) of the axial and lateral mesoderm (orange and green). In zebrafish, somite progenitors (green) internalize at the germ ring (green and light blue belt) and migrate anteriorly (blue arrows) at the early gastrula stage. Cells at the dorsal-most marginal zone internalize and converge (red arrows) to create the physical shield (orange). After 75% epiboly (mid-gasatrula stage), somite progenitors (green) initiate to converge toward the midline, with increasing their speed and directionality (red arrows). The presumptive notochord (orange) elongates and narrows by mediolateral cell intercalation (red and blue arrows). Prechordal plate progenitors (brown) migrate anteriorly as a cohesive cluster in both *Xenopus* and zebrafish. (B) Wild-type and typical mutant phenotypes at the tailbud stage in zebrafish. In PCP mutants, the presumptive notochord (orange) and somites (green) are widened laterally and are shorter along the anterior–posterior axis than in wild-type embryos. This is accompanied with compromised anterior migration of the prospective prechordal plate (brown). (See Color Insert.)

Other than the mesoderm, the prospective hindbrain and spinal cord also narrow mediolaterlly and extend along the body axis during gastrulation/neurulation. The *Xenopus* neural plate consists of the two layers of cells: superficial epithelial cells and deep mesenchyme-like cells. In an isolated neural plate, the deep cells show mediolateral cell intercalation behavior that contributes to elongation of the neural plate independently of the underlying mesoderm (Elul et al., 1997). Although boundary capture is described as for monopolar protrusive activity of neural cells away from the boundary to the prospective floor plate (Elul & Keller, 2000), it remains to be elucidated whether the boundary capture is related to a force-generating process.

2.3. Convergent extension in zebrafish

Different to simultaneous movements of involution and convergence in *Xenopus*, the cells located in the germ ring internalize separately from convergence except within the shield. Indeed, local convergence-like movement mediates the physical appearance of the shield (Montero et al., 2005). Lateral mesoderm cells do not initiate convergence movement until mid-gastrula (75% epiboly). This is coincided with the geometry of the germ ring at this stage of the zebrafish embryo, which is similar to that of the marginal zone at the early gastrula in *Xenopus*. At the onset of mesoderm convergence, cells appear only loosely associated yet to be visibly polarized and show only little coordinated and directed movement (Sepich et al., 2005). However, once these cells get closer to the dorsal side and thus cell density increases, the cells polarize along their mediolateral axis and exhibit highly coordinated and directed convergence movements, which are reminiscent of collective migration (Sepich et al., 2000). When arriving near the prospective notochord, mesoderm cells initiate mediolateral intercalation, thereby contributing to the elongation of the body axis. In contrast, notochord progenitors originated from the shield start showing mediolateral cell intercalation behavior at mid-gastrula stage and exclusively undergo extension movement. Thus, these two different types of cell movements both mediate CE and contribute to the elongation of body in zebrafish (Fig. 4.3).

2.4. Collective migration of head mesendoderm/prechordal plate progenitors

During gastrulation, head mesendoderm cells in *Xenopus* and prechordal plate progenitors in zebrafish migrate as a cohesive sheet of cells toward the animal pole. Cells located at the leading edge form dynamic protrusions

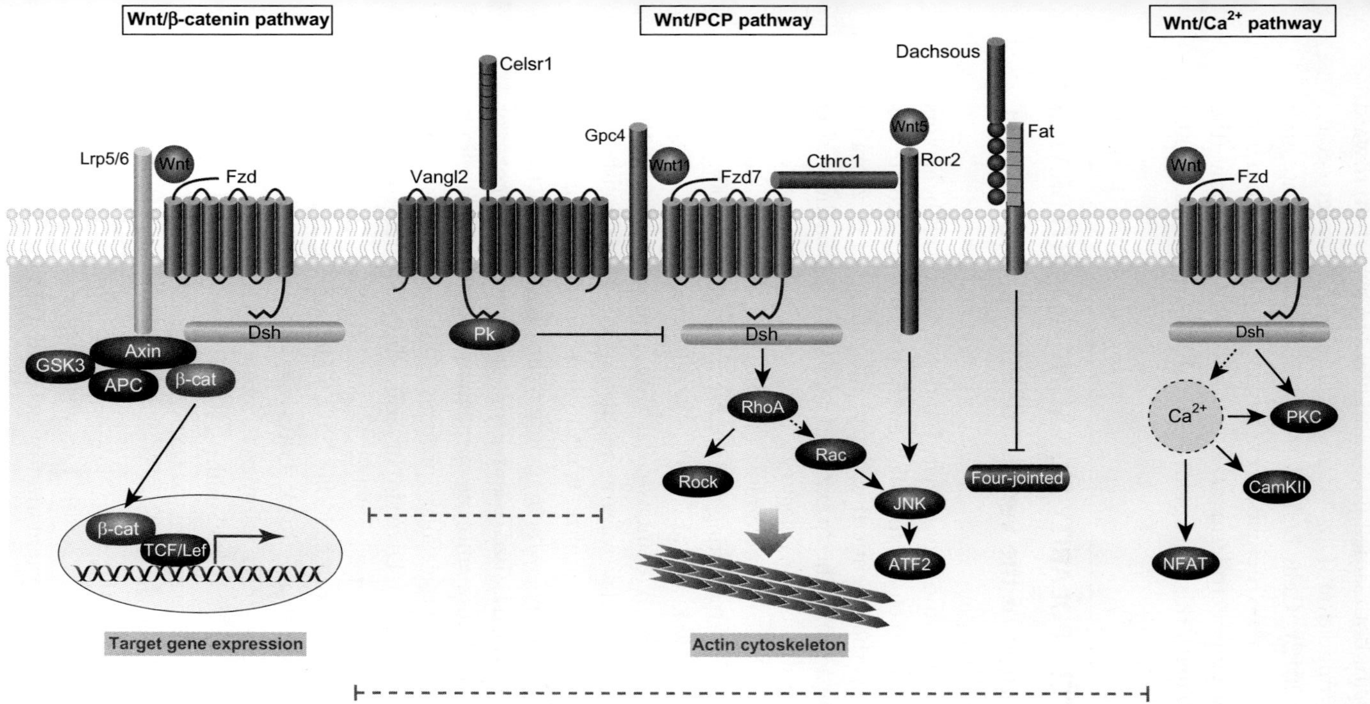

Figure 4.3 Divergence of the Wnt pathways in vertebrates. The Wnt/β-catenin pathway ("canonical" pathway) involves stabilization of β-catenin upon binding of Wnt ligand to Fzd and Lrp5/6 receptors, leading to target gene expression mediated by β-catenin and TCF/Lef. The Wnt/planar cell polarity (PCP) pathway (a "noncanonical" pathway) primarily regulates actin cytoskeleton via several groups of molecules, including PCP core molecules (Vangl2, Celsr1, Prickle), Wnt11 and associated proteins (Glypican4, Fzd7, Cthrc1), Wnt5 and associated proteins (Ror2, Cthrc1), and atypical cadherins (Dachsous, Fat). Wnt11 activates Rock (Rho-associated kinase) via Dsh and RhoA, while Wnt5 functions through JNK to activate ATF2 (activating transcription factor-2). The Wnt/Ca^{2+} pathway (another "noncanonical" pathway) is activated by Wnt11 or Wnt5 and triggers Ca^{2+} signaling through CamKII (calmodulin-dependent protein kinase 2), PKC (protein kinase C), and NFAT (nuclear factor of activated T cells). The Wnt/β-catenin and Wnt/PCP or Wnt/Ca^{2+} pathways are mutually antagonistic (dotted pink lines) in some cases.(For interpretation of the references to color in this figure legend, the reader is referred to the online version of this chapter.)

(lamellipodia in *Xenopus*; both lamellipodia and blebs in zebrafish) toward the direction of their migration (Diz-Munoz et al., 2010; Montero et al., 2003; Weber et al., 2012). Cells behind the leading edge migrate in a coordinated manner with no neighbor exchanges involved. Migrating cells utilize the overlying ectoderm as a substrate for migration (the blastocoel roof in *Xenopus*; the epiblast in zebrafish) (Ulrich et al., 2003; Winklbauer, 1990).

3. PLANAR CELL POLARITY SIGNALING IN VERTEBRATES

3.1. Identification of the vertebrate PCP pathway during gastrulation

Despite the fascinated feature of coordinated and polarized cell behaviors underlying CE, there was no clue as to what genetic pathway(s) mediate this process until the end of the twentieth century. Identification of zebrafish mutants, which exhibit a shorter body axis, and candidate approaches, using dominant-negative constructs in *Xenopus* to test a plausible pathway that inhibits CE movements but not mesoderm specification, has revealed that the PCP pathway is key to mediating CE (Heisenberg et al., 2000; Tada & Smith, 2000; Topczewski et al., 2001; Wallingford et al., 2000). Within a decade, it has been demonstrated that core members of the *Drosophila* PCP pathway all play roles in regulating CE in zebrafish and *Xenopus* except that vertebrates utilize Wnt ligands. Hereafter, this refers to as the Wnt/PCP pathway, which includes the ligands Wnt11 and Wnt5, the receptors Frizzled7 (Fzd7) and Fzd2, the membrane cofactor Glypican4 (Gpc4), the transmembrane protein Vangl2 and Flamingo (Fmi)/Celsr1/2, and the intracellular proteins Disheveled (Dsh/Dvl), Prickle (Pk), and Diego (Dgo)/Diversin/Inversin (Inv) (Carreira-Barbosa et al., 2003, 2009; Darken et al., 2002; Djiane et al. & Shi, 2000; Formstone & Mason, 2005; Goto & Keller, 2002; Heisenberg et al., 2000; Jessen et al., 2002; Kilian et al., 2003; Ohkawara et al., 2003; Park & Moon, 2002; Rauch et al., 1997; Schwarz-Romond et al., 2002; Simons et al., 2005; Sokol, 1996; Tada & Smith, 2000; Takeuchi et al., 2003; Topczewski et al., 2001; Veeman et al., 2003; Wallingford et al., 2000).

The Wnt/PCP pathway is required for both the orientation and polarization of cells undergoing CE (Sepich et al., 2000; Topczewski et al., 2001; Wallingford et al., 2000). Likewise, elevated PCP activity leads to disrupted elongation/orientation of cells, thereby inhibiting CE movements (Wallingford et al., 2000). Moreover, cells with abrogated Wnt/PCP activity cannot undergo mediolateral cell intercalations with neighboring normal cells in *Xenopus* (Kinoshita et al., 2003). Similarly, in zebrafish, embryos with abrogated Wnt/PCP activity exhibit defective cell intercalation behavior, which can be visualized by labeled cells lateral to the shield, distributing as in a string along the AP axis as gastrulation proceeds (Heisenberg et al., 2000; Sepich et al., 2000).

Whereas there is no obvious effect of altered Wnt/PCP activity on collective migration of head mesendoderm cells in *Xenopus*, abrogation of the Wnt/PCP pathway can lead to reduced migration of prechordal plate progenitors in zebrafish (Ulrich et al., 2005). This is primarily due to reduction in coherence of the cells in *wnt11* mutants as Wnt11 is required for E-cadherin-mediated cell cohesion (Ulrich et al., 2005). This suggests that the Wnt/PCP pathway mediates cell cohesion underlying collective migration of a large population of mesenchymal cells.

3.2. Neural tube defects in mice

Similar to zebrafish, identification of mouse mutants that exhibit severe neural tube defects (NTD), reminiscent of craniorachischisis in humans, has uncovered the core PCP genes *vangl2* and *celsr1* (Curtin et al., 2003; Kibar et al., 2001; Murdoch et al., 2001). In addition, double mutants for *dvl1;dvl2* or *fzd3;fzd6* show the severe NTD phenotype (Hamblet et al., 2002; Wang, Guo, et al., 2006; Wang, Hamblet, et al., 2006). Besides the core members of the PCP pathway, mouse genetics has further identified new members of the vertebrate PCP pathway. Despite the fact that *scribble (scrb)* is implicated in regulating basolateral polarity in *Drosophila* (Albertson & Doe, 2003), *scrb* was identified from classical mouse mutants, exhibiting a severe NTD (Murdoch et al., 2003). Further, mutants in the *ptk7* locus, encoding a receptor tyrosine kinase, exhibit severe NTD and genetically interact with *vangl2* (Lu et al., 2004; Paudyal et al., 2010).

Further analysis of the PCP mutant embryos at early-somite stages revealed that NTD arise primarily due to defective CE movements in the notochord and neural plate (Ybot-Gonzalez et al., 2007). Indeed, CE is mediated by mediolateral cell intercalations that contribute to extension of the axial tissue (Yamanaka et al., 2007; Yen et al., 2009). However, it is still unclear whether CE defects in mesenchymal notochord cells secondarily affect CE in overlying neuroepithelial cells or defects in the two layers occur independently (Table 4.1).

Table 4.1 Core PCP genes and their modulators

Vertebrate gene	Species	*Drosophila* PCP gene	Molecular features	Cell movements	References
wnt11	z, X		Secreted Wnt glycoprotein	CE, NC	Heisenberg et al. (2000), Tada and Smith (2000), Carmona-Fontaine et al. (2008)
wnt5b	z		Secreted Wnt glycoprotein	CE, JC	Rauch et al. (1997)
wnt5a	X		Secreted Wnt glycoprotein	CE, NC	Yamanaka et al. (2002), Oishi et al. (2003), Matthews et al. (2008)
vangl2	z, X, m, c	*strabismus/Van Gogh (vang)*	4-pass TM protein	CE, NTD, NM, CO, PS	Darken et al. (2002), Goto and Keller (2002), Park and Moon (2002), Kibar et al. (2001), Murdoch et al. (2001), Jessen et al. (2002), Montcouquiol et al. (2003), Bingham et al. (2002), Wang et al. (2005), Voiculescu et al. (2007)
glypican4 (gpc4)	z, X		Heparan sulfate proteoglycan	CE, JC	Topczewski et al. (2001), Ohkawara et al. (2003), LeClair et al. (2009)
dsh/Dvl1/2	z, X, m, c	*disheveled (dsh)*	DIX, PDZ, and DEP domains	CE, NTD, CO, NM, PS, NC	Sokol (1996), Hamblet et al. (2002), Wang, Hamblet, et al. (2006), Voiculescu et al. (2007), Matthews et al. (2008), Carmona-Fontaine et al. (2008)
fzd3/6	z, m	*frizzled (fz)*	7-pass TM protein, Wnt receptor	NTD, NM	Wang, Guo, et al. (2006), Wada et al. (2006)

fzd7	z, X	*frizzled (fz)*	7-pass TM protein, Wnt receptor	CE	Djiane et al. (2000)Quesada-Hernandez et al. (2010)
prickle1/2	z, X, c	*prickle (pk)*	LIM and PET domains	CE, NM, NC, PS	Carreira-Barbosa et al. (2003), Veeman et al. (2003), Takeuchi et al. (2003), Mapp et al. (2010), Voiculescu et al. (2007), Carmona-Fontaine et al. (2008)
celsr1/2	z, m, c	*flamingo (fmi)/ starry night*	7-pass TM protein, atypical cadherin	CE, Epi, NTD, NM, CO, PS, AVE	Formstone and Mason (2005), Curtin et al. (2003), Wada et al. (2006), Qu et al. (2010), Voiculescu et al. (2007), Trichas et al. (2011)
diversin	z, X	*diego (dgo)*	Ankirin repeat protein	CE	Schwarz-Romond et al. (2002)
inversin (inv)	X, m	*diego (dgo)*	Ciliary-associated protein	CE, NTD	Simons et al. (2005)
scribble	z, m		Scaffolding protein	CE, NTD, NM, CO	Murdoch et al. (2001), Murdoch et al. (2003), Montcouquiol et al. (2003), Wada et al. (2005)
daam1	X		Formin-homology (FH) domain protein	CE	Habas et al. (2001), Habas et al. (2003)
rhoA	X	*dRhoA*	Small GTPase	CE	Habas et al. (2001), Habas et al. (2003)
rac	X	*dRac1*	small GTPase	CE	Habas et al. (2003), Chung et al. (2005)

Continued

Table 4.1 Core PCP genes and their modulators—cont'd

Vertebrate gene	Species	*Drosophila* PCP gene	Molecular features	Cell movements	References
rho kinase 2 (rok2)	z	*Rho kinase (drok)*	Serine/threonine kinase, RhoA effector	CE	Marlow et al. (2002)
jun N-terminal kinase (jnk)	X		Serine/threonine kinase	CE	Yamanaka et al. (2002), Schambony and Wedlich (2007), Oishi et al. (2003)
int	X	*inturned (in)*	Putative PDZ domain	CE†, NTD	Park et al. (2006)
fy	X	*fuzzy (fy)*	4-pass TM protein	CE†, NTD	Park et al. (2006)
ror2	X		Receptor tyrosine kinase, Wnt5 receptor	CE	Schambony and Wedlich (2007), Oishi et al. (2003), Hikasa et al. (2002)
paraxial protocadherin (papc)	X		Protocadherin	CE	Medina et al. (2004), Unterseher et al. (2004), Wang et al. (2008)
neurotrophin receptor homologue 1 (NRH1)	X		Receptor tyrosine kinase	CE	Sasai et al. (2004), Chung et al. (2005)
ptk7	X, m		Receptor tyrosine kinase	NTD	Lu et al. (2004), Paudyal et al. (2010)
Cthrc1	m		Wnt cofactor	NTD*	Yamamoto et al. (2008)

syndecan4	X		Heparan sulfate proteoglycan	CE	Munoz et al. (2006), Ohkawara et al. (2011)
Bardet–Biedl syndrome (bbs)	z, m		Ciliary-associated protein	CE, NTD	Ross et al. (2005), Gerdes et al. (2007), May-Simera et al. (2010)
Fat4	m	*fat (ft)*	Atypical cadherin	NTD*, CO	Saburi et al. (2008), Mao et al. (2011), Saburi et al. (2012)
dachsous (ds)	m	*dachsous (ds)*	Atypical cadherin	CO	Mao et al. (2011)
$G\alpha_{12/13}$	z	*Goα47A/broken heart (bkh)*	Heterotrimeric G-protein subunit	CE	Slusarski et al. (1997), Lin et al. (2005)
Ga_{11}	X	*Goα47A/broken heart (bkh)*	Heterotrimeric G-protein subunit	CE	Iioka et al. (2007)
casein kinase Iε (ckIε)	X	*Casein Kinase Iε (CKIε)*	Serine/threonine kinase	CE	McKay et al. (2001)
nkd1/2	z	*naked cuticle (nkd)*	Wnt antagonist	CE	Van Raay et al. (2007)
LDL receptor related protein 6 (lrp6)	X		TM protein, Wnt coreceptor	CE	Tahinci et al. (2007)
partitioning-defective 1 (par-1)	X		Serine/threonine kinase	CE	Kusakabe and Nishida (2004), Ossipova et al. (2005)
partitioning-defective 6 (par-6)	X		PDZ domain	CE	Kusakabe and Nishida (2004)
atypical protein kinase C (aPKC)	X		Serine/threonine kinase	CE	Kusakabe and Nishida (2004)
ArfGAP	X		GTPase-activating protein	CE	Hyodo-Miura et al. (2006)

Continued

Table 4.1 Core PCP genes and their modulators—cont'd

Vertebrate gene	Species	*Drosophila* PCP gene	Molecular features	Cell movements	References
μ2-adaptin	X		Clathrin adaptor	CE	Yu et al. (2007)
β-arrestin 2 (βarr2)	X		GPCR adaptor	CE	Kim and Han (2007)
fritz	X		WD40 repeat protein	CE	Kim et al. (2010)
Nance–Horan syndrome-like 1b (nhsl1b)	z		Putative actin-binding protein	NM	Walsh et al. (2011)
Smurf1/2	m		Ubiquitin ligase	CO	Narimatsu et al. (2009)
weak-similarity GEF (wgef)	X		Similar to GEF	CE	Tanegashima et al. (2008)
lpp	z		LIM domain protein	CE	Vervenne et al. (2008)
prr	X		Prorenin receptor	CE	Buechling et al. (2010)
membrane matrix metalloproteinase (mmp14)	z		Metalloproteinase	CE	Williams et al. (2012)
rack1	z		Receptor for activated C kinase 1	CE	Li et al. (2011)
R-spondin 3 (rspo3)	X		Secreted Wnt modulator	CE	Ohkawara et al. (2011), Glinka et al. (2011)
lgr4/5	X		G-protein-coupled receptor	CE	Glinka et al. (2011)
mink1	X		Ste20 kinase	CE	Daulat et al. (2012)

z, zebrafish; X, *Xenopus*; m, mouse; c, chick; TM, transmembrane; CE, convergent extension (†very mild phenotype), NTD, neural tube defect (*only manifest in *Vangl2*$^{+/-}$), Epi, epiboly; NM, neuronal migration; CO, cochlea; JC, jaw cartilage; PS, primitive streak; NC, neural crest; AVE, anterior visceral endoderm.

3.3. Divergence of the pathways: Wnt/PCP, Wnt/Ca^{2+}, and Fat/Ds

It has become evident that noncanonical Wnt11 and Wnt5 utilize different receptors and their downstream effectors although *wnt11* and *wnt5b* act redundantly in regulating CE in zebrafish (Kilian et al., 2003). Wnt11 binds to Fzd7 and acts through the core transducer of PCP signal Dsh which in turn activates RhoA and Rok2 to modulate actin cytoskeleton (Djiane et al., 2000; Habas et al., 2003; Marlow et al., 2002; Tada & Smith, 2000). In contrast, it has been shown that Wnt5 can bind to the receptor tyrosine kinase Ror2 and potentially form a complex with Fzd through the secreted glycoprotein Cthrc1 (Yamamoto et al., 2008). Wnt5/Ror2 signal regulates CE to activate JNK then the transcription factor ATF2, mediating activation of target genes, independently of Wnt11 function (Hikasa et al., 2002; Schambony & Wedlich, 2007) and perhaps through Ca^{2+} signal as in cultured cells (Oishi et al., 2003).

There is emerging evidence for mutual antagonistic actions between the Wnt/β-catenin and Wnt/PCP/Wnt/Ca^{2+} pathways. In addition, regulators for Wnt/β-catenin signal can also modulate the Wnt/PCP pathway at different levels. The newly identified module that positively regulates Wnt/β-catenin signal at the cell surface along with the receptors Fzd and Lrp is the secreted glycoprotein R-spondin and its G-protein-coupled receptor Lgr (Carmon et al., 2011; de Lau et al., 2011; Glinka et al., 2011). Interestingly, R-spondin3 via Lgr4/5 modulates CE by positively regulating the Wnt/PCP pathway in the presence of the proteoglycan Syndecan4 (Glinka et al., 2011; Ohkawara et al., 2011), which has been shown to regulate CE through PCP signal in *Xenopus* (Munoz et al., 2006). Conversely, Lrp disrupts CE by negatively regulating the Wnt/PCP pathway in *Xenopus* (Tahinci et al., 2007).

It appears that downstream of Dsh is further divergent, and this can be explained by the presence of a variety of Dsh-binding proteins that bias one branch of the noncanonical Wnt pathway to another. The forming-homology protein Daam1, being identified as a binding protein for Dsh and RhoA, mediates CE through activation of RhoA and Rac based on overexpression of mutant forms of Daam1 (Habas et al., 2001) although it remains to be clarified whether Daam1 fulfills downstream mediator of Dsh based on loss-of-function studies. Casein kinase Iε binds to Dsh and is capable of modulating CE but rather acts by negatively regulating canonical Wnt/β-catenin signal through functional interaction with Wnt5/Ror2 (McKay et al., 2001; Witte et al., 2010). Similarly, Inv and Naked both bind to Dsh to facilitate the Wnt/PCP pathway by antagonizing Wnt/β-catenin signal (Simons et al., 2005; Van Raay et al., 2007). The clathrin AP-2 adaptor μ2-adaptin is associated with Dsh, and mediates Fzd endocytosis, which is required for PCP signaling and thus for the regulation of CE movements in *Xenopus* (Yu et al., 2007). This is supported by the observation in cultured cells that interaction of Wnt5 with Fzd induces internalization of the signaling complex including Dsh and β-arrestin presumably through the clathrin-mediated endocytosis (Chen et al., 2003). Together with the observation that β-arrestin is required for CE and mediates downstream of Dsh (Kim & Han, 2007), these results suggest potential involvement of heterotrimeric G-proteins in Wnt/PCP signaling. Consistent with this notion, the heterotrimeric G-protein a subunit can transduce Wnt/Ca^{2+} signaling in zebrafish embryos (Slusarski et al., 1997), and $G\alpha_{11}$ mediates downstream of Wnt11 and is required for CE in *Xenopus* (Iioka et al., 2007), and $G\alpha_{12/13}$ is required for CE in zebrafish (Lin et al., 2005).

In addition, there are inputs at the level of other core PCP components to regulate CE. Rack1, a Vangl2-interacting protein, is required for CE and is capable of antagonizing Wnt/β-catenin signal (Li et al., 2011). Being identified as a *Drosophila* homologue of misshapen, Mink1 can bind to and phosphorylates Pk, and phosphorylated Pk is able to form a complex with Vangl2 at the membrane while mediating CE in *Xenopus* (Daulat et al., 2012).

It is less clear evidence for homologues of downstream mediators of *Drosophila* PCP, mediating CE in zebrafish and *Xenopus*. Abrogation of *inturned* or *fuzzy* leads to only mild CE phenotypes in *Xenopus* (Park et al., 2006). Consistent with this observation, *fritz* is required for only cell elongation, but not for cell polarization, of chordamesoderm cells, whereas *dsh* is required for both processes (Kim et al., 2010).

There are debatable issues as to whether the Fat/Dachsous (Ds) pathway acts in the Fz/PCP pathway or in a parallel pathway in *Drosophila* (e.g., Lawrence et al., 2008a). In addition, the Fat/Ds regulates growth control through the Hippo pathway. There is no definitive evidence that Fat/Ds acts in the Wnt/PCP pathway in vertebrates although *fat4* genetically interacts with *vangl2* (Saburi et al., 2008; Saburi et al., 2012).

3.4. Localization of core PCP proteins and CE behavior

In the authentic view of PCP in the *Drosophila* wing, the core PCP proteins are asymmetrically localized to establish planar polarization such that the complexes Fz-Dsh and Vang-Pk are localized on distal and proximal sides, respectively, through the mutually antagonistic interactions (Bastock et al., 2003; Jenny et al., 2005; Tree et al., 2002). The long-range cell-nonautonomous effects of PCP are mediated by the Fmi, Fz, and Vang through different mechanisms in *Drosophila* (Lawrence et al., 2008b). During zebrafish gastrulation, similarly, anteriorly localized Pk and posteriorly biased Dsh are observed in both somite and notochord progenitor cells undergoing CE, and the asymmetric localization of these proteins is disrupted in core PCP mutants (Yin et al., 2008). Consistent with the idea that mutually antagonistic activities between Pk and Dsh can establish such asymmetric localization of the two proteins, overexpression of Pk leads to downregulation of Dsh in the zebrafish gastrula (Carreira-Barbosa et al., 2003). Interestingly, the anteriorly biased Pk localization is microtubules (MT) dependent (Sepich et al., 2011), and this is indeed supported by the notion that MT polarity is upstream of the establishment of PCP in the *Drosophila* wing (Shimada et al., 2006). Despite the asymmetric localization using *Drosophila* GFP-Pk and *Xenoups* Dvl2-GFP (Yin et al., 2008), neither zebrafish GFP-Pk1b nor Dvl2-GFP exhibits such asymmetric localization in notochord progenitors (M. Tada, unpublished results). To exclude the possibility of ectopic localization primarily due to gain-of-function effects, this issue needs to be clarified by using antibodies to detect endogenous localization or by knocking-in GFP-tagged constructs into the endogenous locus based on BAC manipulation.

Another interesting feature of lateral mesoderm progenitors undergoing CE is that the positioning of microtubules organizing centre (MTOC) is posteromedially or posterolaterally biased, and its localization is randomized in embryos with compromised PCP function (Sepich et al., 2011). This MTOC localization is different to the authentic view in which MTOC is localized toward the direction of their migration in cultured cells, despite the fact that noncanonical Wnt/PCP signal regulates MT polarity in cultured cells (Schlessinger et al., 2007). In relation to the ability of Wnt/PCP signal to modulate MT polarity, there is increasing evidence that ciliary-associated proteins can regulate CE by interacting with the Wnt/PCP pathway including Inv, Bbs4, and Bbs8 (Gerdes et al., 2007; May-Simera et al., 2010; Ross et al., 2005; Simons et al., 2005).

What establishes AP-biased Pk and Dsh in cells undergoing CE? Presumably, the cue can establish the AP polarity is a gradient of Nodal signal as demonstrated in animal caps treated with different doses of Activin. Only at the interface between high and low Activin, activities confer CE cell behavior (Ninomiya et al., 2004). This hypothesis needs to be clarified *in vivo* in future.

4. PCP REGULATING CE AND COLLECTIVE CELL MIGRATION IN OTHER CONTEXTS

4.1. Collective migration of individual mesenchymal cells—Neural crest cells

Neural crest cells in the cranial region of the embryo are delaminated from the dorsal neural tube and migrate as individual mesenchymal cells in the stream. Migrating neural crest cells are highly polarized with actin-rich processes to orient the direction of their migration (Matthews et al., 2008). Despite their individual migration, there are rare neighbor exchanges involved. There are two mechanisms to achieve collective migration of neural crest cells. First, contact inhibition of locomotion (CIL) allows cells to inhibit from their random movement by collapsing their processes followed by changes in their direction upon encountering between the two cells (Carmona-Fontaine et al., 2008). Second, coattraction by which cells can attract another one at a distance via diffusible molecule(s) maintains their coherence (Carmona-Fontaine et al., 2011). Thus, these two mechanisms counterbalance each other to permit cells to undergo collective migration. Wnt11 signaling through Dsh and RhoA is required for polarization of migrating neural crest cells (Matthews et al., 2008). Moreover, the Wnt/PCP pathway is indispensable for CIL behaviors in that Wnt11, Fzd7, Dsh, and activated RhoA are all accumulated at cell–cell contact in cultured neural crest cells (Carmona-Fontaine et al., 2008). Together, the Wnt/PCP pathway plays a pivotal role in regulating collective migration of neural crest cells.

4.2. Collective migration of a small cluster of mesenchymal cells—Neuronal migration

In contrast to a large population of coherent sheet of prechordal plate progenitors, an example of a small cluster of cohesive cells undergoing collective migration is facial branchiomotor (FBM) neurons in zebrafish and mouse embryos. FBM neurons are born in the rhombomere 4 of the hindbrain and caudally migrate into the rhombomere 7 territory before projecting their axons toward their target tissue branchial arches. This collective migration is PCP dependent such that migration of FBM neurons is totally abolished in the core PCP zebrafish mutants including *vangl2*, *scrb*, *fzd3a*, and *celsr2* (if abrogated with Celsr1a/1b activities) (Jessen et al., 2002; Wada et al., 2006; Wada et al., 2005) and morphants such as *pk1a* and *pk1b* (Carreira-Barbosa et al., 2003; Mapp et al., 2010). Importantly, it appears that this process is Dsh-independent (Bingham et al., 2002). Different to zebrafish, FBM neurons migrate in a sequential manner in mice in that the neurons migrate caudally then laterally. This is evident in *celsr2;celsr3* double mutants or *fzd3* mutants, in which caudal migration of FMB neurons is suppressed while their lateral migration is relatively normal (Qu et al., 2010).

Both cell-autonomous (in migrating neurons) and cell-non-autonomous (in the neuroepithelia) functions of PCP are required for proper collective migration (Jessen et al., 2002; Wada et al., 2005; Wada et al., 2006). Interestingly, Pk1b, whose expression is only in migrating neurons but not in the neuroepithelia, mediates collective migration independently of PCP function, and rather its ability is associated with its nuclear localization (Mapp et al., 2011). In addition to Pk1b, forward genetic approach in zebrafish identified a potential new mediator of PCP in this process. The gene, which encodes Nance–Horan syndrome-like protein 1b with a WAVE homology domain as a potential regulator of actin cytoskeleton, is required only within the migrating neurons (Walsh et al., 2011). Currently, it is unknown whether the PCP pathway controls cell cohesion of migrating FBM neurons. Considering the fact that neurons possess higher cohesive properties than surrounding neuroepithelial cells (Stockinger et al., 2011), it would be plausible since Pk is required for cohesive properties of migrating dorsal forerunner cells during zebrafish gastrulation (Oteiza et al., 2010).

The interpretation for cell-nonautonomous requirement of the PCP pathway in the zebrafish neuroepithelia is a complex issue as to what a primary defect is. In contrast to other vertebrates, zebrafish neurulation involves oriented cell division (crossing-division) that creates the neural tube with a lumen from the neural keel mediated by polarized mesenchymal cells (Tawk et al., 2007). Despite the ability of the cells to stretch and to acquire apicobasal polarity along the mediolateral axis of the embryo, the core PCP mutants exhibit defects in several different aspects of morphogenesis during neurulation such as the orientation and planar polarization of the cells in their AP axis and ectopic lumen formation (Ciruna et al., 2006; Tawk et al., 2007; Zigman et al., 2011). In addition, proper coherent properties of neuroepithelia allow a coherent cluster of FBM neurons to migrate caudally by preventing from their ectopic migration (Stockinger et al., 2011). To clarify the issue of cell-non-autonomous function of PCP during FMB neuron migration, a genetic-based conditional loss-of-function approach in time and in space will be required in future.

4.3. Collective migration of a group of epithelial cells—Mice AVE

Epithelial cells of the mouse AVE migrate as a group from the distal tip of the mouse cylinder proximally and orient the future body axis prior to gastrulation. AVE cells migrate actively with basal protrusions to the direction of their migration while undergoing cell intercalation both among themselves and with the surrounding cells (Migeotte et al., 2010; Srinivas et al., 2004; Trichas et al., 2011). This directional migration is mediated by the formation of rosette, which is defined by five cells or more meeting at the point/vertex (Trichas et al., 2012). As AVE cells migrate anteriorly, the mean rosette density increases in the visceral endoderm, although there is no correlation between the spatial localization of rosettes and the direction of their migration. Based on a mathematical model, the potential role for the rosettes is proposed to ensure the coherent interaction between the AVE cells. Importantly, embryos with disrupted PCP signaling show a reduction in the rosette density, thereby exhibiting defective AVE migration (Trichas et al., 2012). Consistent with the notion that AVE migration is a PCP-dependent process, Dvl2 is enriched in the lateral membrane of migrating AVE cells (Trichas et al., 2011). Despite the fact that this strategy is similar to cells undergoing CE during *Drosophila* germband extension (Bertet et al., 2004; Blankenship et al., 2006), there are differences between the two systems. While mouse AVE cells utilize the rosettes by means of their coordination and coherence rather than being related to their directionality, cells undergoing CE during germband extension mediate to form and resolve the rosettes in a directional manner, thereby being associated with the force generation process underlying directionality. Further, CE during germband extension is a PCP-independent process (Zallen & Wieschaus, 2004).

4.4. PCP mediating CE in tube elongation—Cochlea

Epithelial cells of the mouse cochlear also undergo CE, contributing to the elongation of the cochlear duct in the organ of Corti. This process involves junctional remodeling reminiscent of rosette formation during *Drosophila* germband extension (Bertet et al., 2004; Blankenship et al., 2006).

Consistent with the CE cell behavior underlying germband extension, CE in the mouse cochlea is mediated by Myosin II and planar polarization of E-/N-cadherin, both of which are presumably associated with a force-generated process (Chacon-Heszele et al., 2012; Yamamoto, Okano, Ma, Adelstein, & Kelley, 2009). In contrast to PCP-independent mechanisms during germband extension, core PCP mutants exhibit wider and shorter cochlear ducts (Montcouquiol et al., 2003; Wang, Guo, et al., 2006; Wang, Hamblet, et al., 2006; Wang et al., 2005), suggesting PCP-dependence during cochlear CE. Interestingly, during cochlear CE, polarized localization of Cadherins is altered in the *vangl2* mutants, suggesting core PCP signal and classical Cadherins mediate this process (Chacon-Heszele et al., 2012). Likewise, *fat4* and *ds* mutants show the cochlea CE phenotype (Mao et al., 2011). However, the elongation defect might account for defective oriented cell division as well which has been implicated for PCP. To clarify what is the primary defect during cochlear CE awaits further investigation.

4.5. PCP mediating cartilage elongation

It appears that the elongation of developing cartilages involves morphogenesis of mesenchymal cells, reminiscent of CE during gastrulation. Some of the PCP mutant fish (*gpc4* and *wnt5b*) show elongation defects in jaw cartilages, thereby exhibiting a hammer-like head structure (LeClair et al., 2009; Rauch et al., 1997). However, it is unclear whether this process is dependent largely on CE as cell proliferation significantly accounts for the elongation. Indeed, it seems to be the case for cell proliferation, contributing to the elongation of the mouse limb cartilage. A distal to proximal gradient of Wnt5a regulates the coordinated proliferation and differentiation of chondrocytes in the mouse limb (Yamaguchi et al., 1999). This process is mediated by its receptor Ror2, which in turn phosphorylates Vangl2 and induces asymmetric localization of Vangl2, propagating an activity gradient of Vangl2 in the proximal direction (Gao et al., 2011). In support of this, *ror2* mutants abolish Vangl2 activity gradient and localization, and *ror2;vangl2* double mutants phenocopy the *wnt5a* limb phenotype (Gao et al., 2011).

4.6. PCP mediating CE in primitive streak formation in chick

During chick streak formation prior to gastrulation, global cell flow-like polonaise movements within an epithelial sheet of the epiblast dictate the position of streak formation (Cui et al., 2005). These epithelial cells undergo cell intercalations, reminiscent of CE, contributing to the elongation of the primitive streak (Lawson & Schoenwolf, 2001; Voiculescu et al., 2007).

Importantly, abrogation of core PCP genes interferes with cell intercalations, thereby leading to inhibition of streak formation (Voiculescu et al., 2007). These observations suggest that the PCP pathway plays a pivotal role in regulating elongation of the body axis through CE regardless of prior to or during gastrulation.

5. PCP IN MEDIATING COLLECTIVENESS AND POLARIZED BEHAVIORS

5.1. CE and contact inhibition of locomotion

The separable convergence and extension movements in fish might be due to the density of lateral mesoderm cells lesser than that of the *Xenopus* gastrula. This inspires us to interpret CE at least in zebrafish as part of collective cell migration. How do lateral mesoderm cells read their density while undergoing convergence movement? By analogy to collective migration of neural crest cells (Carmona-Fontaine et al., 2008), one possibility is that cells undergoing CE may involve contact inhibition of locomotion (CIL) in a dispersed location and utilize the Wnt/PCP pathway to ensure this event to occur. In other words, a possible role for the Wnt/PCP pathway is to suppress "randomness" and to maintain proper cell orientation. Consistent with this idea, N-cadherin is required for proper dorsal convergence as well as CIL in neural crest cells (Theveneau et al., 2010; von der Hardt et al., 2007). Moreover, the ability of the Wnt/PCP pathway to modulate actomyosin is correlated with suppression of ectopic blebs, which can be induced by abrogation of myosin phosphatase in cells undergoing CE (Weiser et al., 2009). In other words, if the Wnt/PCP activity is compromised, randomness of lateral mesoderm cells increases as a result of increased blebbing activity or possibly owing to loss-of-CIL behavior of the cells.

During collective migration of prechordal progenitors, if cell cohesion is reduced in *wnt11* mutants, the cells reduce net migration while remaining tightly packed (Ulrich et al., 2005). However, proper cell cohesion can be dispensable for net migration of prechordal plate progenitors, as defective migration of *wnt11*-mutant cells is restored if the cells acquire increased motility while retaining reduced cell cohesion (Kai et al., 2008). Interestingly, in this situation, the cells migrate more randomly, presumably due to reduced CIL behavior with increased motility. This is consistent with the idea that the Wnt/PCP pathway plays a fundamental role in reducing "randomness." This inspires us to hypothesize that mesenchymal cells utilize the common strategy by which to suppress randomness regardless of their density.

5.2. CE and cell–cell adhesion

One unique feature of the ability of the Wnt/PCP pathway to modulate cell–cell adhesion is mediated through the family of atypical Cadherin, which does not utilize the module of Catenin complex linking with actin cytoskeleton. Several lines of evidence support the notion that Fmi/Celsr regulates cell cohesion/adhesion underlying CE behaviors. First, ectopic expression of *celsr2* increases cell–cell contact persistency in cells of the zebrafish blastula and colocalizes with Wnt11-Fzd7 at cell–cell contact (Witzel et al., 2006). Second, dissociated cells with decreased endogenous Celsr activity from zebrafish blastula embryos are segregated out of wild-type cells in hanging drop assays (Carreira-Barbosa et al., 2009). This is consistent with the observation that, when *celsr2* is overexpressed, individually floating S2 cells become cohesive, making a cluster of the cells (Shima et al., 2004). However, loss of function of *celsr1a/1b/2* leads to a defective epiboly phenotype, which arises prior to CE taking place, making it difficult to analyze CE cell behavior properly (Carreira-Barbosa et al., 2009). Rather than the stereotypic CE phenotype in zebrafish, embryos with abrogated Celsr activity exhibit the epiboly phenotype, reminiscent of hypomorphic alleles of *e-cadherin* mutants (Kane et al., 2005). This raises the possibility that Celsr modulates cell adhesion/cohesion through functional interaction with E-cadherin either directly or indirectly.

The other atypical Cadherin mediating CE is paraxial protocadherin (Papc). *papc* is identified as a transcriptional target for the Wnt5/Ror2/JNK pathway in *Xenopus* (Schambony & Wedlich, 2007) and has been shown to be required for CE in *Xenopus* (Medina et al., 2004; Unterseher et al., 2004; Wang et al., 2008). One possible mechanism by which Papc modulates cell adhesion is to act through classical Cadherins (Chen & Gumbiner, 2006). In support of this view, a ventrodorsal gradient of BMP signal mediates convergence movement by modulating a dorsoventral gradient of N-cadherin activity in zebrafish (von der Hardt et al., 2007). However, there is no direct evidence for Papc to act downstream of BMP signal in *Xenopus*.

5.3. Cell–substrate adhesion and integration into apicobasal polarity

The ability of Wnt/PCP signal to modulate cell–substrate adhesion is also fundamental to the regulation of CE. A positive feedback loop, in which Wnt11 regulates Fibronectin assembly (Dzamba et al., 2009) and Fibronectin-Integrin signal in turn modulates Wnt/PCP pathway activities (Davidson et al., 2006; Goto et al., 2005; Munoz et al., 2006), plays an important role in regulating cell polarization underlying CE in *Xenopus*. The ability of Vangl2 to modulate deposition of the extracellular matrix through the regulation of endocytosis of the metalloproteinase MMP14 is associated with polarization of the cells undergoing CE (Williams et al., 2012). Collectively, these suggest that the Wnt/PCP pathway mediates cell polarization underlying CE by organizing extracellular matrices in between mesoderm cells and the overlying ectoderm.

In a sheet of epithelia, basal lamina acts as a cue for apicobasal polarization. However, little is known whether Fibronectin can provide with a basal input into CE behavior of mesenchymal cells and how the apical complex Par3–Par6–aPKC feeds into CE cell behavior. Par6 is required for CE (Kusakabe & Nishida, 2004) and ArfGAP, which can form a complex with Par6 and aPKC, mediates cell polarization underlying CE independently of the Wnt/PCP pathway (Hyodo-Miura et al., 2006). In contrast, Par1, which is known to localize mutually exclusive to the apical complex in several different contexts, regulates CE by mediating Dsh phosphorylation (Kusakabe & Nishida, 2004; Ossipova et al., 2005). Par6 can form a complex with Wnt5-mediated phosphorylated Dvl2 and the ubiquitin ligase Smurf, which in turn targets Pk for degradation (Narimatsu et al., 2009). In addition, *smurf1;smurf2* double mutants and embryos expressing a mutant form of Par6 lacking a Dvl2-binding domain both exhibit severe CE defects in mice (Narimatsu et al., 2009). Together, the mediators of apicobasal polarity and PCP feed in regulating cell polarization independently but can cooperate to mediate CE behavior. It remains to be investigated how these two pathways orchestrate coordinated cell movements.

5.4. Tissue elongation and oriented cell division

In addition to oriented cell interaction (CE), oriented cell division can account for the elongation of a tissue. Indeed, the Wnt/PCP pathway mediates cell division oriented in the AP axis of the epiblast/neural plate in zebrafish (Gong et al., 2004; Quesada-Hernandez et al., 2010). How does the Wnt/PCP pathway regulate oriented cell division? NuMA, a protein associated with nuclear mitotic apparatus, colocalizes with Dvl2 at the cleavage plane of a diving cell and controls mitotic spindle orientation in epiblast cells, and acts downstream of the Wnt/PCP pathway in zebrafish (Segalen et al., 2010). The Fat-Ds pathway has also been implicated in the regulation of oriented cell division, acting as a global cue in the *Drosophila* wing (Mao et al., 2011; Rogulja et al., 2008). Thus, both Wnt/PCP and Fat-Ds pathways can contribute to tissue elongation. Despite its requirement for this process, Wnt/PCP-mediated orientated cell division does not account for the elongation of the body axis, reemphasizing a primary role of the Wnt/PCP pathway in regulating cell intercalation during gastrulation.

Likewise, other processes, in which Wnt/PCP signal has been proposed to mediate CE or collective migration, require reassessment as to whether or not Wnt/PCP signal mediates oriented cell division, accounting for tissue elongation or directionality. For example, during cochlear tube elongation, it appears that significant tissue growth contributes to the elongation and that the Fat-Ds pathway is involved in this process as well as the Wnt/PCP pathway. Therefore, it requires assessment for new regulators as to what is a primary cause and which pathway to interact rather than judging by the terminal phenotype. Similarly, it is plausible that oriented cell division mediated by the Wnt/PCP pathway may in part account for collective migration of mouse AVE cells as significant tissue growth is occurring during their migration.

6. CONCLUDING REMARKS

A variety of different developmental processes underlying collective cell migration and cell intercalation utilize the Wnt/PCP pathway as a conserved genetic module, regardless of the size of cell populations in both mesenchymal and epithelial tissues. The concept of planar polarization applies to understanding of how cells communicate and coordinate in the plane of the tissue. Despite the enormous progress in identification of the modulators of the PCP pathway and of the biological processes involving the PCP pathway, little is known about the cell-nonautonomous mechanisms by which the cells know where they are within the tissue and propagate the signal(s) to their neighbors. Further identification of novel cellular and developmental processes utilizing the PCP pathway and of genetic pathway(s) that interact or counteract with the PCP pathway will provide us with the clues to understanding of the cell-nonautonomous mechanisms underlying the coordination and directionality of cells.

ACKNOWLEDGMENTS

We thank Roberto Mayor and Shankar Srinivas for critical reading of the manuscript. MT is supported by the MRC and Royal Society. MK is supported by KAKENHI 12640066.

REFERENCES

Albertson, R., & Doe, C. (2003). Dlg, Scrib and Lgl regulate neuroblast cell size and mitotic spindle asymmetry. *Nature Cell Biology*, *5*, 166–170.

Bastock, R., Strutt, H., & Strutt, D. (2003). Strabismus is asymmetrically localised and binds to Prickle and Dishevelled during *Drosophila* planar polarity patterning. *Development*, *130*, 3007–3014.

Bertet, C., Sulak, L., & Lecuit, T. (2004). Myosin-dependent junction remodelling controls planar cell intercalation and axis elongation. *Nature*, *429*, 667–671.

Bingham, S., Higashijima, S., Okamoto, H., & Chandrasekhar, A. (2002). The Zebrafish *trilobite* gene is essential for tangential migration of branchiomotor neurons. *Developmental Biology*, *242*, 149–160.

Blankenship, J. T., Backovic, S. T., Sanny, J. S., Weitz, O., & Zallen, J. A. (2006). Multicellular rosette formation links planar cell polarity to tissue morphogenesis. *Developmental Cell*, *11*, 459–470.

Buechling, T., Bartscherer, K., Ohkawara, B., Chaudhary, V., Spirohn, K., Niehrs, C., et al. (2010). Wnt/Frizzled signaling requires dPRR, the Drosophila homolog of the prorenin receptor. *Current Biology*, *20*, 1263–1268.

Carmon, K. S., Gong, X., Lin, Q., Thomas, A., & Liu, Q. (2011). R-spondins function as ligands of the orphan receptors LGR4 and LGR5 to regulate Wnt/beta-catenin signaling. *Proceedings of the National Academy of Sciences of the United States of America*, *108*, 11452–11457.

Carmona-Fontaine, C., Matthews, H. K., Kuriyama, S., Moreno, M., Dunn, G. A., Parsons, M., et al. (2008). Contact inhibition of locomotion in vivo controls neural crest directional migration. *Nature*, *456*, 957–961.

Carmona-Fontaine, C., Theveneau, E., Tzekou, A., Tada, M., Woods, M., Page, K. M., et al. (2011). Complement fragment C3a controls mutual cell attraction during collective cell migration. *Developmental Cell*, *21*, 1026–1037.

Carreira-Barbosa, F., Concha, M. L., Takeuchi, M., Ueno, N., Wilson, S. W., & Tada, M. (2003). Prickle 1 regulates cell movements during gastrulation and neuronal migration in zebrafish. *Development*, *130*, 4037–4046.

Carreira-Barbosa, F., Kajita, M., Morel, V., Wada, H., Okamoto, H., Martinez Arias, A., et al. (2009). Flamingo regulates epiboly and convergence/extension movements through cell cohesive and signalling functions during zebrafish gastrulation. *Development*, *136*, 383–392.

Chacon-Heszele, M. F., Ren, D., Reynolds, A. B., Chi, F., & Chen, P. (2012). Regulation of cochlear convergent extension by the vertebrate planar cell polarity pathway is dependent on p120-catenin. *Development*, *139*, 968–978.

Chen, X., & Gumbiner, B. M. (2006). Paraxial protocadherin mediates cell sorting and tissue morphogenesis by regulating C-cadherin adhesion activity. *The Journal of Cell Biology*, *174*, 301–313.

Chen, W., ten Berge, D., Brown, J., Ahn, S., Hu, L. A., Miller, W. E., et al. (2003). Dishevelled 2 recruits β-arrestin 2 to mediate Wnt5A-stimulated endocytosis of Frizzled 4. *Science*, *301*, 1391–1394.

Chung, H. A., Hyodo-Miura, J., Nagamune, T., & Ueno, N. (2005). FGF signal regulates gastrulation cell movements and morphology through its target NRH. *Developmental Biology*, *282*, 95–110.

Ciruna, B., Jenny, A., Lee, D., Mlodzik, M., & Schier, A. F. (2006). Planar cell polarity signalling couples cell division and morphogenesis during neurulation. *Nature*, *439*, 220–224.

Cui, C., Yang, X., Chuai, M., Glazier, J. A., & Weijer, C. J. (2005). Analysis of tissue flow patterns during primitive streak formation in the chick embryo. *Developmental Biology*, *284*, 37–47.

Curtin, J. A., Quint, E., Tsipouri, V., Arkell, R. M., Cattanach, B., Copp, A. J., et al. (2003). Mutation of Celsr1 disrupts planar polarity of inner ear hair cells and causes severe neural tube defects in the mouse. *Current Biology*, *13*, 1129–1133.

Darken, R. S., Scola, A. M., Rakeman, A. S., Das, G., Mlodzik, M., & Wilson, P. A. (2002). The planar polarity gene *strabismus* regulates convergent extension movements in *Xenopus*. *The EMBO Journal*, *21*, 976–985.

Daulat, A. M., Luu, O., Sing, A., Zhang, L., Wrana, J. L., McNeill, H., et al. (2012). Mink1 regulates beta-catenin-independent Wnt signaling via Prickle phosphorylation. *Molecular and Cellular Biology*, *32*, 173–185.

Davidson, L. A., Marsden, M., Keller, R., & Desimone, D. W. (2006). Integrin α5β1 and fibronectin regulate polarized cell protrusions required for *Xenopus* convergence and extension. *Current Biology*, *16*, 833–844.

de Lau, W., Barker, N., Low, T. Y., Koo, B. K., Li, V. S., Teunissen, H., et al. (2011). Lgr5 homologues associate with Wnt receptors and mediate R-spondin signalling. *Nature*, *476*, 293–297.

Diz-Munoz, A., Krieg, M., Bergert, M., Ibarlucea-Benitez, I., Muller, D. J., Paluch, E., et al. (2010). Control of directed cell migration in vivo by membrane-to-cortex attachment. *PLoS Biology*, *8*, e1000544.

Djiane, A., Riou, J., Umbhauer, M., Boucaut, J., & Shi, D. (2000). Role of *frizzled* 7 in the regulation of convergent extension movements during gastrulation in *Xenopus laevis*. *Development*, *127*, 3091–3100.

Dzamba, B. J., Jakab, K. R., Marsden, M., Schwartz, M. A., & DeSimone, D. W. (2009). Cadherin adhesion, tissue tension, and noncanonical Wnt signaling regulate fibronectin matrix organization. *Developmental Cell*, *16*, 421–432.

Elul, T., & Keller, R. (2000). Monopolar protrusive activity: A new morphogenic cell behavior in the neural plate dependent on vertical interactions with the mesoderm in *Xenopus*. *Developmental Biology*, *224*, 3–19.

Elul, T., Koehl, M. A., & Keller, R. (1997). Cellular mechanism underlying neural convergent extension in *Xenopus laevis* embryos. *Developmental Biology*, *191*, 243–258.

Formstone, C. J., & Mason, I. (2005). Combinatorial activity of Flamingo proteins directs convergence and extension within the early zebrafish embryo via the planar cell polarity pathway. *Developmental Biology*, *282*, 320–335.

Friedl, P., & Gilmour, D. (2009). Collective cell migration in morphogenesis, regeneration and cancer. *Nature Reviews. Molecular Cell Biology*, *10*, 445–457.

Gao, B., Song, H., Bishop, K., Elliot, G., Garrett, L., English, M. A., et al. (2011). Wnt signaling gradients establish planar cell polarity by inducing Vangl2 phosphorylation through Ror2. *Developmental Cell*, *20*, 163–176.

Gerdes, J. M., Liu, Y., Zaghloul, N. A., Leitch, C. C., Lawson, S. S., Kato, M., et al. (2007). Disruption of the basal body compromises proteasomal function and perturbs intracellular Wnt response. *Nature Genetics*, *39*, 1350–1360.

Glinka, A., Dolde, C., Kirsch, N., Huang, Y. L., Kazanskaya, O., Ingelfinger, D., et al. (2011). LGR4 and LGR5 are R-spondin receptors mediating Wnt/β-catenin and Wnt/PCP signalling. *EMBO Reports*, *12*, 1055–1061.

Gong, Y., Mo, C., & Fraser, S. E. (2004). Planar cell polarity signalling controls cell division orientation during zebrafish gastrulation. *Nature*, *430*, 689–693.

Goto, T., Davidson, L., Asashima, M., & Keller, R. (2005). Planar cell polarity genes regulate polarized extracellular matrix deposition during frog gastrulation. *Current Biology*, *15*, 787–793.

Goto, T., & Keller, R. (2002). The planar cell polarity gene strabismus regulates convergence and extension and neural fold closure in *Xenopus*. *Developmental Biology*, *247*, 165–181.

Gray, R. S., Roszko, I., & Solnica-Krezel, L. (2011). Planar cell polarity: Coordinating morphogenetic cell behaviors with embryonic polarity. *Developmental Cell*, *21*, 120–133.

Habas, R., Dawid, I. B., & He, X. (2003). Coactivation of Rac and Rho by Wnt/Frizzled signaling is required for vertebrate gastrulation. *Genes & Development*, *17*, 295–309.

Habas, R., Kato, Y., & He, X. (2001). Wnt/Frizzled activation of rho regulates vertebrate gastrulation and requires a novel formin homology protein daam1. *Cell*, *107*, 843–854.

Hamblet, N. S., Lijam, N., Ruiz-Lozano, P., Wang, J., Yang, Y., Luo, Z., et al. (2002). Dishevelled 2 is essential for cardiac outflow tract development, somite segmentation and neural tube closure. *Development*, *129*, 5827–5838.

Heisenberg, C. P., Tada, M., Rauch, G. J., Saude, L., Concha, M. L., Geisler, R., et al. (2000). Silberblick/Wnt11 mediates convergent extension movements during zebrafish gastrulation. *Nature*, *405*, 76–81.

Hikasa, H., Shibata, M., Hiratani, I., & Taira, M. (2002). The *Xenopus* receptor tyrosine kinase Xror2 modulates morphogenetic movements of the axial mesoderm and neuroectoderm via Wnt signaling. *Development*, *129*, 5227–5239.

Hyodo-Miura, J., Yamamoto, T. S., Hyodo, A. C., Iemura, S., Kusakabe, M., Nishida, E., et al. (2006). XGAP, an ArfGAP, is required for polarized localization of PAR proteins and cell polarity in *Xenopus* gastrulation. *Developmental Cell*, *11*, 69–79.

Iioka, H., Iemura, S., Natsume, T., & Kinoshita, N. (2007). Wnt signalling regulates paxillin ubiquitination essential for mesodermal cell motility. *Nature Cell Biology*, *9*, 813–821.

Jenny, A., Reynolds-Kenneally, J., Das, G., Burnett, M., & Mlodzik, M. (2005). Diego and Prickle regulate Frizzled planar cell polarity signalling by competing for Dishevelled binding. *Nature Cell Biology*, *7*, 691–697.

Jessen, J. R., Topczewski, J., Bingham, S., Sepich, D. S., Marlow, F., Chandrasekhar, A., et al. (2002). Zebrafish trilobite identifies new roles for Strabismus in gastrulation and neuronal movements. *Nature Cell Biology*, *4*, 610–615.

Kai, M., Heisenberg, C. P., & Tada, M. (2008). Sphingosine-1-phosphate receptors regulate individual cell behaviours underlying the directed migration of prechordal plate progenitor cells during zebrafish gastrulation. *Development*, *135*, 3043–3051.

Kane, D. A., McFarland, K. N., & Warga, R. M. (2005). Mutations in *half baked*/E-cadherin block cell behaviors that are necessary for teleost epiboly. *Development*, *132*, 1105–1116.

Keller, R. (2002). Shaping the vertebrate body plan by polarized embryonic cell movements. *Science*, *298*, 1950–1954.

Keller, R., & Danilchik, M. (1988). Regional expression, pattern and timing of convergence and extension during gastrulation of *Xenopus laevis*. *Development*, *103*, 193–209.

Kibar, Z., Vogan, K. J., Groulx, N., Justice, M. J., Underhill, D. A., & Gros, P. (2001). *Ltap*, a mammalian homolog of *Drosophila Strabismus/Van Gogh*, is altered in the mouse neural tube mutant Loop-tail. *Nature Genetics*, *28*, 251–255.

Kilian, B., Mansukoski, H., Barbosa, F. C., Ulrich, F., Tada, M., & Heisenberg, C. P. (2003). The role of Ppt/Wnt5 in regulating cell shape and movement during zebrafish gastrulation. *Mechanisms of Development*, *120*, 467–476.

Kim, G. H., & Han, J. K. (2007). Essential role for β-arrestin 2 in the regulation of *Xenopus* convergent extension movements. *The EMBO Journal*, *26*, 2513–2526.

Kim, S. K., Shindo, A., Park, T. J., Oh, E. C., Ghosh, S., Gray, R. S., et al. (2010). Planar cell polarity acts through septins to control collective cell movement and ciliogenesis. *Science*, *329*, 1337–1340.

Kinoshita, N., Iioka, H., Miyakoshi, A., & Ueno, N. (2003). PKC δ is essential for Dishevelled function in a noncanonical Wnt pathway that regulates *Xenopus* convergent extension movements. *Genes & Development*, *17*, 1663–1676.

Kusakabe, M., & Nishida, E. (2004). The polarity-inducing kinase Par-1 controls *Xenopus* gastrulation in cooperation with 14-3-3 and aPKC. *The EMBO Journal*, *23*, 4190–4201.

Lawrence, P. A., Struhl, G., & Casal, J. (2008a). Do the protocadherins Fat and Dachsous link up to determine both planar cell polarity and the dimensions of organs? *Nature Cell Biology*, *10*, 1379–1382.

Lawrence, P. A., Struhl, G., & Casal, J. (2008b). Planar cell polarity: A bridge too far? *Current Biology*, *18*, R959–R961.

Lawson, A., & Schoenwolf, G. C. (2001). Cell populations and morphogenetic movements underlying formation of the avian primitive streak and organizer. *Genesis*, *29*, 188–195.

LeClair, E. E., Mui, S. R., Huang, A., Topczewska, J. M., & Topczewski, J. (2009). Craniofacial skeletal defects of adult zebrafish Glypican 4 (*knypek*) mutants. *Developmental Dynamics*, *238*, 2550–2563.

Li, S., Esterberg, R., Lachance, V., Ren, D., Radde-Gallwitz, K., Chi, F., et al. (2011). Rack1 is required for Vangl2 membrane localization and planar cell polarity signaling while attenuating canonical Wnt activity. *Proceedings of the National Academy of Sciences of the United States of America*, *108*, 2264–2269.

Lin, F., Sepich, D. S., Chen, S., Topczewski, J., Yin, C., Solnica-Krezel, L., et al. (2005). Essential roles of Gα12/13 signaling in distinct cell behaviors driving zebrafish convergence and extension gastrulation movements. *The Journal of Cell Biology*, *169*, 777–787.

Lu, X., Borchers, A. G., Jolicoeur, C., Rayburn, H., Baker, J. C., & Tessier-Lavigne, M. (2004). PTK7/CCK-4 is a novel regulator of planar cell polarity in vertebrates. *Nature*, *430*, 93–98.

Mao, Y., Mulvaney, J., Zakaria, S., Yu, T., Morgan, K. M., Allen, S., et al. (2011). Characterization of a Dchs1 mutant mouse reveals requirements for Dchs1-Fat4 signaling during mammalian development. *Development*, *138*, 947–957.

Mapp, O. M., Walsh, G. S., Moens, C. B., Tada, M., & Prince, V. E. (2011). Zebrafish Prickle1b mediates facial branchiomotor neuron migration via a farnesylation-dependent nuclear activity. *Development*, *138*, 2121–2132.

Mapp, O. M., Wanner, S. J., Rohrschneider, M. R., & Prince, V. E. (2010). Prickle1b mediates interpretation of migratory cues during zebrafish facial branchiomotor neuron migration. *Developmental Dynamics*, *239*, 1596–1608.

Marlow, F., Topczewski, J., Sepich, D., & Solnica-Krezel, L. (2002). Zebrafish rho kinase 2 acts downstream of wnt11 to mediate cell polarity and effective convergence and extension movements. *Current Biology*, *12*, 876–884.

Matthews, H. K., Marchant, L., Carmona-Fontaine, C., Kuriyama, S., Larrain, J., Holt, M. R., et al. (2008). Directional migration of neural crest cells in vivo is regulated by Syndecan-4/Rac1 and non-canonical Wnt signaling/RhoA. *Development*, *135*, 1771–1780.

May-Simera, H. L., Kai, M., Hernandez, V., Osborn, D. P., Tada, M., & Beales, P. L. (2010). Bbs8, together with the planar cell polarity protein Vangl2, is required to establish left-right asymmetry in zebrafish. *Developmental Biology*, *345*, 215–225.

McKay, R. M., Peters, J. M., & Graff, J. M. (2001). The casein kinase I family: Roles in morphogenesis. *Developmental Biology*, *235*, 378–387.

Medina, A., Swain, R. K., Kuerner, K. M., & Steinbeisser, H. (2004). *Xenopus* paraxial protocadherin has signaling functions and is involved in tissue separation. *The EMBO Journal*, *23*, 3249–3258.

Migeotte, I., Omelchenko, T., Hall, A., & Anderson, K. V. (2010). Rac1-dependent collective cell migration is required for specification of the anterior-posterior body axis of the mouse. *PLoS Biology*, *8*, e1000442.

Montcouquiol, M., Rachel, R. A., Lanford, P. J., Copeland, N. G., Jenkins, N. A., & Kelley, M. W. (2003). Identification of Vangl2 and Scrb1 as planar polarity genes in mammals. *Nature*, *423*, 173–177.

Montero, J. A., Carvalho, L., Wilsch-Brauninger, M., Kilian, B., Mustafa, C., & Heisenberg, C. P. (2005). Shield formation at the onset of zebrafish gastrulation. *Development*, *132*, 1187–1198.

Montero, J. A., Kilian, B., Chan, J., Bayliss, P. E., & Heisenberg, C. P. (2003). Phosphoinositide 3-kinase is required for process outgrowth and cell polarization of gastrulating mesendodermal cells. *Current Biology*, *13*, 1279–1289.

Munoz, R., Moreno, M., Oliva, C., Orbenes, C., & Larrain, J. (2006). Syndecan-4 regulates non-canonical Wnt signalling and is essential for convergent and extension movements in *Xenopus* embryos. *Nature Cell Biology*, *8*, 492–500.

Murdoch, J. N., Henderson, D. J., Doudney, K., Gaston-Massuet, C., Phillips, H. M., Paternotte, C., et al. (2003). Disruption of scribble (Scrb1) causes severe neural tube defects in the circletail mouse. *Human Molecular Genetics*, *12*, 87–98.

Murdoch, J. N., Rachel, R. A., Shah, S., Beermann, F., Stanier, P., Mason, C. A., et al. (2001). Circletail, a new mouse mutant with severe neural tube defects: Chromosomal localization and interaction with the loop-tail mutation. *Genomics*, *78*, 55–63.

Narimatsu, M., Bose, R., Pye, M., Zhang, L., Miller, B., Ching, P., et al. (2009). Regulation of planar cell polarity by Smurf ubiquitin ligases. *Cell*, *137*, 295–307.

Ninomiya, H., Elinson, R. P., & Winklbauer, R. (2004). Antero-posterior tissue polarity links mesoderm convergent extension to axial patterning. *Nature*, *430*, 364–367.

Ohkawara, B., Glinka, A., & Niehrs, C. (2011). Rspo3 binds syndecan 4 and induces Wnt/PCP signaling via clathrin-mediated endocytosis to promote morphogenesis. *Developmental Cell*, *20*, 303–314.

Ohkawara, B., Yamamoto, T. S., Tada, M., & Ueno, N. (2003). Role of glypican 4 in the regulation of convergent extension movements during gastrulation in *Xenopus laevis*. *Development*, *130*, 2129–2138.

Oishi, I., Suzuki, H., Onishi, N., Takada, R., Kani, S., Ohkawara, B., et al. (2003). The receptor tyrosine kinase Ror2 is involved in non-canonical Wnt5a/JNK signalling pathway. *Genes to Cells*, *8*, 645–654.

Ossipova, O., Dhawan, S., Sokol, S., & Green, J. B. (2005). Distinct PAR-1 proteins function in different branches of Wnt signaling during vertebrate development. *Developmental Cell*, *8*, 829–841.

Oteiza, P., Koppen, M., Krieg, M., Pulgar, E., Farias, C., Melo, C., et al. (2010). Planar cell polarity signalling regulates cell adhesion properties in progenitors of the zebrafish laterality organ. *Development*, *137*, 3459–3468.

Park, T. J., Haigo, S. L., & Wallingford, J. B. (2006). Ciliogenesis defects in embryos lacking inturned or fuzzy function are associated with failure of planar cell polarity and Hedgehog signaling. *Nature Genetics*, *38*, 303–311.

Park, M., & Moon, R. T. (2002). The planar cell-polarity gene stbm regulates cell behaviour and cell fate in vertebrate embryos. *Nature Cell Biology*, *4*, 20–25.

Paudyal, A., Damrau, C., Patterson, V. L., Ermakov, A., Formstone, C., Lalanne, Z., et al. (2010). The novel mouse mutant, chuzhoi, has disruption of Ptk7 protein and exhibits defects in neural tube, heart and lung development and abnormal planar cell polarity in the ear. *BMC Developmental Biology*, *10*, 87.

Qu, Y., Glasco, D. M., Zhou, L., Sawant, A., Ravni, A., Fritzsch, B., et al. (2010). Atypical cadherins Celsr1-3 differentially regulate migration of facial branchiomotor neurons in mice. *The Journal of Neuroscience*, *30*, 9392–9401.

Quesada-Hernandez, E., Caneparo, L., Schneider, S., Winkler, S., Liebling, M., Fraser, S. E., et al. (2010). Stereotypical cell division orientation controls neural rod midline formation in zebrafish. *Current Biology*, *20*, 1966–1972.

Rauch, G. J., Hammerschmidt, M., Blader, P., Schauerte, H. E., Strahle, U., Ingham, P. W., et al. (1997). Wnt5 is required for tail formation in the zebrafish embryo. *Cold Spring Harbor Symposia on Quantitative Biology*, *62*, 227–234.

Rogulja, D., Rauskolb, C., & Irvine, K. D. (2008). Morphogen control of wing growth through the Fat signaling pathway. *Developmental Cell*, *15*, 309–321.

Ross, A. J., May-Simera, H., Eichers, E. R., Kai, M., Hill, J., Jagger, D. J., et al. (2005). Disruption of Bardet-Biedl syndrome ciliary proteins perturbs planar cell polarity in vertebrates. *Nature Genetics*, *37*, 1135–1140.

Saburi, S., Hester, I., Fischer, E., Pontoglio, M., Eremina, V., Gessler, M., et al. (2008). Loss of Fat4 disrupts PCP signaling and oriented cell division and leads to cystic kidney disease. *Nature Genetics*, *40*, 1010–1015.

Saburi, S., Hester, I., Goodrich, L., & McNeill, H. (2012). Functional interactions between Fat family cadherins in tissue morphogenesis and planar polarity. *Development*, *139*, 1806–1820.

Sasai, N., Nakazawa, Y., Haraguchi, T., & Sasai, Y. (2004). The neurotrophin-receptor-related protein NRH1 is essential for convergent extension movements. *Nature Cell Biology*, *6*, 741–748.

Schambony, A., & Wedlich, D. (2007). Wnt-5A/Ror2 regulate expression of XPAPC through an alternative noncanonical signaling pathway. *Developmental Cell*, *12*, 779–792.

Schlessinger, K., McManus, E. J., & Hall, A. (2007). Cdc42 and noncanonical Wnt signal transduction pathways cooperate to promote cell polarity. *The Journal of Cell Biology*, *178*, 355–361.

Schwarz-Romond, T., Asbrand, C., Bakkers, J., Kuhl, M., Schaeffer, H. J., Huelsken, J., et al. (2002). The ankyrin repeat protein Diversin recruits Casein kinase Iε to the beta-catenin degradation complex and acts in both canonical Wnt and Wnt/JNK signaling. *Genes & Development*, *16*, 2073–2084.

Segalen, M., Johnston, C. A., Martin, C. A., Dumortier, J. G., Prehoda, K. E., David, N. B., et al. (2010). The Fz-Dsh planar cell polarity pathway induces oriented cell division via Mud/NuMA in *Drosophila* and zebrafish. *Developmental Cell*, *19*, 740–752.

Sepich, D. S., Calmelet, C., Kiskowski, M., & Solnica-Krezel, L. (2005). Initiation of convergence and extension movements of lateral mesoderm during zebrafish gastrulation. *Developmental Dynamics*, *234*, 279–292.

Sepich, D. S., Myers, D. C., Short, R., Topczewski, J., Marlow, F., & Solnica-Krezel, L. (2000). Role of the zebrafish trilobite locus in gastrulation movements of convergence and extension. *Genesis*, *27*, 159–173.

Sepich, D. S., Usmani, M., Pawlicki, S., & Solnica-Krezel, L. (2011). Wnt/PCP signaling controls intracellular position of MTOCs during gastrulation convergence and extension movements. *Development*, *138*, 543–552.

Shih, J., & Keller, R. (1992). Cell motility driving mediolateral intercalation in explants of *Xenopus laevis*. *Development*, *116*, 901–914.

Shima, Y., Kengaku, M., Hirano, T., Takeichi, M., & Uemura, T. (2004). Regulation of dendritic maintenance and growth by a mammalian 7-pass transmembrane cadherin. *Developmental Cell*, *7*, 205–216.

Shimada, Y., Yonemura, S., Ohkura, H., Strutt, D., & Uemura, T. (2006). Polarized transport of Frizzled along the planar microtubule arrays in Drosophila wing epithelium. *Developmental Cell*, *10*, 209–222.

Simons, M., Gloy, J., Ganner, A., Bullerkotte, A., Bashkurov, M., Kronig, C., et al. (2005). Inversin, the gene product mutated in nephronophthisis type II, functions as a molecular switch between Wnt signaling pathways. *Nature Genetics*, *37*, 537–543.

Simons, M., & Mlodzik, M. (2008). Planar cell polarity signaling: From fly development to human disease. *Annual Review of Genetics*, *42*, 517–540.

Slusarski, D. C., Corces, V. G., & Moon, R. T. (1997). Interaction of Wnt and a Frizzled homologue triggers G-protein-linked phosphatidylinositol signalling. *Nature*, *390*, 410–413.

Sokol, S. Y. (1996). Analysis of Dishevelled signalling pathways during *Xenopus* development. *Current Biology*, *6*, 1456–1467.

Srinivas, S., Rodriguez, T., Clements, M., Smith, J. C., & Beddington, R. S. (2004). Active cell migration drives the unilateral movements of the anterior visceral endoderm. *Development*, *131*, 1157–1164.

Stockinger, P., Maitre, J. L., & Heisenberg, C. P. (2011). Defective neuroepithelial cell cohesion affects tangential branchiomotor neuron migration in the zebrafish neural tube. *Development*, *138*, 4673–4683.

Tada, M., & Smith, J. C. (2000). *Xwnt11* is a target of *Xenopus* Brachyury: Regulation of gastrulation movements via Dishevelled, but not through the canonical Wnt pathway. *Development*, *127*, 2227–2238.

Tahinci, E., Thorne, C. A., Franklin, J. L., Salic, A., Christian, K. M., Lee, L. A., et al. (2007). Lrp6 is required for convergent extension during *Xenopus* gastrulation. *Development*, *134*, 4095–4106.

Takeuchi, M., Nakabayashi, J., Sakaguchi, T., Yamamoto, T. S., Takahashi, H., Takeda, H., et al. (2003). The *prickle*-related gene in vertebrates is essential for gastrulation cell movements. *Current Biology*, *13*, 674–679.

Tanegashima, K., Zhao, H., & Dawid, I. B. (2008). WGEF activates Rho in the Wnt-PCP pathway and controls convergent extension in *Xenopus* gastrulation. *The EMBO Journal*, *27*, 606–617.

Tawk, M., Araya, C., Lyons, D. A., Reugels, A. M., Girdler, G. C., Bayley, P. R., et al. (2007). A mirror-symmetric cell division that orchestrates neuroepithelial morphogenesis. *Nature*, *446*, 797–800.

Theveneau, E., Marchant, L., Kuriyama, S., Gull, M., Moepps, B., Parsons, M., et al. (2010). Collective chemotaxis requires contact-dependent cell polarity. *Developmental Cell*, *19*, 39–53.

Topczewski, J., Sepich, D. S., Myers, D. C., Walker, C., Amores, A., Lele, Z., et al. (2001). The zebrafish glypican knypek controls cell polarity during gastrulation movements of convergent extension. *Developmental Cell*, *1*, 251–264.

Tree, D. R., Shulman, J. M., Rousset, R., Scott, M. P., Gubb, D., & Axelrod, J. D. (2002). Prickle mediates feedback amplification to generate asymmetric planar cell polarity signaling. *Cell*, *109*, 371–381.

Trichas, G., Joyce, B., Crompton, L. A., Wilkins, V., Clements, M., Tada, M., et al. (2011). Nodal dependent differential localisation of dishevelled-2 demarcates regions of differing cell behaviour in the visceral endoderm. *PLoS Biology*, *9*, e1001019.

Trichas, G., Smith, A. M., White, N., Wilkins, V., Watanabe, T., Moore, A., et al. (2012). Multi-cellular rosettes in the mouse visceral endoderm facilitate the ordered migration of anterior visceral endoderm cells. *PLoS Biology*, *10*, e1001256.

Ulrich, F., Concha, M. L., Heid, P. J., Voss, E., Witzel, S., Roehl, H., et al. (2003). Slb/Wnt11 controls hypoblast cell migration and morphogenesis at the onset of zebrafish gastrulation. *Development*, *130*, 5375–5384.

Ulrich, F., Krieg, M., Schotz, E. M., Link, V., Castanon, I., Schnabel, V., et al. (2005). Wnt11 functions in gastrulation by controlling cell cohesion through Rab5c and E-cadherin. *Developmental Cell*, *9*, 555–564.

Unterseher, F., Hefele, J. A., Giehl, K., De Robertis, E. M., Wedlich, D., & Schambony, A. (2004). Paraxial protocadherin coordinates cell polarity during convergent extension via Rho A and JNK. *The EMBO Journal*, *23*, 3259–3269.

Van Raay, T. J., Coffey, R. J., & Solnica-Krezel, L. (2007). Zebrafish Naked1 and Naked2 antagonize both canonical and non-canonical Wnt signaling. *Developmental Biology*, *309*, 151–168.

Veeman, M. T., Slusarski, D. C., Kaykas, A., Louie, S. H., & Moon, R. T. (2003). Zebrafish prickle, a modulator of noncanonical Wnt/Fz signaling, regulates gastrulation movements. *Current Biology*, *13*, 680–685.

Vervenne, H. B., Crombez, K. R., Lambaerts, K., Carvalho, L., Koppen, M., Heisenberg, C. P., et al. (2008). Lpp is involved in Wnt/PCP signaling and acts together with Scrib to mediate convergence and extension movements during zebrafish gastrulation. *Developmental Biology*, *320*, 267–277.

Voiculescu, O., Bertocchini, F., Wolpert, L., Keller, R. E., & Stern, C. D. (2007). The amniote primitive streak is defined by epithelial cell intercalation before gastrulation. *Nature, 449*, 1049–1052.

von der Hardt, S., Bakkers, J., Inbal, A., Carvalho, L., Solnica-Krezel, L., Heisenberg, C. P., et al. (2007). The Bmp gradient of the zebrafish gastrula guides migrating lateral cells by regulating cell-cell adhesion. *Current Biology, 17*, 475–487.

Wada, H., Iwasaki, M., Sato, T., Masai, I., Nishiwaki, Y., Tanaka, H., et al. (2005). Dual roles of zygotic and maternal Scribble1 in neural migration and convergent extension movements in zebrafish embryos. *Development, 132*, 2273–2285.

Wada, H., Tanaka, H., Nakayama, S., Iwasaki, M., & Okamoto, H. (2006). Frizzled3a and Celsr2 function in the neuroepithelium to regulate migration of facial motor neurons in the developing zebrafish hindbrain. *Development, 133*, 4749–4759.

Wallingford, J., & Mitchell, B. (2011). Strange as it may seem: The many links between Wnt signaling, planar cell polarity, and cilia. *Genes & Development, 25*, 201–213.

Wallingford, J. B., Rowning, B. A., Vogeli, K. M., Rothbacher, U., Fraser, S. E., & Harland, R. M. (2000). Dishevelled controls cell polarity during *Xenopus* gastrulation. *Nature, 405*, 81–85.

Walsh, G. S., Grant, P. K., Morgan, J. A., & Moens, C. B. (2011). Planar polarity pathway and Nance-Horan syndrome-like 1b have essential cell-autonomous functions in neuronal migration. *Development, 138*, 3033–3042.

Wang, Y., Guo, N., & Nathans, J. (2006). The role of Frizzled3 and Frizzled6 in neural tube closure and in the planar polarity of inner-ear sensory hair cells. *The Journal of Neuroscience, 26*, 2147–2156.

Wang, J., Hamblet, N. S., Mark, S., Dickinson, M. E., Brinkman, B. C., Segil, N., et al. (2006). Dishevelled genes mediate a conserved mammalian PCP pathway to regulate convergent extension during neurulation. *Development, 133*, 1767–1778.

Wang, Y., Janicki, P., Koster, I., Berger, C. D., Wenzl, C., Grosshans, J., et al. (2008). *Xenopus* Paraxial Protocadherin regulates morphogenesis by antagonizing Sprouty. *Genes & Development, 22*, 878–883.

Wang, J., Mark, S., Zhang, X., Qian, D., Yoo, S. J., Radde-Gallwitz, K., et al. (2005). Regulation of polarized extension and planar cell polarity in the cochlea by the vertebrate PCP pathway. *Nature Genetics, 37*, 980–985.

Weber, G. F., Bjerke, M. A., & DeSimone, D. W. (2012). A mechanoresponsive cadherin-keratin complex directs polarized protrusive behavior and collective cell migration. *Developmental Cell, 22*, 104–115.

Weiser, D. C., Row, R. H., & Kimelman, D. (2009). Rho-regulated myosin phosphatase establishes the level of protrusive activity required for cell movements during zebrafish gastrulation. *Development, 136*, 2375–2384.

Williams, B. B., Cantrell, V. A., Mundell, N. A., Bennett, A. C., Quick, R. E., & Jessen, J. R. (2012). VANGL2 regulates membrane trafficking of MMP14 to control cell polarity and migration. *Journal of Cell Science, 125*, 2141–2147.

Winklbauer, R. (1990). Mesodermal cell migration during *Xenopus* gastrulation. *Developmental Biology, 142*, 155–168.

Witte, F., Bernatik, O., Kirchner, K., Masek, J., Mahl, A., Krejci, P., et al. (2010). Negative regulation of Wnt signaling mediated by CK1-phosphorylated Dishevelled via Ror2. *The FASEB Journal, 24*, 2417–2426.

Witzel, S., Zimyanin, V., Carreira-Barbosa, F., Tada, M., & Heisenberg, C. P. (2006). Wnt11 controls cell contact persistence by local accumulation of Frizzled 7 at the plasma membrane. *The Journal of Cell Biology, 175*, 791–802.

Yamaguchi, T. P., Bradley, A., McMahon, A. P., & Jones, S. (1999). A Wnt5a pathway underlies outgrowth of multiple structures in the vertebrate embryo. *Development, 126*, 1211–1223.

Yamamoto, S., Nishimura, O., Misaki, K., Nishita, M., Minami, Y., Yonemura, S., et al. (2008). Cthrc1 selectively activates the planar cell polarity pathway of Wnt signaling by stabilizing the Wnt-receptor complex. *Developmental Cell, 15*, 23–36.

Yamamoto, N., Okano, T., Ma, X., Adelstein, R. S., & Kelley, M. W. (2009). Myosin II regulates extension, growth and patterning in the mammalian cochlear duct. *Development, 136*, 1977–1986.

Yamanaka, H., Moriguchi, T., Masuyama, N., Kusakabe, M., Hanafusa, H., Takada, R., et al. (2002). JNK functions in the non-canonical Wnt pathway to regulate convergent extension movements in vertebrates. *EMBO Reports, 3*, 69–75.

Yamanaka, Y., Tamplin, O. J., Beckers, A., Gossler, A., & Rossant, J. (2007). Live imaging and genetic analysis of mouse notochord formation reveals regional morphogenetic mechanisms. *Developmental Cell, 13*, 884–896.

Ybot-Gonzalez, P., Savery, D., Gerrelli, D., Signore, M., Mitchell, C. E., Faux, C. H., et al. (2007). Convergent extension, planar-cell-polarity signalling and initiation of mouse neural tube closure. *Development, 134*, 789–799.

Yen, W. W., Williams, M., Periasamy, A., Conaway, M., Burdsal, C., Keller, R., et al. (2009). PTK7 is essential for polarized cell motility and convergent extension during mouse gastrulation. *Development, 136*, 2039–2048.

Yin, C., Kiskowski, M., Pouille, P. A., Farge, E., & Solnica-Krezel, L. (2008). Cooperation of polarized cell intercalations drives convergence and extension of presomitic mesoderm during zebrafish gastrulation. *The Journal of Cell Biology, 180*, 221–232.

Yu, A., Rual, J. F., Tamai, K., Harada, Y., Vidal, M., He, X., et al. (2007). Association of Dishevelled with the clathrin AP-2 adaptor is required for Frizzled endocytosis and planar cell polarity signaling. *Developmental Cell, 12*, 129–141.

Zallen, J. A., & Wieschaus, E. (2004). Patterned gene expression directs bipolar planar polarity in *Drosophila*. *Developmental Cell, 6*, 343–355.

Zigman, M., Trinh le, A., Fraser, S. E., & Moens, C. B. (2011). Zebrafish neural tube morphogenesis requires Scribble-dependent oriented cell divisions. *Current Biology, 21*, 79–86.

CHAPTER FIVE

Planar Cell Polarity in the Inner Ear

Helen May-Simera, Matthew W. Kelley[1]

Laboratory of Cochlear Development, NIDCD, NIH, Bethesda, Maryland, USA

[1]Corresponding author: e-mail address: kelleymt@nidcd.nih.gov

Contents

Abstract

The inner ears of vertebrates represent one of the most striking examples of planar cell polarity (PCP). Populations of directionally sensitive mechanosensory hair cells develop actin-based stereociliary bundles that are uniformly oriented. Analysis of perturbations in bundle polarity in mice with mutations in Vangl2 formed the basis for the initial demonstration of conservation of the PCP signaling pathway in vertebrates. Subsequent studies have demonstrated roles for other "core" PCP genes, such as *Frizzled, Disheveled, and Celsr*, and for identifying novel PCP molecules such as Scribble and Ptk7. In addition, the demonstration of hearing deficits in humans with mutations in cilia genes combined with analysis of PCP defects in mice with ciliary deletion has implicated the cilia as an important modulator of hair cell polarization. Finally, the presence of shortened cochleae in many PCP mouse mutants has revealed an additional role for the PCP pathway in the development of the auditory system.

Current Topics in Developmental Biology, Volume 101
ISSN 0070-2153
http://dx.doi.org/10.1016/B978-0-12-394592-1.00006-5
2012, Published by Elsevier Inc.

1. INTRODUCTION TO THE INNER EAR

In vertebrates, the senses of hearing, balance, and acceleration are perceived through discrete sensory structures located within the inner ear. Each of these structures contains an epithelial sheet that comprises mechanosensory hair cells and nonsensory supporting cells. Mechanosensory hair cells act as the primary transducers, detecting movements caused by physical motion or sound pressure waves and generating graded responses in neurotransmitter release. Associated neuronal innervation by branches of the VIIIth nerve conveys those signals to the CNS. The basis for mechanosensitivity in these cells is a group of modified microvilli, referred to as a stereociliary bundle, that project from the lumenal surface. These stereocilia are invariably arranged in a staircase pattern that reflects an inherent asymmetry within the bundle. Tip links, molecular connections that span between adjacent stereocilia, are exclusively oriented along the plane of increasing stereocilia height (Pickles et al., 1989). Each tip link is physically connected to a mechanosensitive channel with an open probability that increases significantly in response to increased tension (reviewed in Stauffer & Holt, 2007). Opening of the channel leads to a rapid influx of calcium and potassium. Therefore, only deflection of the bundle in the direction of the tallest stereocilia leads to increased channel opening and depolarization. Conversely, if tension is decreased along the axis of the staircase, open probability decreases leading to a hyperpolarization of the cell. All of the mechanosensory epithelia within the inner ear contain morphological specializations that take advantage of the directional nature of the stereociliary bundles to achieve increased sensitivity. For instance, in the coiled cochlear sensory epithelium of mammals, all stereociliary bundles are oriented toward the lateral edge of the coil. In response to sound waves, an overlying nonliving membrane, the tectorial membrane is deflected closer and laterally relative to the lumenal surface. This motion applies a lateral force to the stereociliary bundles, leading to channel opening and cellular depolarization. In contrast, some vestibular epithelia contain arrays of hair cells with orientations that are separated by 180°. As a result, deflection of an overlying membrane in one direction leads to depolarization of some cells and hyperpolarization of others. This arrangement is thought to result in heightened sensitivity. Anatomically, this results in a "reversal zone" in which bundle orientation switches by 180°.

The direct link between the orientation of stereociliary bundles and the appropriate function of auditory and vestibular mechanosensory epithelia suggests that PCP must be highly regulated during development of the inner ear. This fact plus the ease with which defects in bundle orientation can be discerned and quantified has made the inner ear one of the best and most heavily studied examples of vertebrate PCP. Moreover, the recent demonstration that elongation of the cochlear duct is also regulated by the PCP pathway has provided an additional aspect of inner ear development that can be examined in terms of the roles of different PCP molecules (Montcouquiol et al., 2003; Wang, Hamblet, et al., 2006). In this review, we provide a general overview of the development of the inner ear, with a primary focus on the mammalian system, followed by a more detailed description of the roles of different PCP molecules and how these factors interact at a molecular level to mediate both elongation and bundle orientation.

2. DEVELOPMENT OF THE INNER EAR

In all vertebrates, the inner ear develops predominantly from the otic placode, a thickening of the dorsal ectoderm located adjacent to the developing hindbrain (reviewed in Driver & Kelley, 2009; Puligilla & Kelley, 2009). The placode undergoes a period of proliferation, sinks inwards toward the hindbrain, and constricts at the ectodermal surface to eventually pinch off and form a fluid-filled otocyst. Formation of the otocyst is followed by a series of elaborate morphogenetic changes that result in the development of the various structures within the inner ear. In mammals, the cochlear duct begins as a ventral out pocketing from the otocyst that extends and coils over time. The epithelial cells that line the otocyst and all of its derivatives give rise to both the sensory patches containing mechanosensory hair cells and supporting cells and the nonsensory cells that comprise the remainder of the inner ear.

In mouse, outgrowth of the cochlea begins around E11 and continues through the early postnatal period (Fig. 5.1). Within the duct, individual cells that will develop as hair cells can first be identified between E13 and E15. As these cells differentiate from surrounding progenitor cells, a period of cellular rearrangement and patterning occurs that leads to the development of a highly ordered pattern of both hair cells and supporting cells that is completed by approximately postnatal day 0 (P0) (McKenzie, Krupin, & Kelley, 2004). Two types of hair cells, inner and outer hair cells, as well as at least six different types of supporting cells are arranged into ordered rows that extend along the basal-to-apical axis of the entire cochlear spiral. During the same developmental time period, several patches of hair cell-containing

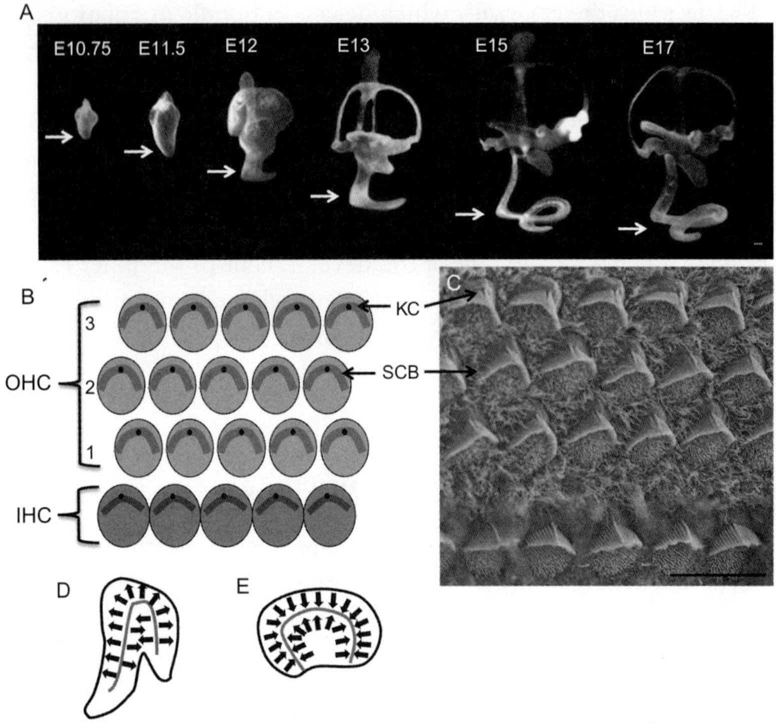

Figure 5.1 Development of the mammalian cochlea. (A) Lateral view of paint-filled membranous labyrinths of embryonic mice at indicating ages between embryonic day 10.75 and embryonic day 17. The cochlear duct (arrows) begins as a ventral out pocketing that begins to extend from the otocyst at E10.75 and continues to extend and coil until the early postnatal period. The mature mouse cochlear duct is between 6 and 10 mm and completes approximately 2.5 coils (scale bar is 100 μm; adapted from Morsli, Choo, Ryan, Johnson, & Wu, 1998). (B) Schematic representation of a surface view of the organ of Corti. The single row of inner hair cells and three rows of outer hair cells (OHC) are depicted, whereas the support cells that interdigitate between the hair cells have been omitted. Each hair cell contains an actin-based stereociliary bundle, represented by a red arch, located on its lumenal surface. The bundles are uniformly oriented toward the lateral edge of each cell. The microtubule-based kinocilium, represented by a black dot, extends from the plasma membrane and is attached to the stereociliary bundles via kinocilia links. (C) Scanning electron micrograph of the organ of Corti (basal turn) at P0. The crescent-shaped stereociliary bundles (SCB) with the single kinocilium (KC) located at the vertex are evident on both inner and outer hair cells. Microvilli on the supporting cells between the hair cells are still present at this stage. Similarly, shorter microvilli located on the medial surface of the hair cells are still present. (D and E) Schematic representations of uniform stereociliary bundle orientations (arrows) in two vestibular sensory epithelial, the saccule (D) and the utricle (E). Note the reversal zones (red lines) present in each epithelium. Scale bar is 10 μm. (For interpretation of the references to color in this figure legend, the reader is referred to the online version of this chapter.)

sensory epithelia develop as specific locations within the vestibular regions of the inner ear (Denman-Johnson & Forge, 1999). The PCP pathway controls at least two crucial steps in the developing inner ear, the orientation of stereociliary bundles, and the outgrowth of the cochlear duct. These events will be discussed separately below.

2.1. Extension of cochlea duct

The extending cochlear duct more than doubles in length between E12.5 and P0, indicating significant growth during this time period (Morsli et al., 1998). However, while nonsensory cells within the duct continue to proliferate during this period, the progenitor pool that will give rise to hair cells and supporting cells within the organ of Corti becomes postmitotic between E13 and E14 (Chen, Johnson, Zoghbi, & Segil, 2002; Chen & Segil, 1999; Ruben, 1967; Wang et al., 2005). Since derivatives from this precursor population (called the prosensory domain) extend along the entire length of the cochlear duct by P0, these cells must undergo a period of extension and rearrangement. Morphological studies in which the distribution of these cells was analyzed at different developmental time points indicated cellular rearrangements that are consistent with the process of convergence and extension (Chen & Segil, 1999; McKenzie et al., 2004). In addition, a by-product of these cell rearrangements is the alignment of both hair cells and supporting cells into the highly ordered rows that extend along the entire long axis of the cochlea. As will be discussed below, analysis of cochleae from PCP mutant mice indicates defects in cochlear extension and patterning, demonstrating a role for PCP signaling in this process (Montcouquiol et al., 2003; Wang et al., 2005; Ybot-Gonzalez et al., 2007).

2.2. Development and orientation of mechanosensory hair cell stereociliary bundles

As discussed above, during extension of the cochlear duct, individual cells within the wall of the duct begin to differentiate into mechanosensory hair cells. Prior to differentiation, as is the case for virtually all epithelial cells post-terminal mitosis, the lumenal surfaces of cells within the duct are covered with a lawn of short actin-based microvilli and one true, microtubule-based, nonmotile cilium. Beginning around E15 in the mouse, the single true cilium located on each developing hair cell begins to elongate and to migrate toward the lateral edge of each cell (Montcouquiol et al., 2003). At the same time, microvilli located adjacent to this cilium also begin to elongate to give rise to individual stereocilia. Specifically in mechanosensory hair cells, this cilium is referred to as the kinocilium. Importantly, the direction of the migration of the cilium/kinocilium is thought to determine the eventual orientation of the stereociliary bundle. While the initial direction of this migration is nonrandom and is biased toward the final orientation of the mature bundle, deviations of as much as 35° from the final orientation are observed (Cotanche & Corwin, 1991; Dabdoub et al., 2003; Denman-Johnson & Forge, 1999). Once the kinocilium reaches the lateral edge of the lumenal surface, deviations are corrected during a period of refinement as developing bundles gradually reorient to ultimately achieve uniform orientation (Fig. 5.2). In the mouse cochlea, stereociliary bundle development and orientation occur along two gradients, a predominantly basal-to-apical gradient that begins in the mid-basal of the duct and a medial-to-lateral gradient that begins in the inner hair cells and extends toward the third row of outer hair cells. Final uniform orientation of all hair cells is not achieved in the mouse until approximately P10 (Dabdoub et al., 2003).

Interestingly, while all stereociliary bundles are arranged in a staircase pattern, the arrangement of the individual stereocilia varies based on location. In the vestibular epithelia, bundles are generally round with the kinocilia located at one edge. In contrast, in the cochlea, the morphology of the bundle is modified such that bundles on inner hair cells are in the shape of an arc, while bundles on outer hair cells appear more like a "W" although with a very shallow central indentation. In both cases, the kinocilium is located at the vertex of the structure. The functional significance of these modifications in morphology is unclear but has been suggested to be linked to the overall structure of the organ of Corti.

While the number of studies on the development of bundle orientation in the vestibular system is more limited by comparison with the cochlea, the overall progression seems similar although apparently less coordinated (Denman-Johnson & Forge, 1999). Developing hair bundles can first be identified around E12.5, approximately 48–72 h prior to the formation of hair cells in the cochlear duct. The single true cilium (also referred to as a kinocilium) begins to elongate and then moves in a nonrandom manner toward one edge of the lumenal surface. As is the case in the cochlea, there is a period of refinement that leads to more uniform orientations over time. However, the overall process of polarization is more difficult to assay in the vestibular epithelia because hair cells are not generated in a discrete gradient as they are in the cochlea. Instead, hair cells seem to arise at random

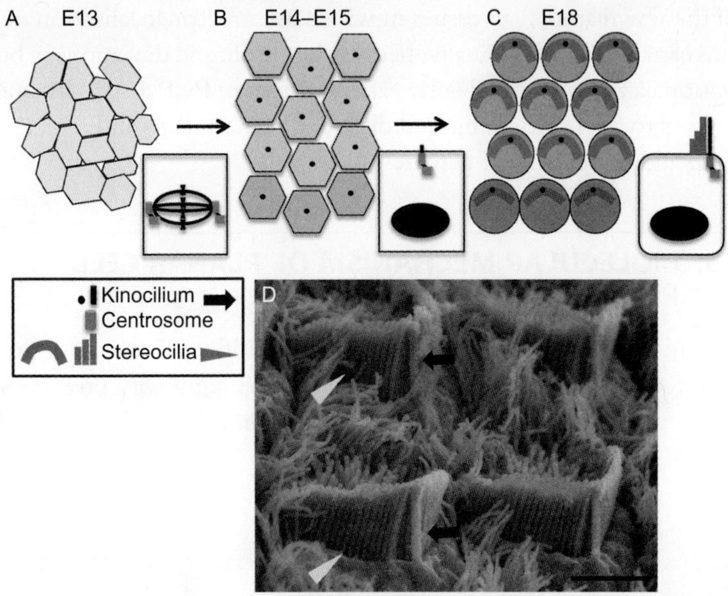

Figure 5.2 The kinocilium during hair cell development. Extension of the cochlea duct occurs concurrently with the polarization of individual hair cells. (A) At E13, the precursors for the organ of Corti are still undergoing mitosis and the cells are still actively dividing. Inset shows lateral view. (B) As the hair cells emerge from terminal mitosis, the basal body/centrosome migrates to the apex of the cell and the kinocilium begins to protrude. Between E15 and E16, the kinocilium begins to migrate toward the lateral edge of the luminal hair cell surface and adjacent microvilli elongate to form stereocilia. Inset shows location of basal body and extending kinocilium in the lateral view. (C) By E18, the initial migration of the kinocilium is complete; a further period of refinement occurs until the developing bundles achieve their ultimate orientation by P10. The kinocilium begins to retract from approximately P3 onward. Inset shows lateral view of hair cell with kinocilium and stereociliary bundle. (D) Scanning electron micrograph of the stereociliary bundles on four outer hair cells at P0 (basal turn of the cochlea). The crescent-shaped stereociliary bundles (red arrowhead) with the single kinocilium (black arrow) located at the vertex are indicated. Microvilli on the supporting cells between the hair cells are still present at this stage. Scale bar is 2.5 μm. (See Color Insert.)

positions within the epithelium over the course of several days. While this pattern makes it more difficult to interpret the process of polarization, it suggests that a unifying polarity may exist within the epithelium prior to the formation of any stereociliary bundles. This possibility will be discussed in greater detail in another section of this chapter.

The clear manifestation of PCP in the orientation of stereociliary bundles coupled with direct effects on function makes the mammalian inner ear an ideal system for the dissection of PCP signaling pathways. In addition, it is one of the few mammalian tissues in which asymmetric localization of PCP proteins can be observed. Finally, because less profound deviations in bundle orientation can be quantified fairly easily, analysis of PCP defects in inner ear can be used to assess the strength of different PCP mutants and to determine potential additive effects of genetic interactions.

3. MOLECULAR MECHANISM OF PLANAR CELL POLARITY

Genetic studies in *Drosophila* initially identified a core set of proteins required for regulation of PCP (Fig. 5.3). These regulatory proteins signal sequentially to integrate directional cues with cellular factors to direct

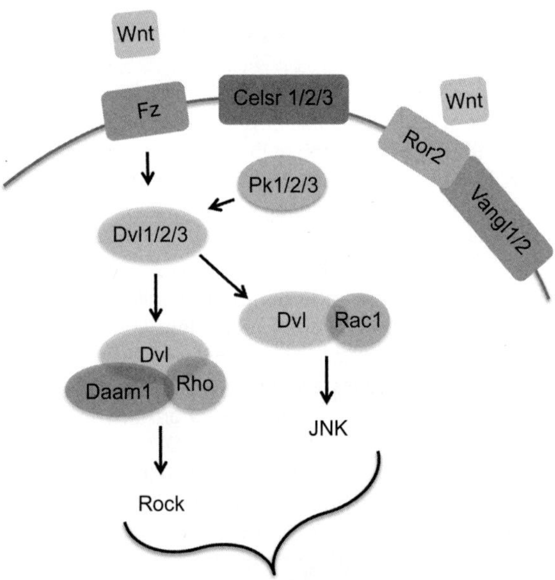

Figure 5.3 Molecular basis of PCP (noncanonical Wnt) signaling. Core components of the planar cell polarity (noncanonical Wnt) signaling pathway. In vertebrates, PCP signaling is initiated by the binding of Wnt ligands to membrane bound receptors (Fz) or through complexes with coreceptors (Vangl1/2, Ror2, and Ptk7). The signal is transduced via the assembly of various Dvl/effector complexes (Dvl1/2/3; Pk1/2/3), which ultimately lead to the activation of various pathways regulating the actin cytoskeleton and cell adhesion. Association of Dvl with Rho and Rac activates Rho Kinase (Rock) and c-Jun N-terminal kinase (JNK). Activation of PLC, PKC, and Cdc42 is mediated via association with intracellular Ca^{2+}. (For color version of this figure, the reader is referred to the online version of this chapter.)

coordinated polarization of cells in a plane of epithelium. While first identified in *Drosophila*, the same core molecules serve a conserved function in the regulation of coordinated uniform cellular activities across a plane of cells in animals as diverse as nematodes to mammals (Semenov, Habas, Macdonald, & He, 2007).

3.1. Core PCP molecules

The first demonstration of a role for one of the core PCP molecules in the orientation of stereociliary bundles was the demonstration of rotated bundles in the cochleae of *Vangl2*$^{Lp/Lp}$ mutants (Montcouquiol et al., 2003). In this spontaneous mouse mutant, a missense mutation in the C-terminal cytoplasmic domain of *Vangl2* is thought to lead to the destabilization and loss of function of the protein (Kibar et al., 2001; Montcouquiol et al., 2006). *Vangl2* is one of two mammalian homologs of *Drosophila van gogh* (vg) (Murdoch, Doudney, Paternotte, Copp, & Stanier, 2001). Stereociliary bundles on hair cells in both cochlear and vestibular epithelia appear phenotypically normal in terms of shape, number, and distribution, but most bundles are misoriented (Montcouquiol et al., 2003). Within the cochlea, analysis of bundle orientation defects based on position within the four rows of hair cells indicated an intriguing gradient in which hair cells located in the single inner hair cell row or in the third row of outer hair cells were most severely affected with orientations in both rows approaching a random distribution. In contrast, the average change in orientation for bundles located in the first row of outer hair cells was significantly less than in the second or third row of outer hair cells and did not significantly differ from control in the basal region of the cochlea. This result emphasizes a common observation for the effects of many PCP mutants within the inner ear: increased sensitivity in the two most peripheral rows of hair cells, and in particular, in the third row of outer hair cells. In contrast, measurements of bundle orientations in the utricle, one of the vestibular epithelia, indicated a near random distribution (Montcouquiol et al., 2006). Finally, overall length of the cochlear duct was decreased and cellular patterning was disrupted in *Vangl2*$^{Lp/Lp}$ mutants, indicating a role for Vangl2 and PCP in cochlear extension. In particular, the apical region of the cochlea, which normally contains three, or in some cases only two, rows of outer hair cells in a wild-type animal, contained five to seven rows of outer hair cells in *Vangl*$^{Lp/Lp}$ mutants (Montcouquiol et al., 2003).

Vangl2 and the related Vangl1 are four-transmembrane domain proteins with a C-terminal PDZ domain. In addition to orientation defects in the inner ear, homozygosity for the *Vangl2*^{Lp} mutation also results in craniorachischisis, a defect that arises from defects in neural tube closure, a process that is also regulated by the PCP pathway (Mullick, Trasler, & Gros, 1995; Ueno & Greene, 2003; Ybot-Gonzalez et al., 2007). Mutations in *Vangl1* have been associated with human neural tube defects (Kibar et al., 2007), but a homozygous *Vangl1* gene trap (Rachel et al., 2012; Rachel, Wellington, Warburton, Mason, & Beermann, 2002) line, encoding a severely truncated nonfunctional version of the protein, displayed no neural tube defects and only mild stereociliary bundle orientation defects (Torban et al., 2008). Double heterozygous animals for *Vangl1*^{gt} and *Vangl2*^{Lp} did develop craniorachischisis and had severely rotated cochlea hair cells, suggesting a degree of functional redundancy between Vangl1 and Vangl2 in the cochlea and neural tube.

In an effort to identify the basis for the orientation defects in *Vangl2*$^{Lp/Lp}$ mice, the migration of developing kinocilia was compared in control and mutant cochleae (Montcouquiol et al., 2003). As described above, developing kinocilia typically undergo a directionally biased migration from the center of the lumenal surface of each developing hair cell toward the lateral edge. While this migration still occurred in hair cells from *Vangl2*$^{Lp/Lp}$ mutants, the directional bias was disrupted. Moreover, a direct correlation between deviation in the directional bias and final bundle orientation was observed, suggesting that this defect underlies the orientation defects observed in these animals.

Subsequent studies have demonstrated similar phenotypes, although with some intriguing caveats, in cochleae from mice with mutations in other core PCP genes. These include targeted double mutants for *Frizzled 3* and *Frizzled 6* (Wang, Guo, & Nathans, 2006), which seem to be functionally redundant in cochlear PCP, missense mutants for *Celsr1* (one of three mammalian homologs of flamingo) (Curtin et al., 2003), and targeted mutations in *Disheveled* (*Dvl*) *1*, *2*, or *3* (Etheridge et al., 2008; Wang, Hamblet, et al., 2006). Intriguingly, the degree of bundle misorientation and the specific hair cell populations that are effected in each of these mutant models are somewhat variable. While the third row of outer hair cells is universally effected, inner hair cells are misoriented in some models (*Fz3/6*) but not in others (*Dvl*). These results presumably indicate differing degrees of functional compensation but may also provide clues regarding the nature of the cell–cell communication that acts to propagate the polarizing signal from one row of hair cells to the next.

3.2. Effects of Wnt ligands

While the core components of the PCP signaling pathway have clearly been conserved between vertebrates and invertebrates, one of the most intriguing differences between the two is an apparent requirement for Wnt molecules in vertebrate PCP signaling (Dabdoub & Kelley, 2005; Ulloa & Marti, 2010) (Fig. 5.2). Wnts are secreted ligands, which can diffuse at a rate of up to 50 μm in 30 min (Strigini & Cohen, 2000) and are known to signal through binding to Frizzled (fz) receptors. Their ability to diffuse allows them to potentially act as long-range signaling molecules in a gradient-dependent manner (Wodarz & Nusse, 1998; Zecca, Basler, & Struhl, 1996). Nineteen *Wnt* genes have been identified in humans and mice, and at least eight are known to be expressed in the developing mouse cochlea (Dabdoub et al., 2003; Daudet, Ripoll, Moles, & Rebillard, 2002). Moreover, *Wnt7a* and *Wnt5a* are expressed in gradients at time points that correlate with bundle orientation. In particular, both are expressed at higher levels on the more medial side of the organ of Corti, suggesting that a gradient could exist across the four rows of hair cells. This hypothesis was initially examined by altering the concentration of Wnt7a in cochlear explant cultures, which resulted in a defect in the orientation of outer hair cells, with more lateral rows more severely affected. However, *Wnt7a* knockout mice do not show any bundle orientation defects (Dabdoub et al., 2003). In contrast, PCP defects are evident in cochleae from *Wnt5a* knockout animals, including shortened and widened cochlear ducts and mild bundle polarity defects, although with incomplete penetrance (Qian et al., 2007). Moreover, mice that are double heterozygous for a *Wnt5a* deletion and *Vangl2*^{Lp} display neural tube defects as well as severe bundle orientation defects. These results strongly implicate Wnt5a as a possible ligand that regulates PCP in the developing cochlea (Qian et al., 2007).

By comparison with *Vangl2* or other core PCP mutants, the cochlear PCP phenotype in the *Wnt5a* mutants is relatively mild, suggesting possible functional redundancy, a hypothesis that is supported by the high number of *Wnts* that have been reported to be expressed in the inner ear (Dabdoub et al., 2003). For this reason, it has been helpful to examine the effects of Wnt antagonists *in vitro*. Using cochlear explant cultures, investigators have examined the effects of various Wnt antagonists. Members of the secreted fz-related proteins (Sfrp1–5) and Wnt inhibitory factor 1 (WIF1) bind directly to Wnt thereby inhibiting its binding to fz (Kawano & Kypta,

2003). Cochlea explants cultured in the presence of Sfrp1, Frzb (Sfrp3), or WIF1 displayed disrupted bundle orientation (Dabdoub & Kelley, 2005; Qian et al., 2007). Moreover, at least in the case of Frzb, the effects could be partially rescued by increasing the concentration of Wnt5a, indicating that Wnt signaling is disrupted by Frzb. Similarly, application of sodium chlorate, which disrupts sulfation of polysaccharides, and therefore disrupts HSPGs, which have a high affinity for Wnt proteins (Baeg, Selva, Goodman, Dasgupta, & Perrimon, 2004; Bradley & Brown, 1990), also disrupted outer hair cell orientation *in vitro* (Dabdoub & Kelley, 2005). Taken together, these results support the idea that endogenous Wnts probably play a role in cochlear PCP; however, whether a gradient of any Wnt is actually instructive for bundle orientation has not yet been demonstrated.

Interestingly, several Wnt antagonists are expressed endogenously in the cochlea. In particular, *Frzb* is expressed in a counter gradient to *Wnt5a* (Qian et al., 2007). Since expression of Wnt antagonists has been shown to play a role in refining and/or amplifying Wnt gradients (Dabdoub & Kelley, 2005; Kawano & Kypta, 2003; Leimeister, Bach, & Gessler, 1998), these results are consistent with a role for a Wnt gradient in cochlear PCP. Overall, these results strongly support a role for Wnt signaling in the generation of inner ear PCP, at least in the cochlea. However, the lack of a mouse mutant with a strong inner ear PCP phenotype is clearly a crucial requirement to confirm this hypothesis. Unfortunately, because of functional redundancy, multiple *Wnts* may need to be deleted to achieve this goal, but with the advent of tissue-specific deletion techniques, the necessary mouse models may not be too far off.

3.3. Asymmetric protein localization

One of the more insightful discoveries regarding the cellular mechanisms underlying PCP was the demonstration of asymmetric protein localization for several of the core genes in *Drosophila* (Jenny & Mlodzik, 2006; Strutt, Johnson, Cooper, & Bray, 2002; Uemura & Shimada, 2003). Within the developing fly wing, fz and disheveled become localized to the distal portion of each cell, while vg and prickle (pk) become localized to the proximal region of each cell. Flamingo colocalizes on both the distal and the proximal regions of each cell but is excluded from cellular membranes oriented parallel with the axis of extension. While similarly intriguing localization of PCP proteins in mammalian tissues has proved challenging to visualize, the organ of Corti in the cochlear duct represents one of the few tissues in which asymmetric localization of PCP proteins has been demonstrated. Prior to the migration of the kinocilium, Vangl2 becomes localized to cell–cell junctions between the proximal regions of developing hair cells and the distal regions of developing supporting cells (Montcouquiol et al., 2006). While it has not been possible to determine whether Vangl2 is present in hair cells, support cells, or both, the predominant expression at these boundaries seems consistent with results from *Drosophila*. Similarly, consistent observations were made for Dvl1, and Dvl2, both of which appeared to localize in the distal regions of developing hair cells (Wang, Hamblet, et al., 2006; Wang et al., 2005).

While the results described above are consistent with a conservation of PCP signaling mechanisms between fly wing and vertebrate inner ear, several recent studies have suggested either significant deviations or the existence of more complex patterns of polarization within the inner ear. As discussed, codeletion of *Fz3* and *Fz6* leads to bundle orientation defects, and by analogy with *Drosophila*, Fz3 and Fz6 would be expected to localize on the same side of the cell as Dvl. However, localization of Fz3 and Fz6 in the developing cochlea indicated colocalization with Vangl2 on the medial side of developing hair cells (Wang, Guo, et al., 2006). This result was somewhat surprising and was made more so by the observation that in developing vestibular epithelia, Fz3 and Fz6 localized to the distal side of hair cells. Subsequently, a study examining localization of Prickle2 (Pk2), a vertebrate homolog of pk, in the vestibular system, took advantage of the reversal zone that exists within the utricle to determine whether localization of Pk2 changes with orientation (Deans et al., 2007). Results indicated an asymmetric localization of Pk2 to one side of each hair cell and supporting cell, but this asymmetry was unchanged on either side of the reversal zone even though hair cell bundle orientation rotated by 180°. Finally, immunolocalization for Vangl2 through the reversal zone of the chicken utricle indicated no obvious switch in subcellular distribution (Warchol & Montcouquiol, 2010).

These results have several intriguing implications. First, the asymmetric distribution of multiple core PCP molecules is consistent with strong polarization of both auditory and vestibular epithelia, as well as a general conservation of the PCP signaling mechanism. But, the observation that asymmetric distribution does not parallel stereociliary bundle orientation suggests that bundle orientation may not be determined directly through cellular polarization. Rather, the underlying polarity of the epithelium may dictate other aspects of cellular identity with bundle orientation determined based on as-yet-unknown facets of that identity. Second, the results also suggest fundamental differences between PCP in the auditory and vestibular systems despite the use of the same core PCP components. This suggestion is supported by the observation of significantly different polarity defects in response to similar mutations. For instance, deletion of *Fz3* and *Fz6* has a minimal effect on bundle orientation in the utricle but leads to a nearly 180° reversal in the orientation of inner hair cell bundles. While these observations do not minimize the importance of the PCP pathway in stereociliary bundle orientation or the validity of bundle orientation as a read out of PCP, they do represent important considerations for the examination of other PCP mutants in terms of inner ear defects.

3.4. Novel coreceptors

As discussed, Wnt signaling plays a role in vertebrate PCP. Therefore, it is not unexpected that some Wnt coreceptors might also influence bundle orientation. In particular, the Ror kinases are a family of transmembrane receptors that have been proposed to mediate the role of Wnt ligands in PCP signaling (Minami, Oishi, Endo, & Nishita, 2010). Ror kinases physically interact with Wnt5a homologs in *Caenorhabditis elegans* and *Xenopus* and are required for cell polarization and convergence extension (Forrester, Dell, Perens, & Garriga, 1999; Oishi et al., 2003). *Ror2* knockout mice display a similar phenotype to *Wnt5a* mutants, including convergence extension defects and misoriented cochlear hair cells (Forrester et al., 1999). Recently, Wnt5a was shown to act through Ror2 to induce Vangl2 phosphorylation, establishing a PCP activity gradient of Vangl2 in the developing vertebrate limb. Vangl2 and Ror2 were shown to form a Wnt-induced receptor complex resulting in the dosage-dependent phosphorylation of Vangl2 depending on Wnt5a dosages (Gao et al., 2011). Similarly, the secreted collagen triple helix repeat-containing 1 glycoprotein (Cthrc1) has been shown to stabilize Wnt, Fz, Ror2 receptor complexes, and *Cthrc1*-null mice genetically interact with *Vangl2*^{Lp} to cause cochlear PCP phenotypes (Yamamoto et al., 2008).

3.5. Cytoplasmic mediators

In addition to the core PCP cytoplasmic mediators Dvl and Pk, several novel cytoplasmic proteins have also been shown to mediate PCP in vertebrates and, in particular, in the inner ear. The mammalian *Scribble* (*Scrib*) gene, a homolog of *Drosophila scribble* (*scrib*), encodes a leucine-rich protein with four PDZ domains. In *Drosophila scrib* mediates both apical–basal and planar polarity (Bilder, 2004; Courbard et al., 2009; Macara, 2004; Nelson, 2003). However, *Scrib* was first identified as a PCP gene in the *circletail* mouse mutant. This spontaneous point mutation results in a severely truncated Scrib protein, craniorachischisis, and misoriented cochlea hair cells (Montcouquiol et al., 2003; Murdoch et al., 2003; Rachel et al., 2002). Further, *Scrib* interacts genetically with *Vangl2* and Scrib protein has been shown to bind to Vangl2 via its PDZ domain. Scrib is thought to function in the asymmetric targeting of Vangl2-containing PCP complexes (Montcouquiol et al., 2006). Asymmetric localization of Vangl2 is lost in Scrib^{crc} mutants, confirming a role for Scrib in Vangl2 localization.

An additional cytoplasmic protein that can also modulate inner ear PCP is *protein tyrosine kinase* 7 (*Ptk7*) (Lu et al., 2004), which encodes a tyrosine kinase signal peptide, seven immunoglobulin domains, a single transmembrane domain, and an intracellular tyrosine kinase domain. Ptk7 has been shown to interact directly with Wnt ligands as well as with β-catenin (Peradziryi et al., 2011; Puppo et al., 2011); however, its role in PCP signaling is likely via its interactions with Rack1 and Pkcδ1, known effectors of Dvl membrane translocation (Wehner, Shnitsar, Urlaub, & Borchers, 2011). Rack1 is needed for Ptk7-mediated membrane localization of Dvl, one of the earlier events during PCP signaling. The *Xenopus* homolog of Ptk7 has been implicated in convergence and extension and fz-dependent localization of disheveled (Lu et al., 2004; Shnitsar & Borchers, 2008). In mouse, Ptk7 has been shown to be essential for polarized cell motility and convergence and extension (Yen et al., 2009). Various mouse mutants of *Ptk7* display craniorachischisis and disrupted hair cell orientation and interact genetically with *Vangl2* and *Celsr1* mutants (Lu et al., 2004; Paudyal et al., 2010). In addition, *Ptk7* has recently been shown to regulate the activity of Myosin II which plays a role in both cochlear convergence and extension and bundle orientation (Lee et al., 2012).

3.6. Downstream targets

Downstream effects of PCP signaling result in the regulation of the actin cytoskeleton via the family of Rho GTPases. Rho family members include Rho, Rac, and Cdc42 molecules, central regulators of actin dynamics, which influence fundamental cell processes including morphogenesis, polarity, movement, and cell division (Jaffe & Hall, 2005). The specific roles of these downstream PCP targets have not yet been extensively characterized in the mammalian inner ear, but a few studies into the role of Rac have begun to yield some insights.

Rac1 was previously shown to be a downstream effector of PCP signaling during convergence extension movements in mammalian cells (Habas, Dawid, & He, 2003). Based on these results, inner ear-specific *Rac1* conditional mutants were generated (Grimsley-Myers, Sipe, Geleoc, & Lu, 2009). Resulting cochleae were severely shortened with disrupted cellular patterning and misoriented stereociliary bundles in all hair cell rows. Initial establishment of PCP appeared normal, as indicated by the asymmetric migration of the kinocilium. Yet subsequent asymmetric membrane localization of the core PCP protein Fz3 was disrupted, implicating Rac1 in the maintenance but not initial establishment of PCP in the mammalian cochlea (Grimsley-Myers et al., 2009). Rac1 was also shown to be crucial for continued stereociliary bundle and accompanying kinocilium morphogenesis. Functional redundancy among the GTPases has been suggested, as not all hair cells in *Rac1* mutants are disrupted. Homozygous *Rac3* knockouts showed no morphological cochlear defects, yet double homozygous conditional knockout mutants for *Rac3* and *Rac1* displayed more severe cochlea defects than *Rac1* mutants alone (Grimsley-Myers et al., 2009). Along with *Rac1* and *Rac3*, *Cdc42* RNA expression has also been detected in the developing cochlea (Grimsley-Myers et al., 2009), although a *Cdc42* cochlea mutant has yet to be reported.

The p21-activated kinases (PAKs) are downstream effectors of Rac and are known to regulate cytoskeletal dynamics (Bokoch, 2003). Immunohistochemistry using an antibody against activated phospho-PAK in developing cochlea tissue showed asymmetric lateral membrane localization in inner and outer hair cells in a developmental gradient concomitant with the wave of hair cell differentiation (Grimsley-Myers et al., 2009). In $Vangl2^{Lp}$ mutants, pPAK was present but mislocalized in a pattern that strongly correlated with hair bundle misorientation. Additionally, *in vitro* cultures treated with IPA-3, a specific chemical inhibitor of PAK, displayed bundle defects that were similar to those observed in *Rac1* mutants (Grimsley-Myers et al., 2009). Combined, these data suggest that PAK acts downstream of Rac1 to regulate bundle orientation and morphogenesis.

Finally, c-Jun N-terminal kinases (JNKs) have also been shown to act as cytoplasmic mediators of PCP signaling. In particular, JNKs have been shown to be required for convergence and extension as well as newly identified PCP-dependent processes such as neuronal migration and axon guidance. While effects of inhibition of JNKs during cochlear development have not been examined, *in vitro* inhibition of JNKs during hair cell regeneration in the chicken utricle has been shown to lead to an inhibition of uniform bundle orientation (Warchol & Montcouquiol, 2010).

3.7. The fat/dachsous pathway

There is an ongoing debate as to the degree of interaction between the core fz–Vangl2 pathway and the fat/dachsous signaling pathway in PCP. In *Drosophila*, fat/dachsous signaling involves two protocadherins, fat and dachsous, and a transmembrane protein four-jointed, which was initially characterized as a PCP protein (Zeidler, Perrimon, & Strutt, 1999). Four-jointed, a Golgi protein, acts to phosphorylate both fat and dachsous. Fat suppresses transcription of four-jointed, while dachsous in turn represses fat activity (Simons & Mlodzik, 2008). While still not completely clear, fat/dachsous signaling appears to work in parallel with the core PCP pathway.

Recent work in mammals has demonstrated roles for the murine homologs of these three genes that are consistent with a role in PCP signaling (Rock, Schrauth, & Gessler, 2005). A single homolog for four-jointed, *Fjx1*, has been identified, and a mutant mouse has been generated, although no PCP phenotype was observed in the ear or elsewhere (Probst, Rock, Gessler, Vortkamp, & Puschel, 2007). In contrast, clear PCP phenotypes have been observed in one of the murine *Fat* mutants. Four mouse homologs of *fat* have been identified, *Fat1–4* with *Fat4* most closely related structurally to *Drosophila fat* (Saburi, Hester, Goodrich, & McNeill, 2012). Targeted deletion of *Fat4* leads to PCP defects in mutant mice including disrupted neural tube closure, shortened cochlea ducts, and disrupted hair bundles, although the latter two defects are relatively mild (Saburi et al., 2008). Expression of *Fjx1* was increased in *Fat4* mutants, indicating a conserved role for *Fat* in regulation of *Fjx1* in vertebrates. *Fat4* was also shown to interact genetically with *Fjx1* and *Vangl2* in kidney, yet localization of Vangl2 in the kidney in *Fat4* mutants was not disrupted, consistent with parallel pathways. Similar studies in the inner ear have not been conducted as yet.

Similarly, targeted disruption of *Dchs1*, the most closely related mammalian homolog of *dachsous*, leads to similarly mild cochlear PCP defects (Mao et al., 2011). Cochlear PCP defects were not exacerbated in *Dchs1/Fat4* double mutants, suggesting disruption of a common pathway. As is the case in *Drosophila*, the precise interactions between Fjx1–dachsous–fat proteins and the core PCP proteins remain unclear. For a large part, functional redundancy between various members of the fat and dachsous families may mask the role of individual proteins. For example, *Fat1* mutant mice show no PCP phenotype (Ciani, Patel, Allen, & ffrench-Constant, 2003). It may be necessary to generate multiple gene knockout animals in order to tease apart the molecular mechanisms, which may also be tissue and time dependent.

4. UPSTREAM REGULATORS OF PCP

As discussed, appropriate localization of PCP proteins represents a key step in the generation of polarity and mislocalization of PCP proteins can result in phenotypic defects that are indistinguishable from phenotypes observed as a result of the complete deletion of a PCP gene. These observations have led to an emerging understanding of the importance of highly specific synthesis, trafficking, and targeting of proteins to particular regions of the cell. One such process is the transport of proteins from the endoplasmic reticulum (ER) to the Golgi apparatus. Surprisingly, "classic" PCP defects including craniorachischisis and misorientation of cochlear hair cells have been observed in *Sec24b* mutants (Merte et al., 2010; Wansleeben et al., 2010). *Sec24b* encodes a cargo-sorting coat protein that forms the Cop II-coated vesicles required for ER to Golgi transport (Wendeler, Paccaud, & Hauri, 2007). Both *Vangl2* and *Scrib* have been shown to genetically interact with *Sec24b*, and evidence suggests that Sec24b selectively transports Vangl2 from the ER to the Golgi (Merte et al., 2010; Wansleeben et al., 2010). Of note is that the genes encoding Sec23 and Sec24 proteins in *Drosophila* have been shown to be required for apical–basal cell polarity (Norum et al., 2010). As such, these data implicate a high degree of selective protein sorting and trafficking upstream of protein localization.

The concept of regulated protein transport of PCP proteins to the cell membrane has been studied in the *Drosophila* wing epithelium. Polarized localization of fz was shown to be regulated via vesicular transport along proximodistally oriented microtubules and disruption of microtubules perturbed fz localization (Shimada, Yonemura, Ohkura, Strutt, & Uemura, 2006). Based on these results, a similar mechanism for the targeting of other PCP proteins in invertebrates as well as in vertebrates seems likely. Protein degradation is another mechanism by which subcellular protein levels can be regulated. Intriguingly, PCP phenotypes have been observed in *Smurf1* and *Smurf2* mutant mice (Narimatsu et al., 2009). *Smurf* genes encode ubiquitin ligases that are thought to spatiotemporally regulate ubiquitination of diverse substrates. The PCP phenotype in *Smurf* mutants has been attributed to loss of asymmetric subcellular localization of Prickle1 due to defective protein degradation via a Smurf, Par6, and Dvl complex (Narimatsu et al., 2009).

Dact1 (Dapper, antagonist of beta-catenin, homolog 1) has been shown to be important in regulating cytoplasmic PCP components upstream of PCP signaling. It is thought to modulate PCP signaling via interaction with the PDZ domain of Dvl (Wen et al., 2010). *Dact1* homozygous null mice are perinatal lethal and exhibit a variety of developmental defects including neural tube closure defects and shortened tails (Wen et al., 2010); however, cochleae were not examined in these mice. Dact1 is thought to control localization and levels of cellular Dvl, disrupting Rho GTPase and JNK. Dact1 has also been shown to regulate Vangl2 localization during gastrulation via direct protein interaction (Suriben, Kivimae, Fisher, Moon, & Cheyette, 2009). Bicaudal C is another potential candidate for modulation of Dvl during PCP signaling in the mammalian cochlea as it has been shown to do so in *Xenopus* and *Drosophila* (Bouvrette, Sittaramane, Heidel, Chandrasekhar, & Bryda, 2010; Maisonneuve et al., 2009).

Finally, it is important to consider that PCP defects have also been reported in mice carrying mutations in genes that act at time points considerably earlier than the formation of stereociliary bundles. For instance, the homeodomain protein *Emx2* has been shown to regulate early developmental events within the inner ear including balancing cell proliferation and differentiation. Homozygous *Emx2* knockouts survive until birth but display shorter cochlear ducts and reduced hair cell numbers with rotated stereociliary bundles (Holley et al., 2010). Classical PCP signaling is not thought to be directly affected by Emx2, rather the polarity defects observed in these mice are more likely to be a result of the developmental defects related to the extension and patterning of the sensory precursor population that secondarily result in rotated stereociliary bundles. However, considering that many of the PCP mutations also result in defects in cochlear extension, separating direct and indirect roles of individual genes in PCP signaling within the cochlea can be challenging.

5. CILIA AND THEIR ROLE IN PCP

No discussion of PCP in the mammalian inner ear would be complete without an examination of the role of the kinocilium during cochlear and inner ear development and bundle orientation. As discussed, kinocilia are not required for mechanotransduction (Hudspeth & Jacobs, 1979), and while they persist in most hair cells, they are actually lost in the mammalian cochlea shortly before the onset of hearing. These results, and earlier discussed data on kinocilia movements during bundle formation, have led to the suggestion that they play an important role in the development and orientation of the stereociliary bundles.

One of the first demonstrations that developing kinocilia are important in PCP signaling was the identification of PCP defects in ciliary mutants. Ciliopathies refer to a range of human disease syndromes that are caused by defects in primary cilia function. One of these syndromes, Bardet–Biedl syndrome (BBS), is caused by mutations in several genes including *MKKS* and *BBS4* (Forsythe & Beales, 2012), which have been shown to localize to the base of the cilium. Analysis of the effects of loss or inhibition of either gene indicated convergence extension defects in zebrafish and rotated stereociliary bundles in mice (Ross et al., 2005; Simons et al., 2005). *Bbs4*-null mice also displayed craniorachischisis and open eyelids, phenotypes that are also associated with PCP defects. Further, in addition to abnormal rotation, in *Mkks*$^{-/-}$ cochleae, some stereociliary bundles displayed kinocilia that were detached from the rest of the bundle. These mice also had abnormal distortion product otoacoustic emissions (DPOAEs) indicating defects in auditory function, possibly as a result of abnormal stereociliary bundles. Finally, *Mkks* and *Bbs4* were shown to interact genetically with *Vangl2* in both mice and zebrafish (Ross et al., 2005; Simons et al., 2005). More recently, Vangl2 has also been shown to interact biochemically with another BBS protein, *bbs8*, in zebrafish (May-Simera et al., 2010). Consistent with this finding, knockdown of *bbs8* in combination with a mutation in *vangl2* results in cilia-based defects in fluid flow in Kuppfer's vesicle (May-Simera et al., 2010). A potential mechanism for some of these defects was recently presented in a study that examined interactions between *Mkks* and *Cep290*, another ciliopathy gene (Rachel et al., 2012). While animals that were homozygous for either *Cep290*^{rd16}, the causative mutation for LCA-like (ciliopathy) defects in mice, or an *Mkks* deletion both displayed significant hearing loss and some degree of PCP defect, combinations of the two mutant genes actually displayed a significant degree of rescue. Further, examination of kinocilia indicated a significant increase in length in *Cep290*^{rd16} mice, a significant shortening in *Mkks* mutants and a rescue in some genetic combinations, suggesting that regulation of kinociliary length is important for normal PCP and auditory function.

PCP phenotypes in other ciliopathy models strengthen the association of PCP with cilia. In particular, mispositioned and abnormally rotated stereociliary bundles are present on outer hair cells in *Mks1* mutant mice (a mouse model for the ciliopathy Meckel–Gruber syndrome) (Cui et al., 2011). Similar to *Mkks*$^{-/-}$, kinocilia were also abnormally localized in relation to the stereociliary bundles in these mutants. Mutations in *Inversin*, a ciliary protein that localizes to the basal body, a modified centriole anchoring the cilium, cause nephronophthisis, which is also considered a ciliopathy-related disease (Lienkamp, Ganner, & Walz, 2012; Wolf & Hildebrandt, 2011). Knockdown of *Inversin* in *Xenopus* or zebrafish leads to convergence and extension defects, and Inversin has been shown to act as a molecular switch between canonical and noncanonical (PCP) signaling, by targeting cytoplasmic disheveled for degradation (Simons et al., 2005). However, a role for Inversin within the inner ear has not been examined.

Alstrom syndrome is the only ciliopathy in which hearing loss is a component of the clinical phenotype (Marshall, Maffei, Collin, & Naggert, 2011). Cochlear analysis of *Alms1* knockout mice, the only known causative gene for Alstrom syndrome thus far, identified flattened and rotated stereociliary bundles, yet no detachment of the kinocilium from the bundles was reported (Jagger et al., 2011). Auditory brainstem responses and DPOAEs were normal in juvenile mutants but did become disrupted in older animals. Similar to other ciliopathy proteins, Alms1 localizes to basal bodies in both hair cells and supporting cells.

Complete ablation of the kinocilium, as occurs in *Ift88* and *Kif3a* conditional mutants, results in more severe cochlear PCP phenotypes (Jones et al., 2008; Sipe & Lu, 2011) that include shortened cochleae, and rotated and misshapen bundles, some of which resemble donuts with stereocilia forming a circle on the surface of the hair cell. Both of these proteins are crucial for intraflagellar transport (IFT), which involves the bidirectional transport of proteins along the microtubule-based ciliary axoneme. As the cilia lacks the ability to generate its own proteins, loss of IFT leads to cilia agenesis. Although *Ift88* has been shown to genetically interact with *Vangl2* mutants resulting in more extreme cochlear phenotypes, the asymmetric distribution of both Vangl2 and Fz3 within developing hair cells is undisturbed in *Ift88* and *Kif3a* mutants, suggesting that the role of the kinocilium is downstream from the core PCP molecules. However, results in other model systems suggest that cilia mediate cell polarity via positioning of the centrosome (Park et al., 2008; Tissir et al., 2010), placing cilia and basal body proteins upstream of core

PCP signaling. These conflicting or incomplete results will hopefully be resolved through further examination.

Recently, several core PCP proteins have been shown to localize to the cilium or basal body, adding a further layer of complexity to this relationship. The membrane covering the protruding cilium is continuous with the cell membrane, and several recent studies have demonstrated that numerous molecules are trafficked onto the ciliary membrane (Bloodgood, 2012). In particular, Fat4 and Vangl2 have been detected on some ciliary membranes and, as discussed, mutations in both genes cause PCP phenotypes, including defects in stereociliary bundle orientation (Saburi et al., 2008). In addition, Vangl2 and Dvl have also been reported to associate with basal bodies. Dvl acts to dock basal bodies to the apical membrane preceding ciliogenesis (Park et al., 2008), while Vangl2 has been shown to be responsible for asymmetric positioning of motile cilia (Borovina, Superina, Voskas, & Ciruna, 2010). Moreover, a recent proteomic screen found several PCP-associated proteins in the human centrosome such as PRICKLE3, SCRIB, CCDC66, and ALBATROSS (Jakobsen et al., 2011). Finally, though not a core PCP gene, Fritz controls localization of the cytoskeletal protein Septin, crucial for collective cell movement and ciliogenesis in *Xenopus* embryos. Although not a cilia protein *per se*, mutations in human *FRITZ* have been identified in two ciliopathies, BBS and Meckel–Gruber syndrome (Kim et al., 2010).

Overall, these findings clearly implicate cilia and cilia-related proteins in PCP signaling within the auditory system. However, the specific nature of the interactions between cilia and the core PCP signaling cassette remains unclear. Analysis of *Ift88* and *Kif3a* mutants would suggest that positioning of the kinocilium and its basal body is a downstream effect of PCP signaling. But other data potentially place ciliary signaling upstream of, or parallel with, the PCP cassette. The lumenal location of developing kinocilia puts them in an excellent position to act as receptive antennae for extracellular signaling cues (May-Simera & Kelley, 2012), a role that has been confirmed for other cilia by the recent localization of multiple receptor molecules on their surfaces (Bloodgood, 2012). Further, the basal body, located at the base of the kinocilium, is emerging as a regional signaling hot spot. Not only do numerous signaling molecules localize to this region, but also activation and/or degradation of signaling proteins occurs in this region. Additionally, the basal body, which is derived from the mother centriole, acts as a microtubule organizing center, which can regulate the nucleation and outgrowth of intracellular microtubules (Kim et al., 2004; May-Simera & Kelley, 2012;

May-Simera et al., 2009). Potentially correct basal body/centrosome positioning could be required for subsequent regulation of cytoskeletal architecture.

Therefore, at this point, it seems unreasonable to try to definitively place cilia-related signaling either upstream, downstream, or in parallel with PCP. One could easily envisage a possible feedback loop where cilia-related signaling mediates cytoskeletal architecture/function which then plays a role in PCP signaling, which, in turn, regulates cilia position and function. Another key issue to be considered here is whether all of the defects observed in "ciliopathies" are actually a result of disruption in cilia function. Many of the ciliopathy proteins are now being reported to be expressed in other locations within the cell, suggesting that they could be mediating other aspects of cellular biology and development.

6. LOOKING FORWARD: PCP SIGNALING IN THE INNER EAR, WHERE DO WE GO FROM HERE?

As discussed, the very obvious uniform alignment of hair cell stereociliary bundles and the direct link of this alignment with inner ear function have made the inner ear one of the best model systems for studying PCP in vertebrates. Bundle orientation has demonstrated a high degree of sensitivity to perturbations in PCP signaling, in particular, in terms of genetic interactions between key PCP molecules, such as Vangl2, and other candidates that might not be sufficient to produce obvious results in simple deletion models. Further, the very clear morphological manifestation of PCP defects in terms of bundle rotation allows for straightforward quantification of orientation defects which can be used to confirm and compare effects of different mutations. However, more recent studies have yielded results which suggest that our understanding of polarization within the auditory system is far more limited than once thought. The fact that asymmetric localization of core PCP proteins does not directly correlate with bundle orientation suggests that different types of polarization may occur at multiple time points during the development of a given auditory epithelium. Moreover, if asymmetric localization of core PCP proteins is a hallmark of PCP, then the observation that bundle orientation is not dictated by this asymmetry suggests the possibility that bundle rotation should not be considered as a readout of polarization. Needless to say, this would have a significant impact on our understanding of how PCP functions within the auditory system and on our ability to analyze PCP defects.

With these caveats in mind, the auditory system still has considerable potential as a model system for studying PCP. As demonstrated by the Sienknecht, Anderson, Parodi, Fantetti, and Fekete (2011) study, the existence of broad fields of hair cells in both avian and mammalian inner ears provides a unique opportunity to test hypotheses related to autonomous versus nonautonomous signaling by PCP proteins as well as to examine the possible roles of Wnt morphogens in the generation of polarity. However, while effects on bundle orientation cannot be ignored, it will be equally important to determine changes in asymmetric protein localization and to consider the degree of correlation between the two. Further, as the ability to generate tissue- and cell-specific mutations increases, it will be extremely enlightening to determine the effects of bundle rotations and other PCP defects on the specific functions of the inner ear. The rotation of even a small number of bundles has been shown to lead to specific defects in the perception of some frequencies (Yoshida & Liberman, 1999), but the effects of similar rotations on vestibular function or other aspects of auditory perception remain to be determined.

Finally, the additional role of PCP signaling in the extension and patterning of the cochlear duct provides both a new and intriguing model system for the study of PCP signaling and, unfortunately, a confounding factor in the analysis of PCP-mediated effects. Recent work has demonstrated that the alignment of auditory hair cells is dependent on the successful outgrowth of the duct (Wang et al., 2005; Yamamoto, Okano, Ma, Adelstein, & Kelley, 2009), suggesting that PCP signaling plays a key regulatory role in the morphogenetic movements required to generate these ordered rows. Further studies should be able to take advantage of this developmental process to better understand how cells become precisely aligned. However, outgrowth defects that result in misaligned hair cells, often manifesting in the formation of fourth or fifth rows of outer hair cells, have the potential to lead to bundle misorientation as a secondary effect. As discussed, hair cells located in the third row demonstrate heightened sensitivity to PCP defects. Therefore, it seems possible that hair cells located in ectopic fourth or fifth rows might demonstrate disruptions in bundle alignment, not because of a direct role of a particular gene in bundle orientation but because of a role of that gene in cellular patterning. As is the case for many future studies, the ability to target genetic deletions to individual cell types at specific development time points will be required in order to be able to address these possibilities.

REFERENCES

Baeg, G. H., Selva, E. M., Goodman, R. M., Dasgupta, R., & Perrimon, N. (2004). The Wingless morphogen gradient is established by the cooperative action of Frizzled and Heparan Sulfate Proteoglycan receptors. *Developmental Biology*, *276*, 89–100.

Bilder, D. (2004). Epithelial polarity and proliferation control: Links from the Drosophila neoplastic tumor suppressors. *Genes & Development*, *18*, 1909–1925.

Bloodgood, R. A. (2012). The future of ciliary and flagellar membrane research. *Molecular Biology of the Cell*, *23*, 2407–2411.

Bokoch, G. M. (2003). Biology of the p21-activated kinases. *Annual Review of Biochemistry*, *72*, 743–781.

Borovina, A., Superina, S., Voskas, D., & Ciruna, B. (2010). Vangl2 directs the posterior tilting and asymmetric localization of motile primary cilia. *Nature Cell Biology*, *12*, 407–412.

Bouvrette, D. J., Sittaramane, V., Heidel, J. R., Chandrasekhar, A., & Bryda, E. C. (2010). Knockdown of bicaudal C in zebrafish (Danio rerio) causes cystic kidneys: A nonmammalian model of polycystic kidney disease. *Comparative Medicine*, *60*, 96–106.

Bradley, R. S., & Brown, A. M. (1990). The proto-oncogene int-1 encodes a secreted protein associated with the extracellular matrix. *The EMBO Journal*, *9*, 1569–1575.

Chen, P., Johnson, J. E., Zoghbi, H. Y., & Segil, N. (2002). The role of Math1 in inner ear development: Uncoupling the establishment of the sensory primordium from hair cell fate determination. *Development*, *129*, 2495–2505.

Chen, P., & Segil, N. (1999). p27(Kip1) links cell proliferation to morphogenesis in the developing organ of Corti. *Development*, *126*, 1581–1590.

Ciani, L., Patel, A., Allen, N. D., & ffrench-Constant, C. (2003). Mice lacking the giant protocadherin mFAT1 exhibit renal slit junction abnormalities and a partially penetrant cyclopia and anophthalmia phenotype. *Molecular and Cellular Biology*, *23*, 3575–3582.

Cotanche, D. A., & Corwin, J. T. (1991). Stereociliary bundles reorient during hair cell development and regeneration in the chick cochlea. *Hearing Research*, *52*, 379–402.

Courbard, J. R., Djiane, A., Wu, J., & Mlodzik, M. (2009). The apical/basal-polarity determinant Scribble cooperates with the PCP core factor Stbm/Vang and functions as one of its effectors. *Developmental Biology*, *333*, 67–77.

Cui, C., Chatterjee, B., Francis, D., Yu, Q., SanAgustin, J. T., Francis, R., et al. (2011). Disruption of Mks1 localization to the mother centriole causes cilia defects and developmental malformations in Meckel-Gruber syndrome. *Disease Models & Mechanisms*, *4*, 43–56.

Curtin, J. A., Quint, E., Tsipouri, V., Arkell, R. M., Cattanach, B., Copp, A. J., et al. (2003). Mutation of Celsr1 disrupts planar polarity of inner ear hair cells and causes severe neural tube defects in the mouse. *Current Biology*, *13*, 1129–1133.

Dabdoub, A., Donohue, M. J., Brennan, A., Wolf, V., Montcouquiol, M., Sassoon, D. A., et al. (2003). Wnt signaling mediates reorientation of outer hair cell stereociliary bundles in the mammalian cochlea. *Development*, *130*, 2375–2384.

Dabdoub, A., & Kelley, M. W. (2005). Planar cell polarity and a potential role for a Wnt morphogen gradient in stereociliary bundle orientation in the mammalian inner ear. *Journal of Neurobiology*, *64*, 446–457.

Daudet, N., Ripoll, C., Moles, J. P., & Rebillard, G. (2002). Expression of members of Wnt and Frizzled gene families in the postnatal rat cochlea. *Brain Research. Molecular Brain Research*, *105*, 98–107.

Deans, M. R., Antic, D., Suyama, K., Scott, M. P., Axelrod, J. D., & Goodrich, L. V. (2007). Asymmetric distribution of prickle-like 2 reveals an early underlying polarization of vestibular sensory epithelia in the inner ear. *The Journal of Neuroscience*, *27*, 3139–3147.

Denman-Johnson, K., & Forge, A. (1999). Establishment of hair bundle polarity and orientation in the developing vestibular system of the mouse. *Journal of Neurocytology, 28*, 821–835.

Driver, E. C., & Kelley, M. W. (2009). Specification of cell fate in the mammalian cochlea. *Birth Defects Research. Part C, Embryo Today, 87*, 212–221.

Etheridge, S. L., Ray, S., Li, S., Hamblet, N. S., Lijam, N., Tsang, M., et al. (2008). Murine dishevelled 3 functions in redundant pathways with dishevelled 1 and 2 in normal cardiac outflow tract, cochlea, and neural tube development. *PLoS Genetics, 4*, e1000259.

Forrester, W. C., Dell, M., Perens, E., & Garriga, G. (1999). A C. elegans Ror receptor tyrosine kinase regulates cell motility and asymmetric cell division. *Nature, 400*, 881–885.

Forsythe, E., & Beales, P. L. (2012). Bardet-Biedl syndrome. *European Journal of Human Genetics*. doi: 10.1038/ejhg.2012.115. [Jun 20, Epub ahead of print].

Gao, B., Song, H., Bishop, K., Elliot, G., Garrett, L., English, M. A., et al. (2011). Wnt signaling gradients establish planar cell polarity by inducing Vangl2 phosphorylation through Ror2. *Developmental Cell, 20*, 163–176.

Grimsley-Myers, C. M., Sipe, C. W., Geleoc, G. S., & Lu, X. (2009). The small GTPase Rac1 regulates auditory hair cell morphogenesis. *The Journal of Neuroscience, 29*, 15859–15869.

Habas, R., Dawid, I. B., & He, X. (2003). Coactivation of Rac and Rho by Wnt/Frizzled signaling is required for vertebrate gastrulation. *Genes & Development, 17*, 295–309.

Holley, M., Rhodes, C., Kneebone, A., Herde, M. K., Fleming, M., & Steel, K. P. (2010). Emx2 and early hair cell development in the mouse inner ear. *Developmental Biology, 340*, 547–556.

Hudspeth, A. J., & Jacobs, R. (1979). Stereocilia mediate transduction in vertebrate hair cells (auditory system/cilium/vestibular system). *Proceedings of the National Academy of Sciences of the United States of America, 76*, 1506–1509.

Jaffe, A. B., & Hall, A. (2005). Rho GTPases: Biochemistry and biology. *Annual Review of Cell and Developmental Biology, 21*, 247–269.

Jagger, D., Collin, G., Kelly, J., Towers, E., Nevill, G., Longo-Guess, C., et al. (2011). Alstrom Syndrome protein ALMS1 localizes to basal bodies of cochlear hair cells and regulates cilium-dependent planar cell polarity. *Human Molecular Genetics, 20*, 466–481.

Jakobsen, L., Vanselow, K., Skogs, M., Toyoda, Y., Lundberg, E., Poser, I., et al. (2011). Novel asymmetrically localizing components of human centrosomes identified by complementary proteomics methods. *The EMBO Journal, 30*, 1520–1535.

Jenny, A., & Mlodzik, M. (2006). Planar cell polarity signaling: A common mechanism for cellular polarization. *The Mount Sinai Journal of Medicine, New York, 73*, 738–750.

Jones, C., Roper, V. C., Foucher, I., Qian, D., Banizs, B., Petit, C., et al. (2008). Ciliary proteins link basal body polarization to planar cell polarity regulation. *Nature Genetics, 40*, 69–77.

Kawano, Y., & Kypta, R. (2003). Secreted antagonists of the Wnt signalling pathway. *Journal of Cell Science, 116*, 2627–2634.

Kibar, Z., Torban, E., McDearmid, J. R., Reynolds, A., Berghout, J., Mathieu, M., et al. (2007). Mutations in VANGL1 associated with neural-tube defects. *The New England Journal of Medicine, 356*, 1432–1437.

Kibar, Z., Underhill, D. A., Canonne-Hergaux, F., Gauthier, S., Justice, M. J., & Gros, P. (2001). Identification of a new chemically induced allele (Lp(m1Jus)) at the loop-tail locus: Morphology, histology, and genetic mapping. *Genomics, 72*, 331–337.

Kim, J. C., Badano, J. L., Sibold, S., Esmail, M. A., Hill, J., Hoskins, B. E., et al. (2004). The Bardet-Biedl protein BBS4 targets cargo to the pericentriolar region and is required for microtubule anchoring and cell cycle progression. *Nature Genetics, 36*, 462–470.

Kim, S. K., Shindo, A., Park, T. J., Oh, E. C., Ghosh, S., Gray, R. S., et al. (2010). Planar cell polarity acts through septins to control collective cell movement and ciliogenesis. *Science*, *329*, 1337–1340.

Lee, J., Andreeva, A., Sipe, C. W., Liu, L., Cheng, A., & Lu, X. (2012). PTK7 regulates myosin II activity to orient planar polarity in the mammalian auditory epithelium. *Current Biology*, *22*, 956–966.

Leimeister, C., Bach, A., & Gessler, M. (1998). Developmental expression patterns of mouse sFRP genes encoding members of the secreted frizzled related protein family. *Mechanisms of Development*, *75*, 29–42.

Lienkamp, S., Ganner, A., & Walz, G. (2012). Inversin, Wnt signaling and primary cilia. *Differentiation; Research in Biological Diversity*, *83*, S49–S55.

Lu, X., Borchers, A. G., Jolicoeur, C., Rayburn, H., Baker, J. C., & Tessier-Lavigne, M. (2004). PTK7/CCK-4 is a novel regulator of planar cell polarity in vertebrates. *Nature*, *430*, 93–98.

Macara, I. G. (2004). Parsing the polarity code. *Nature Reviews. Molecular Cell Biology*, *5*, 220–231.

Maisonneuve, C., Guilleret, I., Vick, P., Weber, T., Andre, P., Beyer, T., et al. (2009). Bicaudal C, a novel regulator of Dvl signaling abutting RNA-processing bodies, controls cilia orientation and leftward flow. *Development*, *136*, 3019–3030.

Mao, Y., Mulvaney, J., Zakaria, S., Yu, T., Morgan, K. M., Allen, S., et al. (2011). Characterization of a Dchs1 mutant mouse reveals requirements for Dchs1-Fat4 signaling during mammalian development. *Development*, *138*, 947–957.

Marshall, J. D., Maffei, P., Collin, G. B., & Naggert, J. K. (2011). Alstrom syndrome: Genetics and clinical overview. *Current Genomics*, *12*, 225–235.

May-Simera, H. L., Kai, M., Hernandez, V., Osborn, D. P., Tada, M., & Beales, P. L. (2010). Bbs8, together with the planar cell polarity protein Vangl2, is required to establish left-right asymmetry in zebrafish. *Developmental Biology*, *345*, 215–225.

May-Simera, H. L., & Kelley, M. W. (2012). Cilia, Wnt signaling, and the cytoskeleton. *Cilia*, *1*.

May-Simera, H. L., Ross, A., Rix, S., Forge, A., Beales, P. L., & Jagger, D. J. (2009). Patterns of expression of Bardet-Biedl syndrome proteins in the mammalian cochlea suggest noncentrosomal functions. *The Journal of Comparative Neurology*, *514*, 174–188.

McKenzie, E., Krupin, A., & Kelley, M. W. (2004). Cellular growth and rearrangement during the development of the mammalian organ of Corti. *Developmental Dynamics*, *229*, 802–812.

Merte, J., Jensen, D., Wright, K., Sarsfield, S., Wang, Y., Schekman, R., et al. (2010). Sec24b selectively sorts Vangl2 to regulate planar cell polarity during neural tube closure. *Nature Cell Biology*, *12*, 41–46 sup pp. 1–8.

Minami, Y., Oishi, I., Endo, M., & Nishita, M. (2010). Ror-family receptor tyrosine kinases in noncanonical Wnt signaling: Their implications in developmental morphogenesis and human diseases. *Developmental Dynamics*, *239*, 1–15.

Montcouquiol, M., Rachel, R. A., Lanford, P. J., Copeland, N. G., Jenkins, N. A., & Kelley, M. W. (2003). Identification of Vangl2 and Scrb1 as planar polarity genes in mammals. *Nature*, *423*, 173–177.

Montcouquiol, M., Sans, N., Huss, D., Kach, J., Dickman, J. D., Forge, A., et al. (2006). Asymmetric localization of Vangl2 and Fz3 indicate novel mechanisms for planar cell polarity in mammals. *The Journal of Neuroscience*, *26*, 5265–5275.

Morsli, H., Choo, D., Ryan, A., Johnson, R., & Wu, D. K. (1998). Development of the mouse inner ear and origin of its sensory organs. *The Journal of Neuroscience*, *18*, 3327–3335.

Mullick, A., Trasler, D., & Gros, P. (1995). High-resolution linkage map in the vicinity of the Lp locus. *Genomics*, *26*, 479–488.

Murdoch, J. N., Doudney, K., Paternotte, C., Copp, A. J., & Stanier, P. (2001). Severe neural tube defects in the loop-tail mouse result from mutation of Lpp1, a novel gene involved in floor plate specification. *Human Molecular Genetics*, *10*, 2593–2601.

Murdoch, J. N., Henderson, D. J., Doudney, K., Gaston-Massuet, C., Phillips, H. M., Paternotte, C., et al. (2003). Disruption of scribble (Scrb1) causes severe neural tube defects in the circletail mouse. *Human Molecular Genetics*, *12*, 87–98.

Narimatsu, M., Bose, R., Pye, M., Zhang, L., Miller, B., Ching, P., et al. (2009). Regulation of planar cell polarity by Smurf ubiquitin ligases. *Cell*, *137*, 295–307.

Nelson, W. J. (2003). Adaptation of core mechanisms to generate cell polarity. *Nature*, *422*, 766–774.

Norum, M., Tang, E., Chavoshi, T., Schwarz, H., Linke, D., Uv, A., et al. (2010). Trafficking through COPII stabilises cell polarity and drives secretion during Drosophila epidermal differentiation. *PLoS One*, *5*, e10802.

Oishi, I., Suzuki, H., Onishi, N., Takada, R., Kani, S., Ohkawara, B., et al. (2003). The receptor tyrosine kinase Ror2 is involved in non-canonical Wnt5a/JNK signalling pathway. *Genes to Cells*, *8*, 645–654.

Park, C. C., Ahn, S., Bloom, J. S., Lin, A., Wang, R. T., Wu, T., et al. (2008). Fine mapping of regulatory loci for mammalian gene expression using radiation hybrids. *Nature Genetics*, *40*, 421–429.

Paudyal, A., Damrau, C., Patterson, V. L., Ermakov, A., Formstone, C., Lalanne, Z., et al. (2010). The novel mouse mutant, chuzhoi, has disruption of Ptk7 protein and exhibits defects in neural tube, heart and lung development and abnormal planar cell polarity in the ear. *BMC Developmental Biology*, *10*, 87.

Peradziryi, H., Kaplan, N. A., Podleschny, M., Liu, X., Wehner, P., Borchers, A., et al. (2011). PTK7/Otk interacts with Wnts and inhibits canonical Wnt signalling. *The EMBO Journal*, *30*, 3729–3740.

Pickles, J. O., Brix, J., Comis, S. D., Gleich, O., Koppl, C., Manley, G. A., et al. (1989). The organization of tip links and stereocilia on hair cells of bird and lizard basilar papillae. *Hearing Research*, *41*, 31–41.

Probst, B., Rock, R., Gessler, M., Vortkamp, A., & Puschel, A. W. (2007). The rodent Four-jointed ortholog Fjx1 regulates dendrite extension. *Developmental Biology*, *312*, 461–470.

Puligilla, C., & Kelley, M. W. (2009). Building the world's best hearing aid; regulation of cell fate in the cochlea. *Current Opinion in Genetics and Development*, *19*, 368–373.

Puppo, F., Thome, V., Lhoumeau, A. C., Cibois, M., Gangar, A., Lembo, F., et al. (2011). Protein tyrosine kinase 7 has a conserved role in Wnt/beta-catenin canonical signalling. *EMBO Reports*, *12*, 43–49.

Qian, D., Jones, C., Rzadzinska, A., Mark, S., Zhang, X., Steel, K. P., et al. (2007). Wnt5a functions in planar cell polarity regulation in mice. *Developmental Biology*, *306*, 121–133.

Rachel, R. A., May-Simera, H. L., Veleri, S., Gotoh, N., Choi, B. Y., Murga-Zamalloa, C., et al. (2012). Combining Cep290 and Mkks ciliopathy alleles in mice rescues sensory defects and restores ciliogenesis. *The Journal of Clinical Investigation*, *122*, 1233–1245.

Rachel, R. A., Wellington, S. J., Warburton, D., Mason, C. A., & Beermann, F. (2002). A new allele of Gli3 and a new mutation, circletail (Crc), resulting from a single transgenic experiment. *Genesis*, *33*, 55–61.

Rock, R., Schrauth, S., & Gessler, M. (2005). Expression of mouse dchs1, fjx1, and fat-j suggests conservation of the planar cell polarity pathway identified in Drosophila. *Developmental Dynamics*, *234*, 747–755.

Ross, A. J., May-Simera, H., Eichers, E. R., Kai, M., Hill, J., Jagger, D. J., et al. (2005). Disruption of Bardet-Biedl syndrome ciliary proteins perturbs planar cell polarity in vertebrates. *Nature Genetics*, *37*, 1135–1140.

Ruben, R. J. (1967). Development of the inner ear of the mouse: A radioautographic study of terminal mitoses. *Acta Oto-Laryngologica. Supplementum*, *220*, 1–44.

Saburi, S., Hester, I., Fischer, E., Pontoglio, M., Eremina, V., Gessler, M., et al. (2008). Loss of Fat4 disrupts PCP signaling and oriented cell division and leads to cystic kidney disease. *Nature Genetics*, *40*, 1010–1015.

Saburi, S., Hester, I., Goodrich, L., & McNeill, H. (2012). Functional interactions between Fat family cadherins in tissue morphogenesis and planar polarity. *Development*, *139*, 1806–1820.

Semenov, M. V., Habas, R., Macdonald, B. T., & He, X. (2007). SnapShot: Noncanonical Wnt signaling pathways. *Cell*, *131*, 1378.

Shimada, Y., Yonemura, S., Ohkura, H., Strutt, D., & Uemura, T. (2006). Polarized transport of Frizzled along the planar microtubule arrays in Drosophila wing epithelium. *Developmental Cell*, *10*, 209–222.

Shnitsar, I., & Borchers, A. (2008). PTK7 recruits dsh to regulate neural crest migration. *Development*, *135*, 4015–4024.

Sienknecht, U. J., Anderson, B. K., Parodi, R. M., Fantetti, K. N., & Fekete, D. M. (2011). Non-cell-autonomous planar cell polarity propagation in the auditory sensory epithelium of vertebrates. *Developmental Biology*, *352*, 27–39.

Simons, M., Gloy, J., Ganner, A., Bullerkotte, A., Bashkurov, M., Kronig, C., et al. (2005). Inversin, the gene product mutated in nephronophthisis type II, functions as a molecular switch between Wnt signaling pathways. *Nature Genetics*, *37*, 537–543.

Simons, M., & Mlodzik, M. (2008). Planar cell polarity signaling: From fly development to human disease. *Annual Review of Genetics*, *42*, 517–540.

Sipe, C. W., & Lu, X. (2011). Kif3a regulates planar polarization of auditory hair cells through both ciliary and non-ciliary mechanisms. *Development*, *138*, 3441–3449.

Stauffer, E. A., & Holt, J. R. (2007). Sensory transduction and adaptation in inner and outer hair cells of the mouse auditory system. *Journal of Neurophysiology*, *98*, 3360–3369.

Strigini, M., & Cohen, S. M. (2000). Wingless gradient formation in the Drosophila wing. *Current Biology*, *10*, 293–300.

Strutt, D., Johnson, R., Cooper, K., & Bray, S. (2002). Asymmetric localization of frizzled and the determination of notch-dependent cell fate in the Drosophila eye. *Current Biology: CB*, *12*, 813–824.

Suriben, R., Kivimae, S., Fisher, D. A., Moon, R. T., & Cheyette, B. N. (2009). Posterior malformations in Dact1 mutant mice arise through misregulated Vangl2 at the primitive streak. *Nature Genetics*, *41*, 977–985.

Tissir, F., Qu, Y., Montcouquiol, M., Zhou, L., Komatsu, K., Shi, D., et al. (2010). Lack of cadherins Celsr2 and Celsr3 impairs ependymal ciliogenesis, leading to fatal hydrocephalus. *Nature Neuroscience*, *13*, 700–707.

Torban, E., Patenaude, A. M., Leclerc, S., Rakowiecki, S., Gauthier, S., Andelfinger, G., et al. (2008). Genetic interaction between members of the Vangl family causes neural tube defects in mice. *Proceedings of the National Academy of Sciences of the United States of America*, *105*, 3449–3454.

Uemura, T., & Shimada, Y. (2003). Breaking cellular symmetry along planar axes in Drosophila and vertebrates. *Journal of Biochemistry*, *134*, 625–630.

Ueno, N., & Greene, N. D. (2003). Planar cell polarity genes and neural tube closure. *Birth Defects Research. Part C, Embryo Today: Reviews*, *69*, 318–324.

Ulloa, F., & Marti, E. (2010). Wnt won the war: Antagonistic role of Wnt over Shh controls dorso-ventral patterning of the vertebrate neural tube. *Developmental Dynamics*, *239*, 69–76.

Wang, Y., Guo, N., & Nathans, J. (2006). The role of Frizzled3 and Frizzled6 in neural tube closure and in the planar polarity of inner-ear sensory hair cells. *The Journal of Neuroscience*, *26*, 2147–2156.

Wang, J., Hamblet, N. S., Mark, S., Dickinson, M. E., Brinkman, B. C., Segil, N., et al. (2006). Dishevelled genes mediate a conserved mammalian PCP pathway to regulate convergent extension during neurulation. *Development*, *133*, 1767–1778.

Wang, J., Mark, S., Zhang, X., Qian, D., Yoo, S. J., Radde-Gallwitz, K., et al. (2005). Regulation of polarized extension and planar cell polarity in the cochlea by the vertebrate PCP pathway. *Nature Genetics*, *37*, 980–985.

Wansleeben, C., Feitsma, H., Montcouquiol, M., Kroon, C., Cuppen, E., & Meijlink, F. (2010). Planar cell polarity defects and defective Vangl2 trafficking in mutants for the COPII gene Sec24b. *Development*, *137*, 1067–1073.

Warchol, M. E., & Montcouquiol, M. (2010). Maintained expression of the planar cell polarity molecule Vangl2 and reformation of hair cell orientation in the regenerating inner ear. *Journal of the Association for Research in Otolaryngology*, *11*, 395–406.

Wehner, P., Shnitsar, I., Urlaub, H., & Borchers, A. (2011). RACK1 is a novel interaction partner of PTK7 that is required for neural tube closure. *Development*, *138*, 1321–1327.

Wen, J., Chiang, Y. J., Gao, C., Xue, H., Xu, J., Ning, Y., et al. (2010). Loss of Dact1 disrupts planar cell polarity signaling by altering dishevelled activity and leads to posterior malformation in mice. *The Journal of Biological Chemistry*, *285*, 11023–11030.

Wendeler, M. W., Paccaud, J. P., & Hauri, H. P. (2007). Role of Sec24 isoforms in selective export of membrane proteins from the endoplasmic reticulum. *EMBO Reports*, *8*, 258–264.

Wodarz, A., & Nusse, R. (1998). Mechanisms of Wnt signaling in development. *Annual Review of Cell and Developmental Biology*, *14*, 59–88.

Wolf, M. T., & Hildebrandt, F. (2011). Nephronophthisis. *Pediatric Nephrology*, *26*, 181–194.

Yamamoto, S., Nishimura, O., Misaki, K., Nishita, M., Minami, Y., Yonemura, S., et al. (2008). Cthrc1 selectively activates the planar cell polarity pathway of Wnt signaling by stabilizing the Wnt-receptor complex. *Developmental Cell*, *15*, 23–36.

Yamamoto, N., Okano, T., Ma, X., Adelstein, R. S., & Kelley, M. W. (2009). Myosin II regulates extension, growth and patterning in the mammalian cochlear duct. *Development*, *136*, 1977–1986.

Ybot-Gonzalez, P., Savery, D., Gerrelli, D., Signore, M., Mitchell, C. E., Faux, C. H., et al. (2007). Convergent extension, planar-cell-polarity signalling and initiation of mouse neural tube closure. *Development*, *134*, 789–799.

Yen, W. W., Williams, M., Periasamy, A., Conaway, M., Burdsal, C., Keller, R., et al. (2009). PTK7 is essential for polarized cell motility and convergent extension during mouse gastrulation. *Development*, *136*, 2039–2048.

Yoshida, N., & Liberman, M. C. (1999). Stereociliary anomaly in the guinea pig: Effects of hair bundle rotation on cochlear sensitivity. *Hearing Research*, *131*, 29–38.

Zecca, M., Basler, K., & Struhl, G. (1996). Direct and long-range action of a wingless morphogen gradient. *Cell*, *87*, 833–844.

Zeidler, M. P., Perrimon, N., & Strutt, D. I. (1999). The four-jointed gene is required in the Drosophila eye for ommatidial polarity specification. *Current Biology: CB*, *9*, 1363–1372.

CHAPTER SIX

Does Planar Cell Polarity Signaling Steer Growth Cones?

Yimin Zou[1]

Neurobiology Section, Biological Sciences Division, University of California, San Diego, La Jolla, California, USA

[1]Corresponding author: e-mail address: yzou@ucsd.edu

Contents

Abstract

Recent studies established the role of planar cell polarity signaling in axon guidance. Signaling mechanisms controlling the direction of axon growth are poorly understood. The possibility that conserved and robust cell polarity signaling pathways may be reused as a key mechanism to convey asymmetric signaling in growth cones will provide insights to solving this long-standing mystery. Insights gained from growth cones can also shed light on general principles of cell polarity signaling. This review also discusses the possibility that this cell polarity signaling-based mechanism may be a general mechanism for mediating directional control by many, if not all, axon guidance molecules.

Current Topics in Developmental Biology, Volume 101
ISSN 0070-2153
http://dx.doi.org/10.1016/B978-0-12-394592-1.00009-0
© 2012 Elsevier Inc.
All rights reserved.

1. OUTLINE OF REVIEW

This review summarizes the role of planar cell polarity (PCP) signaling components in axon guidance and whether and how growth cones may use cell polarity pathways (planar and apical–basal) or a subset of cell polarity pathways to detect guidance cues and to turn up or down along gradients of guidance molecules. This review also explores, for the first time, the exciting possibility that this cell polarity signaling-based mechanism (or module) may be a general principle in growth cone guidance.

In stark contrast to our rich knowledge of the identity of axon guidance molecules, how signaling conveys directionality is poorly understood. The possibility that cell polarity signaling may be a key mediator of turning will provide answers to fundamental questions in signaling and cell biological mechanisms of growth cone steering. The systemic and global feature of planar polarity (also referred to as tissue polarity) and apical–basal polarity (A-BP) suggests that this cell polarity-based signaling mechanism may be responsible for establishing the highly organized axonal and dendritic wiring patterns throughout the nervous system, a striking feature of neural circuitry.

Even though the growth cone is a motile structure, apparently distinct from nonmotile epithelial sheets, there are potentially common links at the molecular and cellular levels because the molecular and cellular components are not "stationary" in nonmotile epithelial cells. In fact, recent work showed that the key component in the adherence junction, E-cadherin, is actively turned over while setting up polarity. Ongoing polarized endocytosis and exocytosis in stationary epithelial cells are also widely recognized. Insights gained from cell polarity signaling pathways in growth cones can also shed light on general principles of cell polarity signaling. This review focuses on PCP components, although A-BP will be included because of the intimate interactions of these two cell polarity pathways.

2. PCP SIGNALING COMPONENTS MEDIATE AXON GUIDANCE

Two independent studies published in 2003 converged on the surprising discovery that the Wnt family morphogens are conserved axon guidance molecules (Lyuksyutova et al., 2003; Yoshikawa, McKinnon, Kokel, & Thomas, 2003). In the vertebrate spinal cord, commissural axons that ascend after midline crossing are attracted by Wnts, which are expressed in an anterior-high–posterior-low (rostral high–caudal low) graded fashion

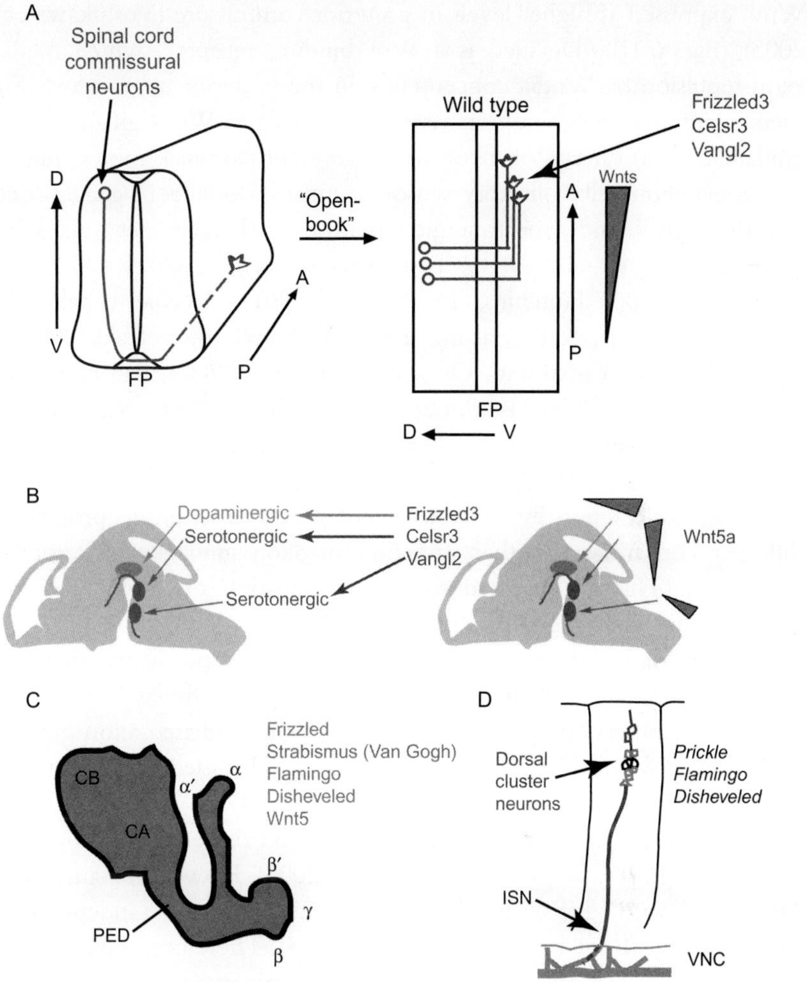

Figure 6.1 Evidence of PCP signaling in axon guidance. (A) Spinal cord commissural axons reply on PCP signaling components for faithful anterior turning after crossing the midline (Lyuksyutova et al., 2003; Shafer, Onishi, Lo, Colakoglu, & Zou, 2011). (B) Dopaminergic and serotonergic axons in the midbrain and hindbrain are guided to project along the anterior–posterior axis by PCP components (Fenstermaker et al., 2010). (C) *Drosophila* mushroom body axon projection patterns are organized by PCP components (Shimizu, Sato, & Tabata, 2011). (D) *Drosophila* sensory axons rely on core PCP components for outgrowth (Mrkusich, Flanagan, & Whitington, 2011). (For color version of this figure, the reader is referred to the online version of this chapter.)

along the length of the spinal cord and detected by the receptor, Frizzled3, in the commissural neurons (Lyuksyutova et al., 2003) (Fig. 6.1A). Studies in *Drosophila* showed that Derailed-expressing commissural axons only cross the midline along the anterior commissure due to the repulsive function of

Wnt5 expressed at higher levels in posterior commissure (Yoshikawa et al., 2003) (Fig. 6.1B). Derailed is a Wnt binding receptor, which mediates axon repulsion by Wnt5a concentrated in the posterior commissure. Since then, a number of publications reported the role of Wnt signaling in axon guidance in a variety of neuronal types (corticospinal motor neurons, retinal ganglion cells, olfactory sensory neurons, dorsal root ganglion cells, dopaminergic, and serotonergic neurons) (Blakely et al., 2011; Domanitskaya et al., 2010; Fenstermaker et al., 2010; Hilliard & Bargmann, 2006; Hutchins, Li, & Kalil, 2011; Keeble et al., 2006; Kennerdell, Fetter, & Bargmann, 2009; Li, Hutchins, & Kalil, 2009; Liu et al., 2005; Lu, Yamamoto, Ortega, & Baltimore, 2004; Pan et al., 2006; Prasad & Clark, 2006; Rodriguez-Gil & Greer, 2008; Sato, Umetsu, Murakami, Yasugi, & Tabata, 2006; Schmitt et al., 2006; Shimizu et al., 2011; Song et al., 2010). Because morphogens are known for their role in specifying cell types by activating different transcription programs at different concentrations, their function in axon guidance had not been anticipated. This has also led to the investigation on how Wnt family morphogens signal in growth cones to guide axons (Zou, 2004).

In an earlier review in 2004, it was cautiously postulated that "PCP signaling pathway," a "variation of PCP signaling pathway," or a "subset of PCP machinery" might be involved in Wnt-mediated axon guidance. The caution was motivated by the fact that only Frizzled3 had been shown to mediate axon guidance at that time (Zou, 2004). In 2005, another PCP component, Ceslr3, showed similar function to Frizzled3 in axon guidance (Tissir, Bar, Jossin, De Backer, & Goffinet, 2005). However, Flamingo, the *Drosophila* homologue of the Celsrs, has additional function independent of PCP signaling (Berger-Muller & Suzuki, 2011). Therefore, the evidence of PCP signaling in axon guidance was still preliminary.

In more recent years, new work has begun to reveal that cell polarity signaling components may play a central role in Wnt-mediated axon guidance. In 2008, components of A-BP signaling, atypical PKC/Par3/Par6 complex, were first shown to directly mediate Wnt attraction and anterior turning of spinal cord commissural axons (Wolf et al., 2008). In 2010 and 2011, multiple PCP components, Frizzled3, Ceslr3, and Vangl2, were directly tested and found to be required in anterior–posterior guidance of brainstem serotonergic and dopaminergic axons and the spinal cord commissural axons (Fenstermaker et al., 2010; Shafer et al., 2011) (Fig. 6.1A and B). In addition, Vangl2 was found localized on tips of extending filopodia and promoted commissural axon growth in response to Wnt5a

(Shafer et al., 2011). The involvement of multiple PCP signaling components in axon guidance appears to be evolutionarily conserved because in *Drosophila*, frizzled (fz), strabismus (stbm/Van Gogh), flamingo (fmi), and disheveled (dsh) are cooperatively required for axonal targeting and branching of the *Drosophila* mushroom body neurons and Wnt5 was implicated as the ligand (Shimizu et al., 2011) (Fig. 6.1C). In the dorsal cluster neurons, *prickle*, *flamingo*, and *disheveled* promote sensory axon advance in *Drosophila* (Mrkusich et al., 2011) (Fig. 6.1D). These recent studies strongly favor the view that the PCP signaling pathway (and together with A-BP signaling) may provide a major axon steering mechanism in response to Wnts.

In each set of published study, a subset of PCP components, three to four of six, were analyzed. However, if one pools all studies together, five of six PCP components have been shown required for axon guidance, with the exception of Diego. Given the conservation of the PCP pathways in other systems, it is likely that the same set of PCP proteins is involved in axons guidance. Nonetheless, analyses of complete set of "core" PCP genes in each of the systems will be necessary to conclude whether the entire complement of PCP pathway is engaged in the growth cones or only a subset of PCP signaling pathway is sufficient. In addition, analyzing the role of Fat/Dachsous set of PCP genes in axon guidance will further test the idea how similar the PCP signaling in growth cone is to other examples of PCP signaling.

3. CAN PCP SIGNALING BE USED IN MOTILE GROWTH CONES?

PCP is a common structural feature of tissues throughout the animal kingdom, although most of our knowledge of PCP has been derived from studies in *Drosophila* (hairs of wing and abdomen, bristles on the surface of the body, and the ommatidia in the eye) (Goodrich & Strutt, 2011). Elegant fly genetic studies led to the discovery of the key regulatory system, especially the "core" PCP components that control planar polarity. In more recent years, orientation of hair follicles in mammalian skins, polarized stereocilia of inner ear hair cells, and asymmetric position of primary cilia in the ependymal lining of mammalian brain have emerged as examples of planar polarity in vertebrates. Strikingly, the same set of "core" PCP components controls planar polarity in a highly conserved manner, suggesting that these "core" components and their robust interactions are part of the "universal" code for planar polarity. For more complete description of

PCP signaling, please consult other articles in the same issue. So far, only the "core" PCP components have been analyzed in axon guidance. Therefore, this review focuses on the "core" components.

Planar polarity has mostly been studied in stationary cells that form two-dimensional tissue sheets. Therefore, it is surprising that PCP signaling components are involved in directional control of the motile axonal growth cones. In addition to axon guidance, the same conserved genes that control the "classic" PCP events are also important in many types of moving cells or cellular structures, such as in convergent extension and neuronal migration (Fenstermaker et al., 2010; Mrkusich et al., 2011; Shafer et al., 2011; Shimizu et al., 2011; Wada & Okamoto, 2009). The function of PCP components is required for the precise anterior turning of commissural axon growth cones after midline crossing in the spinal cord and proper anterior–posterior guidance of dopaminergic and serotonergic axons in the midbrain and hindbrain. The anterior–posterior guidance of the spinal cord and brainstem neurons, as well as the anterior–posterior migration of *zebrafish* facial motor neurons and mouse branchinomotor neurons, coincides with the anterior–posterior polarity observed in the intercalating cells in *zebrafish* convergent extension (Lyuksyutova et al., 2003; Qu et al., 2010; Shafer et al., 2011; Wada & Okamoto, 2009). In addition to axon guidance, there has been an explosion of papers demonstrating the function of PCP genes in various developmental and disease processes in different tissues (Barrow, 2011; Carroll & Das, 2011; Happe, de Heer, & Peters, 2011; Heinonen, Vanegas, Lew, Krosl, & Perreault, 2011; Ng, 2012; Sugiyama, Lovicu, & McAvoy, 2011; Sundberg et al., 2011; Wu, Ge, Huang, Hua, & Mu, 2011; Yates & Dean, 2011; Zou, 2011).

The rapid expansion of the roles of these PCP genes automatically invites the following questions: Is the term "planar cell polarity" being used too loosely? Are some of the functions separate from the real PCP signaling? Alternatively, this may suggest that a common signaling mechanism which can impart polarity drives many morphogenesis processes, some of which are obviously analogous to the classic PCP described in *Drosophila* and others are not immediately similar at least on the surface. Given the broad nature of planar polarity in tissues from all germ layers (Zou, 2011), it is a formal possibility that the core PCP signaling system, being first characterized in fly wing and eye, may be a widely used signaling module conveying directionality in many cell types during various events of morphogenesis. "PCP signaling" or "PCP-like signaling" or "PCP signaling module" refers to this common molecular and cellular signaling mechanism in general, including the classic PCP events in the fly. Needless to say, the incredible versatility of the PCP signaling module will need to be achieved by many different upstream input and downstream output in different morphogenesis events, which will be interesting topics to study.

4. IS THE GROWTH CONE POLARIZED?

Axonal growth cones have been studied for many years. They are known to be very sensitive to concentration differences and are able to turn to areas with higher or lower concentration of molecular guidance cues. However, whether growth cones are polarized or asymmetric has never been established or at least there has not been a widely accepted consensus. The highly motile membrane and cytoskeletal structures tend to lead one to think that growth cones are highly "fluid" and "dynamic" and not polarized. However, when looking inside the growth cone, there is ample evidence of polarity (Fig. 6.2).

First, microtubules are polarized with plus ends pointing toward the distal end of axons and represent the "forward" direction of the growth cone. This polarized organization of microtubule structure is established as early as the axon is formed. This "proximal–distal" growth cone axis controls not only the direction of microtubule polymerization (and depolymerization) but also the direction of vesicular trafficking.

Second, actin filaments also show polarity. The plus ends of actin filaments (barbed ends) point to the tips of filopodia and the minus ends point to the inside of the growth cone. Actin filaments undergo retrograde flow with monomers moving inside the growth cone due to treadmilling (Yang, Zhang, Pollard, & Forscher, 2012).

Third, endo- and exocytosis are found polarized in growth cones and involved in navigation (Itofusa & Kamiguchi, 2011). Whether endo- and exocytosis can provide sufficient membrane translocation to turn growth cones still needs experimental evidence. However, the existence of polarized membrane trafficking suggests that this could at least be a mechanism of setting up an asymmetric signaling gradient within the growth cone.

Based on the aforementioned polarized microtubule and actin organization and membrane trafficking (Etienne-Manneville, 2011; Itofusa & Kamiguchi, 2011), the growth cone is a highly polarized structure.

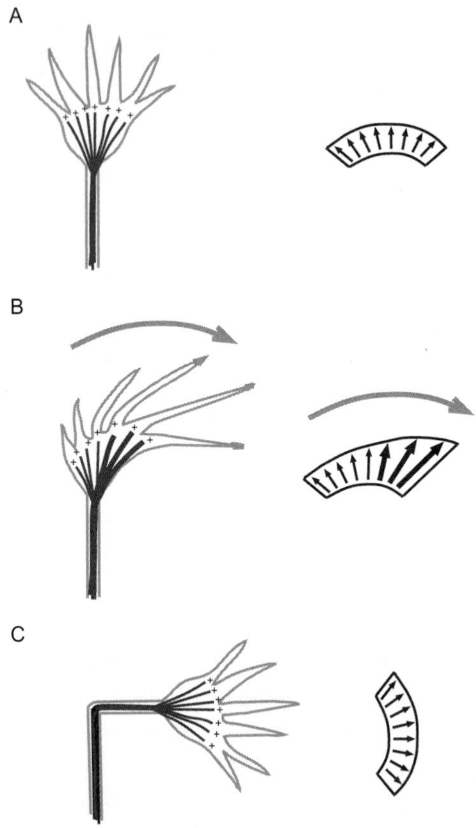

Figure 6.2 A model for growth cone polarity and growth cone turning. Blue lines represent microtubules, which have plus ends pointing toward the distal end of the growth cone. The thickness of microtubules represents their stability. Green indicates the localization of some signaling molecules, which may be asymmetrically localized, such as Vangl2 (Shafer et al., 2011). (For interpretation of the references to color in this figure legend, the reader is referred to the online version of this chapter.)

Perhaps a unique feature about growth cone polarity is that polarity is much more dynamic than stationary cells and the direction of polarity can change, for example, during turning. Recent studies show that both apical–basal and planar polarity signaling components in epithelial polarity are involved in Wnt-mediated turning (Fenstermaker et al., 2010; Shafer et al., 2011; Wolf et al., 2008). This suggests that the growth cone may engage these potent cell polarity regulators to change its polarity in response to guidance cues to achieve turning.

5. BIOCHEMISTRY AND CELL BIOLOGY OF PCP SIGNALING

PCP components have mostly been identified by genetic analyses. Much less is understood in terms of biochemical and cell biological mechanisms. In addition, neither the upstream regulators of PCP nor the downstream effectors that put out the asymmetry in most cases are well understood. Consistent with that, there has not been reliable or relevant biochemical readout specific for PCP signaling. JNK and Rac1 activations are often used in many studies, and they are definitely involved in PCP signaling. However, their exact roles in PCP signaling are unknown because JNK and Rac1 also respond to many other signaling pathways. Asymmetric localization of PCP components has been shown to be essential to PCP signaling. However, how such asymmetric localization is established and what this asymmetric localization encodes are not clear. The ubiquitin proteosome system is a key mechanism of asymmetric localization of some PCP components, suggesting that selective degradation could be a way to introduce asymmetry (Narimatsu et al., 2009). Recent studies established that endocytosis is required for PCP signaling (Sato, Yamamoto, Sakane, Koyama, & Kikuchi, 2010; Yu et al., 2007). Based on all these findings, it is possible that multiple signaling events may take place in different parts of the cell during PCP signaling at the same time or in sequence. Therefore, it is necessary to understand all the biochemical interactions among PCP components before the complete picture of PCP mechanisms can emerge.

A recent study on the biochemical interactions of the core PCP components led to a possibly general mechanism for setting up and/or maintaining the asymmetric localization of PCP components (Shafer et al., 2011) (Fig. 6.3). The distribution of Frizzled3 protein appears to depend on its state of phosphorylation. Frizzled3 protein is mostly localized in intracellular vesicles, and hyperphosphorylation of Frizzled3, induced by Disheveled1, causes Frizzled3 to be targeted to the plasma membrane (Fig. 6.3B). Vangl2, which antagonizes Disheveled1, reduces Frizzled3 phosphorylation and membrane localization on the cell surface (Shafer et al., 2011) (Fig. 6.3C). These findings are consistent with the observations that Van Gogh and Prickle tend to have opposite functions from Frizzled and Disheveled in PCP signaling and suggest that the antagonism may be achieved by opposite effects on Frizzled phosphorylation/membrane localization

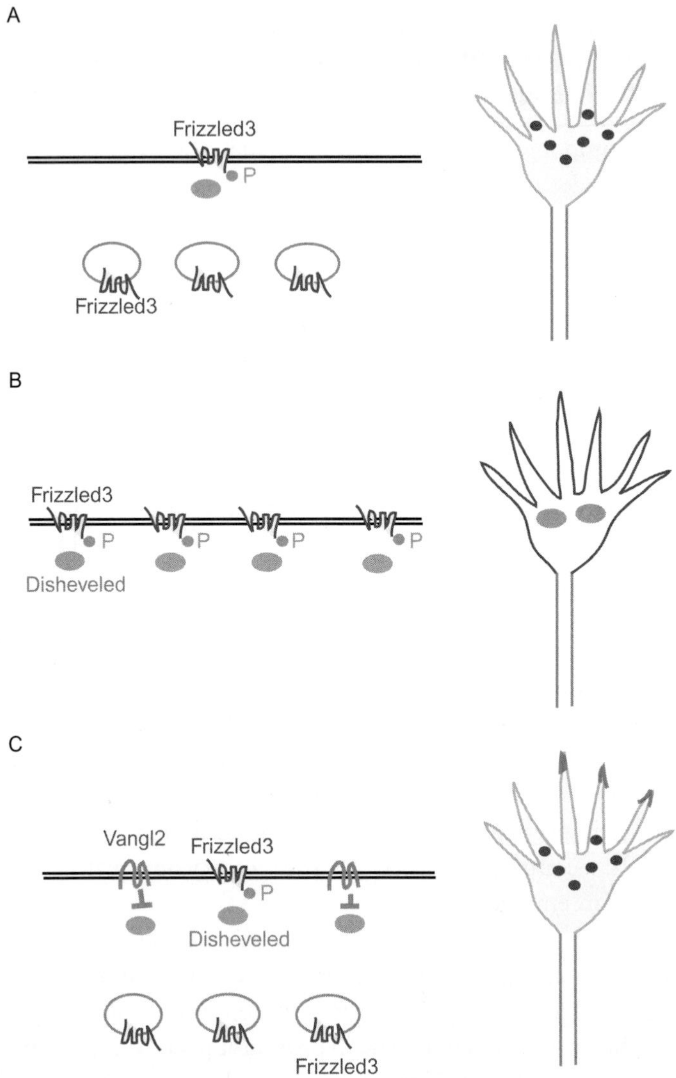

Figure 6.3 Regulation of Frizzled3 membrane localization by phosphorylation. (A) Frizzled3 exists mostly in intracellular vesicles. (B) Disheveled1 can cause Frizzled3 hyperphosphorylation and accumulation on plasma membrane, which stops PCP signaling. (C) Vangl2 antagonizes Disheveled1-mediated Frizzled3 inactivation by promoting Frizzled3 endocyotosis. Vangl2 is enriched on tips of extending filopodia not in retracting filopodia, suggesting Vangl2 activates PCP signaling in a subset of filopodia (Shafer et al., 2011). (For color version of this figure, the reader is referred to the online version of this chapter.)

mediated by Disheveled and Van Gogh. In addition, Frizzled and Disheveled are localized on the distal membrane of fly wing epithelial cells and Van Gogh and Prickle are localized on the proximal membrane shortly before the appearance of morphological asymmetry along the proximal–distal axis. It is possible that Van Gogh may be activated on the proximal membrane only and therefore Frizzled is removed from the cell surface on the proximal membrane. It is also possible that Disheveled is highly activated on the distal membrane such that it is able to keep Frizzled on the plasma membrane. Vangl2 itself undergoes complex phosphorylation as well (Gao et al., 2011), which may represent an input of another regulatory signal. However, it is currently unknown whether Frizzled3 hyperphosphrylation induced by Disheveled inhibits endocytosis or promotes exocytosis. Both cases are consistent with the current findings, and determining which case is true will certainly shed more light on PCP signaling.

6. LOCALIZATION OF VANGL2 PUNCTA ON FILOPODIA TIPS SUGGESTS THAT PCP SIGNALING MAY STEER GROWTH CONES

In live growth cones, Vangl2 protein is highly enriched on the tips of extending filopodia but not the shrinking filopodia. This suggests that at least one aspect of PCP-like signaling is selectively activated on those filopodia tips and not in the rest of the growth cones, which is inactivated by Disheveled1 (Shafer et al., 2011) (Fig. 6.3C). In other words, the asymmetry in growth cones may be manifested by differences among filopodia, the ones with Vangl2 versus those without. This opens up the opportunity to understand how PCP signaling takes place in growth cones. First, this provides a clue that the filopodia tips, not the entire filopodia, are sensors of the growth cone; second, this also suggests that Frizzled3, whose phosphorylation is reduced by Vangl2, may be endocytosed from the tips, carrying signal into the growth cone. It will be very informative to follow Frizzled3 trafficking and decipher the information Frizzled3 may bring into the growth cone. On the other hand, following the question on how Vangl2 is localized to the tips can potentially lead to the answers of how upstream activators of PCP signaling regulate polarity.

In a previous study, aPKC/Par3/Par6 complex, a key component of the A-BP signaling, was found to mediate Wnt attraction during anterior turning of commissural axons (Wolf et al., 2008). There has been evidence that A-BP and PCP signaling pathways interact with each other intimately. First, aPKC can inhibit Frizzled/PCP signaling by directly phosphorylating the intracellular domain of Frizzled (Djiane, Yogev, & Mlodzik, 2005); second, Disheveled directly binds to Lgl, a substrate of aPKC, and regulates the localization of Lgl (Dollar, Weber, Mlodzik, & Sokol, 2005). Therefore, there may exist a connection that integrates these two polarity signaling pathways in epithelial cells and in axonal growth cones. It is now possible to study the mechanisms of how A-BP signaling components may interact with PCP components in commissural axon growth cones. How, for example, does aPKC affect Frizzled3 phosphorylation and trafficking will provide clues to how signaling events take place in growth cones. Understanding what effect these signaling events have on cytoskeletal structures (the microtubules and the actin system) during growth cone turning will eventually solve this century-old mystery of axon guidance.

Growth cones are also repelled by Wnts via a different receptor Ryk (Keeble et al., 2006; Liu et al., 2005). Ryk signaling is still relatively unclear although it is thought to involve the src family kinases in *Drosophila* (Wouda, Bansraj, de Jong, Noordermeer, & Fradkin, 2008). In vertebrate, the Wnt/Calcium signaling has been implicated in Ryk signaling (Hutchins et al., 2011; Li et al., 2009; Li, Hutchins, & Kalil, 2010). The relationship of Ryk with PCP signaling components has been unknown. However, a recent study suggests that Ryk signaling may converge with Frizzled/PCP via interacting with Vangl2 (Macheda et al., 2012). This raises the interesting question whether a common core mechanism causes both attraction and repulsion or whether attraction and repulsion are mediated by totally different mechanisms. The remarkable observation that growth cone attraction and repulsion can be switched by the ratio of cAMP and cGMP suggests a common core may exist (Song et al., 1998; Song, Ming, & Poo, 1997).

7. ARE THERE GROWTH CONE–GROWTH CONE INTERACTIONS DURING PCP-MEDIATED TURNING?

Cell–cell interaction is an essential component in PCP signaling. Several PCP components have both cell-autonomous and cell-nonautonomous functions. If PCP signaling module functions in the growth cones, are there any similar cell–cell interactions?

Neurons of the same type are often born at the same time and have similar time course in their developmental program. Neurons and axons certainly have the opportunities to interact with each other during pathfinding. In the spinal cord, commissural neurons that are born at the

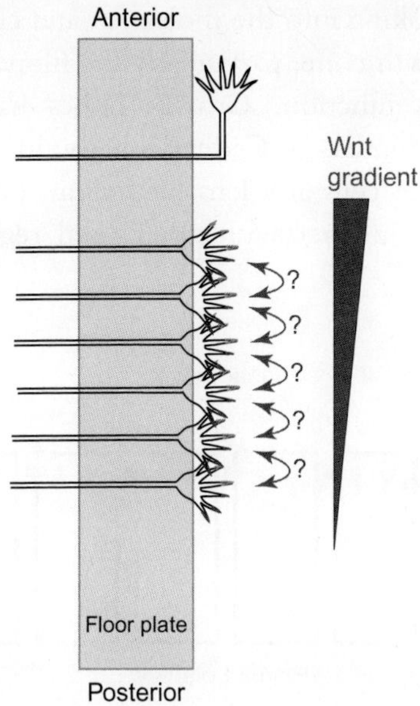

Figure 6.4 Possible growth cone–growth cone interactions during midline crossing and anterior turning of commissural axons. (See Color Insert.)

same time cross the midline at the same time and turn at the same time (Fig. 6.4). It takes many hours (8–9 h) to cross the midline and 1–2 h to turn anteriorly (Y. Zou lab unpublished results). There is plenty of time for commissural neurons and their axons to interact with each other, although there have been no studies addressing this possibility so far. If commissural axon growth cones interact with each other, the global anterior turning of this "sheet" of commissural axon growth cones is highly reminiscent of planar polarity. Furthermore, a sheet of axons may detect greater concentration drop than individual growth cones if they interact with each other. Alternatively, growth cones may not talk to each other. But rather, they work individually because they have sensitive filopodia to detect large enough concentration drops. These intriguing possibilities deserve further investigation.

8. IS THIS CELL POLARITY-BASED SIGNALING SYSTEM A GENERAL GROWTH CONE STEERING MECHANISM?

By studying how Wnts signal in growth cones, we learned that the growth cone uses both A-BP and PCP signaling pathways that are essential for cell and tissue polarity in epithelia to control the direction of turning

(Fig. 6.5A and B). Looking into the molecular and cellular anatomy of the growth cone, the growth cone, particularly the filopodia, has similar compositions as adherens junction. Growth cones have both aPKC and N-cadherin in the filopodia. N-Cadherin is found in adherens junctions in chick cardiac muscle cells and lens epithelium (Volk & Geiger, 1984, 1986). Cadherins also interact with aPKC and regulate its activity and

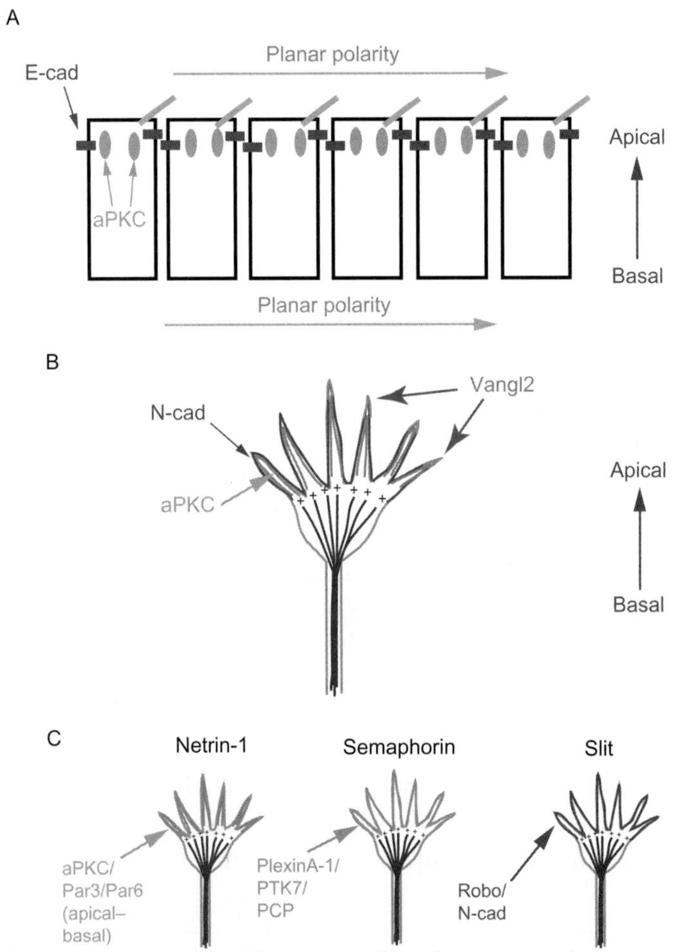

Figure 6.5 A cell polarity-signaling-based machinery for growth cone turning. (A) In epithelia, E-cadherin and aPKC defines the apical basal polarity and planar polarity is perpendicular to apical–basal axis. (B) In neuronal growth cones, N-cadherin and aPKC may define the distal–proximal (apical–basal) axis and PCP signaling may be asymmetrically activated in a subset of filopodia and steer turning in axis perpendicular to the distal–proximal (apical–basal) axis. (C) Other axon guidance signaling may access this

therefore may regulate cell polarity signaling during gastrulation (Seifert, Ibrahim, Stodtmeister, Winklbauer, & Niessen, 2009). I would like to propose that the growth cone filopodia can be viewed as "mobile adherens junction" searching for the missing half and can respond to many cues (Fig. 6.5B).

The cell polarity signaling pathways offer an opportunity to understand signaling mechanisms for growth cone steering. Because of the intimate interactions of A-BP and PCP signaling, it is likely that one may depend on the other. For example, one possibility is that A-BP specifies the proximal–distal axis of the growth cone and PCP components may polarize growth cones perpendicular to the proximal–distal axis (Fig. 6.5B). An alternative to this model is that PCP and A-BP are not organized along these perpendicular axes in growth cones but rather may function in collaboration to amplify each other's signaling level in certain selected filopodia, causing a massive turning signal.

Because A-BP and PCP components are expressed in all neurons and these signaling components can regulate actin, microtubule and membrane trafficking, a spontaneous question is whether this mechanism, or "growth cone compass," is universal. Is there any evidence that other axon guidance signaling system than the Wnt system can also access this "compass?" The following findings are beginning to provide encouraging clues:

1. Par3, Par6, and aPKC are required for axon outgrowth-promoting effects of Netrin-1 and NGF, and Par3 and Par6 are required for ventrally directed growth cone commissural axons to the rat spinal cord midline (Hengst et al., 2009). This study suggests that Netrin-1 can access this potentially universal machinery via A-BP components (Fig. 6.5C).
2. PTK7 is a newly identified Wnt coreceptor in PCP signaling (Peradziryi et al., 2012). Its *Drosophila* orthologue, OTK, forms a complex with

---

machinery. aPKC/Par3/Par6 is required for Netrin-1-stimulated axon outgrowth (Hengst, Deglincerti, Kim, Jeon, & Jaffrey, 2009). PTK7, a Wnt coreceptor in PCP signaling is a coreceptor for PlexinA1, which mediates semaphorin signaling (Peradziryi, Tolwinski, & Borchers, 2012; Toyofuku et al., 2004; Wagner, Peradziryi, Wehner, & Borchers, 2010). N-Cadherin, a potent stimulator for axon out growth, is found in the adherens junction of chick cardiac muscle cells and lens epithelial cells. Slit–Robo can inhibit adhesion by both E-cadherins and N-cadherins and regulate retinal neurite outgrowth, adhesion, or retinal ganglion cell apical process retraction (Rhee, Buchan, Zukerberg, Lilien, & Balsamo, 2007; Rhee et al., 2002; Santiago-Martinez, Soplop, Patel, & Kramer, 2008; Wong, Baudet, Norden, Leung, & Harris, 2012). (For color version of this figure, the reader is referred to the online version of this chapter.)

PlexinA1 to mediate repulsion by Sema 1a. PTK7 also interacts with PlexinA1 to regulate cranial neural crest migration in *Xenopus* (Wagner et al., 2010). In chick, KLG/Otk is a coreceptor with PlexinA1 and VEGFR2 to respond to Sema6D in cell migration (Toyofuku et al., 2004). These studies suggest that Semaphorin signaling may access this machinery via PTK7 through the PCP side (Fig. 6.5C).

3. N-Cadherin is long known to be a potent stimulator for axon outgrowth. N-Cadherin was found in the adherens junction of chick cardiac muscle cells and lens epithelial cells and therefore can exert similar adhesion and cell polarization functions as E-cadherin (Volk & Geiger, 1984, 1986). Recent studies show that a classic guidance system, Slit–Robo system, can inhibit adhesion by both E-cadherins and N-cadherins and regulate retinal neurite outgrowth, adhesion, or retinal ganglion cell apical process retraction (Rhee et al., 2007, 2002; Santiago-Martinez et al., 2008; Wong et al., 2012). Cadherin can interact with aPKC and regulate its function (Seifert et al., 2009). Therefore, Slit may access the cell polarity-based steering machinery via the cadherin complex, which may affect both A-BP and PCP signaling (Etienne-Manneville, 2011) (Fig. 6.5C).
4. In addition to adherens junction, focal adhesion can also affect aPKC (Itoh et al., 2010). Therefore, axon guidance signaling which affects focal adhesion kinase may also access this cell polarity-signaling-based turning machinery.

Therefore, it is possible that growth cone signaling can be unified under a common cell polarity-based machinery. If this is true, this common cell polarity-based compass in growth cone is analogous to the core cell cycle mechanism (CDKs and cyclins), which controls cell cycle of all cell types but can be regulated by many different factors and at different checkpoints in different cell types.

9. SUMMARY

It is somewhat surprising that dynamic growth cones can utilize a signaling system that establishes and maintains polarity in nonmoving cell sheets such as in epithelia. However, at the molecular and cellular level, cell polarity signaling may be similarly dynamic in both motile and nonmotile cells. Recent studies suggest that components of adherens junctions, even E-cadherin, are actively turned over even in these nonmoving cells (Baum & Georgiou, 2011). The robustness of cell polarity signaling pathways imparting asymmetry throughout the entire cells makes them ideally suitable for building the stunningly organized neuronal morphology that is essential for neural circuit function. PCP and A-BP signaling, which intimately interact with each other, are used for axon wiring, and they can, in principle, also be used to regulate designs such as dendrites, axonal and dendritic branches, and even spines (Moreau et al., 2010). These discoveries will not only provide insights of brain development but also identify the components that may be affected in developmental disorders causing nervous system disorders. When there is a need to steer growth cones to repair brain circuits, the knowledge of how PCP signaling may guide growth cones will be instrumental.

REFERENCES

Barrow, J. (2011). Wnt/planar cell polarity signaling: An important mechanism to coordinate growth and patterning in the limb. *Organogenesis*, 7, 260–266.

Baum, B., & Georgiou, M. (2011). Dynamics of adherens junctions in epithelial establishment, maintenance, and remodeling. *The Journal of Cell Biology*, *192*, 907–917.

Berger-Muller, S., & Suzuki, T. (2011). Seven-pass transmembrane cadherins: Roles and emerging mechanisms in axonal and dendritic patterning. *Molecular Neurobiology*, *44*, 313–320.

Blakely, B. D., Bye, C. R., Fernando, C. V., Horne, M. K., Macheda, M. L., Stacker, S. A., et al. (2011). Wnt5a regulates midbrain dopaminergic axon growth and guidance. *PloS One*, *6*, e18373.

Carroll, T. J., & Das, A. (2011). Planar cell polarity in kidney development and disease. *Organogenesis*, 7, 180–190.

Djiane, A., Yogev, S., & Mlodzik, M. (2005). The apical determinants aPKC and dPatj regulate Frizzled-dependent planar cell polarity in the Drosophila eye. *Cell*, *121*, 621–631.

Dollar, G. L., Weber, U., Mlodzik, M., & Sokol, S. Y. (2005). Regulation of Lethal giant larvae by Dishevelled. *Nature*, *437*, 1376–1380.

Domanitskaya, E., Wacker, A., Mauti, O., Baeriswyl, T., Esteve, P., Bovolenta, P., et al. (2010). Sonic hedgehog guides post-crossing commissural axons both directly and indirectly by regulating Wnt activity. *The Journal of Neuroscience*, *30*, 11167–11176.

Etienne-Manneville, S. (2011). Control of polarized cell morphology and motility by adherens junctions. *Seminars in Cell & Developmental Biology*, *22*, 850–857.

Fenstermaker, A. G., Prasad, A. A., Bechara, A., Adolfs, Y., Tissir, F., Goffinet, A., et al. (2010). Wnt/planar cell polarity signaling controls the anterior–posterior organization of monoaminergic axons in the brainstem. *The Journal of Neuroscience*, *30*, 16053–16064.

Gao, B., Song, H., Bishop, K., Elliot, G., Garrett, L., English, M. A., et al. (2011). Wnt signaling gradients establish planar cell polarity by inducing Vangl2 phosphorylation through Ror2. *Developmental Cell*, *20*, 163–176.

Goodrich, L. V., & Strutt, D. (2011). Principles of planar polarity in animal development. *Development*, *138*, 1877–1892.

Happe, H., de Heer, E., & Peters, D. J. (2011). Polycystic kidney disease: The complexity of planar cell polarity and signaling during tissue regeneration and cyst formation. *Biochimica et Biophysica Acta*, *1812*, 1249–1255.

Heinonen, K. M., Vanegas, J. R., Lew, D., Krosl, J., & Perreault, C. (2011). Wnt4 enhances murine hematopoietic progenitor cell expansion through a planar cell polarity-like pathway. *PloS One*, *6*, e19279.

Hengst, U., Deglincerti, A., Kim, H. J., Jeon, N. L., & Jaffrey, S. R. (2009). Axonal elongation triggered by stimulus-induced local translation of a polarity complex protein. *Nature Cell Biology, 11*, 1024–1030.

Hilliard, M. A., & Bargmann, C. I. (2006). Wnt signals and frizzled activity orient anterior–posterior axon outgrowth in C. elegans. *Developmental Cell, 10*, 379–390.

Hutchins, B. I., Li, L., & Kalil, K. (2011). Wnt/calcium signaling mediates axon growth and guidance in the developing corpus callosum. *Developmental Neurobiology, 71*, 269–283.

Itofusa, R., & Kamiguchi, H. (2011). Polarizing membrane dynamics and adhesion for growth cone navigation. *Molecular and Cellular Neurosciences, 48*, 332–338.

Itoh, N., Nakayama, M., Nishimura, T., Fujisue, S., Nishioka, T., Watanabe, T., et al. (2010). Identification of focal adhesion kinase (FAK) and phosphatidylinositol 3-kinase (PI3-kinase) as Par3 partners by proteomic analysis. *Cytoskeleton (Hoboken, N.J.), 67*, 297–308.

Keeble, T. R., Halford, M. M., Seaman, C., Kee, N., Macheda, M., Anderson, R. B., et al. (2006). The Wnt receptor Ryk is required for Wnt5a-mediated axon guidance on the contralateral side of the corpus callosum. *The Journal of Neuroscience, 26*, 5840–5848.

Kennerdell, J. R., Fetter, R. D., & Bargmann, C. I. (2009). Wnt-Ror signaling to SIA and SIB neurons directs anterior axon guidance and nerve ring placement in C. elegans. *Development, 136*, 3801–3810.

Li, L., Hutchins, B. I., & Kalil, K. (2009). Wnt5a induces simultaneous cortical axon outgrowth and repulsive axon guidance through distinct signaling mechanisms. *The Journal of Neuroscience, 29*, 5873–5883.

Li, L., Hutchins, B. I., & Kalil, K. (2010). Wnt5a induces simultaneous cortical axon outgrowth and repulsive turning through distinct signaling mechanisms. *Science Signaling, 3*, pt2.

Liu, Y., Shi, J., Lu, C. C., Wang, Z. B., Lyuksyutova, A. I., Song, X., et al. (2005). Ryk-mediated Wnt repulsion regulates posterior-directed growth of corticospinal tract. *Nature Neuroscience, 8*, 1151–1159.

Lu, W., Yamamoto, V., Ortega, B., & Baltimore, D. (2004). Mammalian Ryk is a Wnt coreceptor required for stimulation of neurite outgrowth. *Cell, 119*, 97–108.

Lyuksyutova, A. I., Lu, C. C., Milanesio, N., King, L. A., Guo, N., Wang, Y., et al. (2003). Anterior–posterior guidance of commissural axons by Wnt-frizzled signaling. *Science, 302*, 1984–1988.

Macheda, M. L., Sun, W. W., Kugathasan, K., Hogan, B. M., Bower, N. I., Halford, M. M., et al. (2012). The Wnt receptor Ryk plays a role in mammalian planar cell polarity signaling. *The Journal of Biological Chemistry, 287*(35), 29312–29323.

Moreau, M. M., Piguel, N., Papouin, T., Koehl, M., Durand, C. M., Rubio, M. E., et al. (2010). The planar polarity protein Scribble1 is essential for neuronal plasticity and brain function. *The Journal of Neuroscience, 30*, 9738–9752.

Mrkusich, E. M., Flanagan, D. J., & Whitington, P. M. (2011). The core planar cell polarity gene prickle interacts with flamingo to promote sensory axon advance in the Drosophila embryo. *Developmental Biology, 358*, 224–230.

Narimatsu, M., Bose, R., Pye, M., Zhang, L., Miller, B., Ching, P., et al. (2009). Regulation of planar cell polarity by Smurf ubiquitin ligases. *Cell, 137*, 295–307.

Ng, J. (2012). Wnt/PCP proteins regulate stereotyped axon branch extension in Drosophila. *Development, 139*, 165–177.

Pan, C. L., Howell, J. E., Clark, S. G., Hilliard, M., Cordes, S., Bargmann, C. I., et al. (2006). Multiple Wnts and frizzled receptors regulate anteriorly directed cell and growth cone migrations in Caenorhabditis elegans. *Developmental Cell, 10*, 367–377.

Peradziryi, H., Tolwinski, N. S., & Borchers, A. (2012). The many roles of PTK7: A versatile regulator of cell–cell communication. *Archives of Biochemistry and Biophysics, 524*, 71–76.

Prasad, B. C., & Clark, S. G. (2006). Wnt signaling establishes anteroposterior neuronal polarity and requires retromer in C. elegans. *Development, 133*, 1757–1766.

Qu, Y., Glasco, D. M., Zhou, L., Sawant, A., Ravni, A., Fritzsch, B., et al. (2010). Atypical cadherins Celsr1-3 differentially regulate migration of facial branchiomotor neurons in mice. *The Journal of Neuroscience*, *30*, 9392–9401.

Rhee, J., Buchan, T., Zukerberg, L., Lilien, J., & Balsamo, J. (2007). Cables links Robo-bound Abl kinase to N-cadherin-bound beta-catenin to mediate Slit-induced modulation of adhesion and transcription. *Nature Cell Biology*, *9*, 883–892.

Rhee, J., Mahfooz, N. S., Arregui, C., Lilien, J., Balsamo, J., & VanBerkum, M. F. (2002). Activation of the repulsive receptor Roundabout inhibits N-cadherin-mediated cell adhesion. *Nature Cell Biology*, *4*, 798–805.

Rodriguez-Gil, D. J., & Greer, C. A. (2008). Wnt/Frizzled family members mediate olfactory sensory neuron axon extension. *The Journal of Comparative Neurology*, *511*, 301–317.

Santiago-Martinez, E., Soplop, N. H., Patel, R., & Kramer, S. G. (2008). Repulsion by Slit and Roundabout prevents Shotgun/E-cadherin-mediated cell adhesion during Drosophila heart tube lumen formation. *The Journal of Cell Biology*, *182*, 241–248.

Sato, M., Umetsu, D., Murakami, S., Yasugi, T., & Tabata, T. (2006). DWnt4 regulates the dorsoventral specificity of retinal projections in the Drosophila melanogaster visual system. *Nature Neuroscience*, *9*, 67–75.

Sato, A., Yamamoto, H., Sakane, H., Koyama, H., & Kikuchi, A. (2010). Wnt5a regulates distinct signalling pathways by binding to Frizzled2. *The EMBO Journal*, *29*, 41–54.

Schmitt, A. M., Shi, J., Wolf, A. M., Lu, C. C., King, L. A., & Zou, Y. (2006). Wnt-Ryk signalling mediates medial-lateral retinotectal topographic mapping. *Nature*, *439*, 31–37.

Seifert, K., Ibrahim, H., Stodtmeister, T., Winklbauer, R., & Niessen, C. M. (2009). An adhesion-independent, aPKC-dependent function for cadherins in morphogenetic movements. *Journal of Cell Science*, *122*, 2514–2523.

Shafer, B., Onishi, K., Lo, C., Colakoglu, G., & Zou, Y. (2011). Vangl2 promotes Wnt/planar cell polarity-like signaling by antagonizing Dvl1-mediated feedback inhibition in growth cone guidance. *Developmental Cell*, *20*, 177–191.

Shimizu, K., Sato, M., & Tabata, T. (2011). The Wnt5/planar cell polarity pathway regulates axonal development of the Drosophila mushroom body neuron. *The Journal of Neuroscience*, *31*, 4944–4954.

Song, H., Ming, G., He, Z., Lehmann, M., McKerracher, L., Tessier-Lavigne, M., et al. (1998). Conversion of neuronal growth cone responses from repulsion to attraction by cyclic nucleotides. *Science*, *281*, 1515–1518.

Song, H. J., Ming, G. L., & Poo, M. M. (1997). cAMP-induced switching in turning direction of nerve growth cones. *Nature*, *388*, 275–279.

Song, S., Zhang, B., Sun, H., Li, X., Xiang, Y., Liu, Z., et al. (2010). A Wnt-Frz/Ror-Dsh pathway regulates neurite outgrowth in Caenorhabditis elegans. *PLoS Genetics*, *6*.

Sugiyama, Y., Lovicu, F. J., & McAvoy, J. W. (2011). Planar cell polarity in the mammalian eye lens. *Organogenesis*, 7, 191–201.

Sundberg, T. B., Darricarrere, N., Cirone, P., Li, X., McDonald, L., Mei, X., et al. (2011). Disruption of Wnt planar cell polarity signaling by aberrant accumulation of the MetAP-2 substrate Rab37. *Chemistry & Biology*, *18*, 1300–1311.

Tissir, F., Bar, I., Jossin, Y., De Backer, O., & Goffinet, A. M. (2005). Protocadherin Celsr3 is crucial in axonal tract development. *Nature Neuroscience*, *8*, 451–457.

Toyofuku, T., Zhang, H., Kumanogoh, A., Takegahara, N., Suto, F., Kamei, J., et al. (2004). Dual roles of Sema6D in cardiac morphogenesis through region-specific association of its receptor, Plexin-A1, with off-track and vascular endothelial growth factor receptor type 2. *Genes & Development*, *18*, 435–447.

Volk, T., & Geiger, B. (1984). A 135-kd membrane protein of intercellular adherens junctions. *The EMBO Journal*, *3*, 2249–2260.

Volk, T., & Geiger, B. (1986). A-CAM: A 135-kD receptor of intercellular adherens junctions. I. Immunoelectron microscopic localization and biochemical studies. *The Journal of Cell Biology*, *103*, 1441–1450.

Wada, H., & Okamoto, H. (2009). Roles of planar cell polarity pathway genes for neural migration and differentiation. *Development, Growth & Differentiation*, *51*, 233–240.

Wagner, G., Peradziryi, H., Wehner, P., & Borchers, A. (2010). PlexinA1 interacts with PTK7 and is required for neural crest migration. *Biochemical and Biophysical Research Communications*, *402*, 402–407.

Wolf, A. M., Lyuksyutova, A. I., Fenstermaker, A. G., Shafer, B., Lo, C. G., & Zou, Y. (2008). Phosphatidylinositol-3-kinase-atypical protein kinase C signaling is required for Wnt attraction and anterior–posterior axon guidance. *The Journal of Neuroscience*, *28*, 3456–3467.

Wong, G. K., Baudet, M. L., Norden, C., Leung, L., & Harris, W. A. (2012). Slit1b-Robo3 signaling and N-cadherin regulate apical process retraction in developing retinal ganglion cells. *The Journal of Neuroscience*, *32*, 223–228.

Wouda, R. R., Bansraj, M. R., de Jong, A. W., Noordermeer, J. N., & Fradkin, L. G. (2008). Src family kinases are required for WNT5 signaling through the Derailed/RYK receptor in the Drosophila embryonic central nervous system. *Development*, *135*, 2277–2287.

Wu, G., Ge, J., Huang, X., Hua, Y., & Mu, D. (2011). Planar cell polarity signaling pathway in congenital heart diseases. *Journal of Biomedicine & Biotechnology*, *2011*, 589414.

Yang, Q., Zhang, X. F., Pollard, T. D., & Forscher, P. (2012). Arp2/3 complex-dependent actin networks constrain myosin II function in driving retrograde actin flow. *The Journal of Cell Biology*, *197*, 939–956.

Yates, L. L., & Dean, C. H. (2011). Planar polarity: A new player in both lung development and disease. *Organogenesis*, *7*, 209–216.

Yoshikawa, S., McKinnon, R. D., Kokel, M., & Thomas, J. B. (2003). Wnt-mediated axon guidance via the Drosophila Derailed receptor. *Nature*, *422*, 583–588.

Yu, A., Rual, J. F., Tamai, K., Harada, Y., Vidal, M., He, X., et al. (2007). Association of Dishevelled with the clathrin AP-2 adaptor is required for Frizzled endocytosis and planar cell polarity signaling. *Developmental Cell*, *12*, 129–141.

Zou, Y. (2004). Wnt signaling in axon guidance. *Trends in Neurosciences*, *27*, 528–532.

Zou, Y. (2011). When it is hard to get to with genetics—Planar cell polarity under a chemical scalpel. *Chemical Biology*, *18*, 1350–1351.

CHAPTER SEVEN

Celsr1–3 Cadherins in PCP and Brain Development

Camille Boutin, André M. Goffinet[1], Fadel Tissir[1]
Institute of Neuroscience, Developmental Neurobiology, Université Catholique de Louvain, Brussels, Belgium
[1]Corresponding authors: Equal contribution. e-mail address: fadel.tissir@uclouvain.be; andre.goffinet@uclouvain.be

Contents

Abstract

Cadherin EGF LAG seven-pass G-type receptors 1, 2, and 3 (Celsr1–3) form a family of three atypical cadherins with multiple functions in epithelia and in the nervous system. During the past decade, evidence has accumulated for important and distinct roles of Celsr1–3 in planar cell polarity (PCP) and brain development and maintenance. Although the role of Celsr in PCP is conserved from flies to mammals, other functions may be more distantly related, with Celsr working only with one or a subset of the classical PCP partners. Here, we review the literature on Celsr in PCP and neural development, point to several remaining questions, and consider future challenges and possible research trends.

Celsr1–3 genes encode atypical cadherins of more than 3000 amino acids (Fig. 7.1). Their large ectodomain is composed of nine N-terminal cadherin repeats (typical cadherins have five repeats), six epidermal growth factor (EGF)-like domains, two laminin G repeats, one hormone receptor motif (HRM), and a G-protein-coupled receptor proteolytic site (GPS). This is followed by seven transmembrane domains (classic cadherins are single-pass type I proteins) and a cytoplasmic tail (CT) that varies in size and, in contrast to the ectodomain, is poorly conserved among the three members. Celsr1–3 belong to the "adhesion receptor" family of G-protein-coupled receptors (GPCRs), also referred to as long N-terminal group B (LNB) GPCRs.

Current Topics in Developmental Biology, Volume 101
ISSN 0070-2153
http://dx.doi.org/10.1016/B978-0-12-394592-1.00010-7
© 2012 Elsevier Inc.
All rights reserved.

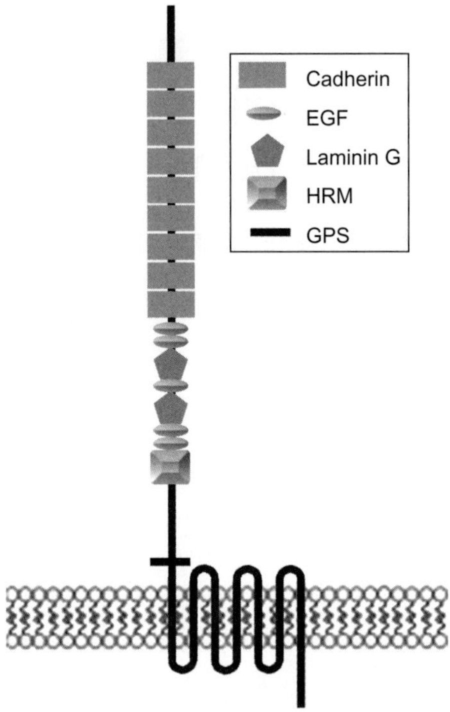

Figure 7.1 Schematic representation of the Celsr1 protein. The extracellular domain includes nine cadherin repeats, six epidermal growth factor (EGF)-like domains, two laminin G repeats, one hormone receptor motif (HRM), and a G-protein-coupled receptor proteolytic site (GPS). In contrast to typical cadherins which are single-pass proteins, Celsr1–3 cadherins are anchored to the plasma membrane by seven transmembrane domains. The cytoplasmic tail varies in size and is poorly conserved among the three Celsrs. (For color version of this figure, the reader is referred to the online version of this chapter.)

These receptors are thought to be natural chimeras of cell adhesion proteins and signaling receptors, able to convert cell–cell communication cues into intracellular signals. Studies of the latrotoxin receptor latrophilin, a member of LNB–GPCR, showed that cleavage at the GPS occurs intracellularly. However, the N-terminal (cell adhesion domain) and C-terminal (GPCR) fragments remain associated by noncovalent bonds at the plasma membrane (Krasnoperov et al., 2002; Volynski et al., 2004). It is not clear yet whether cleavage at the GPS occurs in Celsr1–3 proteins (see below). In addition to cadherin and EGF-like repeats which confer adhesive properties and facilitate cell–cell communication, the extracellular domain of Celsr1–3 contains a HRM which may bind putative ligands, although none has been identified thus far.

Celsr genes were first identified in mice (Hadjantonakis et al., 1997). They form a highly conserved family with homologs in ascidians, worms, flies, and vertebrates. The fruit fly Celsr ortholog was cloned independently by two groups and is known as *Flamingo* (*Fmi*) or *Starry night* (*Stan*) (Chae et al., 1999; Usui et al., 1999). Genetic studies have established a critical role for *Fmi/Stan* in planar cell polarity (PCP) events, particularly in the stereotypic organization of wing hairs (trichomes) and sensory bristles (Chae et al., 1999; Usui et al., 1999), the alignment of ommatidia in the eye (Mlodzik, 1999), and the asymmetric division of sensory organ precursors (Bellaiche, Gho, Kaltschmidt, Brand, & Schweisguth, 2001; Segalen & Bellaiche, 2009). In these processes, Fmi/Stan interacts genetically and functionally with so-called core PCP proteins. Drosophila core PCP proteins include serpentine receptors Frizzled (fz and fz2 are partially redundant in PCP); the tetraspannin Van Gogh (vang, also named strabismus); and the three cytoplasmic proteins Disheveled (dsh), Prickle (pk), and Diego (dgo) (Adler, 2002; Feiguin, Hannus, Mlodzik, & Eaton, 2001; Gubb et al., 1999; Strutt, Johnson, Cooper, & Bray, 2002; Taylor, Abramova, Charlton, & Adler, 1998; Wolff & Rubin, 1998). In the insect wing, the distribution of core PCP proteins is tightly and dynamically regulated. Initially distributed in all apical junctions, they adopt a transient polarized partition along the proximal–distal axis shortly before the growth of wing hairs, and this partition is essential to the establishment of polarity and the proper orientation of hairs. fmi/stan, fzd, dsh, and dgo are enriched at the distal border of wing cells, whereas fmi/stan, vang, and pk are enriched at the proximal border (Bastock, Strutt, & Strutt, 2003; Jenny, Darken, Wilson, & Mlodzik, 2003; Usui et al., 1999). Given that fmi/stan localizes to both proximal and distal junctions, it was considered a permissive molecule with no active effect on polarity: fmi/stan homodimers would act as a scaffold, promoting cell adhesion and bridging the distal (fz-expressing) side of a cell and the proximal (vang-expressing) side of the adjacent cell. However, more recent data provide evidence that fmi/stan plays an instructive role: its central portion containing the HRM and TM domains interacts physically with fz. Furthermore, fmi/stan selectively recruits fz and vang to opposite cell boundaries, thereby initiating bidirectional polarity signals (Chen et al., 2008). In line with this, genetic analyses in flies demonstrated a mutual requirement for fmi/stan, fz, and dsh to achieve a correct partition of polarity complexes (Das, Reynolds-Kenneally, & Mlodzik, 2002). In zebrafish, a serine acidic amino acid-rich domain (SE/D) in the CT of Celsr is required for membrane localization of the Frizzled–Disheveled complex, PCP signaling, and convergent extension (Carreira-Barbosa et al., 2009).

1. CELSR1–3 EXPRESSION PATTERNS

In mammals, the *Celsr* family is composed of three members, whereas birds apparently lack *Celsr2* (Formstone, 2010) and fish have two, *Celsr1a* and *1b* (Wada, Tanaka, Nakayama, Iwasaki, & Okamoto, 2006). In the mouse, *Celsr1–3* genes have similar genomic organizations, with 35 (*Celsr1* & *Celsr3*) or 34 exons (*Celsr2*). Apart from 3′ alternative exons in *Celsr2*, alternative splicing has not been described. *Celsr1–3* expression is regulated spatially and temporally, indicating that they are important for development. A striking feature is the complementary pattern of *Celsr1* and *Celsr3* expressions in different developing systems (Formstone & Little, 2001; Shima et al., 2002; Tissir, De Backer, Goffinet, & Lambert de Rouvroit, 2002). In the nervous system, *Celsr1* mRNA is heavily expressed in zones of neural stem cell (NSC) proliferation, namely, all ventricular zones during embryonic and early postnatal development, and telencephalic ependymal zones as well as subgranular layer of the dentate gyrus in the mature brain. In contrast, *Celsr3* mRNA is absent from NSC and associated with most postmitotic neural cells, whereas *Celsr2* mRNA is found in both NSC and postmitotic cells. Expression of *Celsr1* abates during early postnatal development, in parallel to decreasing numbers of NSC. That of *Celsr3* is downregulated postnatally but persists in the cerebellar granular layer, the hilus of the dentate gyrus, the rostral migratory stream, and the central region of the olfactory bulb. By contrast, *Celsr2* expression remains stable throughout life. These expression patterns hint to functions of *Celsr1* in NSC, *Celsr3* in neural cell maturation, and *Celsr2* in development and maintenance of the nervous system. *Celsr1–3* mRNAs are also variably expressed in nonneural tissue, such as the skin, lungs, kidney, and digestive and reproductive systems. In the rodent testis, the spatiotemporal pattern of *Celsr1–3* expression is somewhat reminiscent of that in the brain. *Celsr3* is expressed exclusively in postmeiotic germ cells, and *Celsr1* and *Celsr2* in Sertoli cells, with a postnatal downregulation of *Celsr1* and persistence of

Celsr2 in the adult (Beall, Boekelheide, & Johnson, 2005; Hadjantonakis, Formstone, & Little, 1998; Johnson, Patel, & Boekelheide, 2000).

Polyclonal antibodies against the extra- and intracellular segments of Celsr1 allowed detection of two forms, full length (p400) and cleaved (p85 kD), generated by as yet unidentified proteolysis events not involving the GPS (Formstone, Moxon, Murdoch, Little, & Mason, 2010). In the hindbrain and spinal cord, Celsr1 protein immunoreactivity was detected in the floor and roof plates, as well as in radial neuroepithelial progenitors, not only in the apical but also in their basolateral domain, particularly in end-feet abutting the pial surface (Formstone et al., 2010). In the embryonic skin, Celsr1 is asymmetrically expressed in hair germ cells and in basal layer epidermal cells, a pattern evocative of that of fmi/stan in the Drosophila wing (Devenport & Fuchs, 2008). The cellular localization of Celsr2 and Celsr3 could not be investigated thus far due to lack of antibodies suitable in immunohistochemistry.

2. CELSR1: A MAJOR PLAYER IN VERTEBRATE PCP

Two *Celsr1* mutant alleles, *Crash* (*Celsr1*^{Crsh}) and *Spin Cycle* (*Celsr1*^{Scy}), were identified in an ENU screen (Curtin et al., 2003). In *Celsr1*^{Crsh}, a G-to-A mutation at nucleotide 3126 results in an aspartate-to-glycine substitution in codon 1040, within the eighth cadherin repeat. In *Celsr1*^{Scy}, a T-to-A point mutation at nucleotide 3337 results in an asparagine-to-lysine substitution in codon 1110, in a region connecting cadherin repeats. Heterozygous animals show abnormal head-shaking behavior. Both heterozygous and homozygous mice have defective organization of stereocilia bundles in inner ear hair cells. Normally, the apical surface of each cochlear hair cell is decorated with actin-filled stereocilia forming a "V" centered on a microtubular kinocilium, with all "Vs" pointing to the external aspect of the cochlear canal. In *Celsr*^{Crsh} and *Celsr1*^{Scy} mutant mice, this typical organization is altered. Stereociliary bundles are randomly oriented, some displaying up to 180° rotation relative to the normal orientation (Curtin et al., 2003). The precise orientation of hair bundles in the cochlea is a classical hallmark of PCP and parallels the polarized distribution of core PCP proteins such as Fzd3, Fzd6, Vangl2, and Prickle2 (Deans et al., 2007; Montcouquiol et al., 2006; Wang, Guo, & Nathans, 2006). PCP proteins localize asymmetrically to one edge of the apical cortex of the cells by mechanisms involving selective targeting and protein stabilization and degradation. Intriguingly, studies of the distribution of PCP proteins were carried out in late embryogenesis, starting from embryonic day 18 when the kinocilium has already migrated from the center to the lateral edge of the cell, raising the question whether the asymmetric partition of PCP proteins is the initial event that sets up the polarity and organizes the epithelium, or simply a molecular readout of polarity signals induced by as yet unidentified cues.

Celsr1^{Crsh} and *Celsr1*^{Scy} homozygotes as well as compound heterozygotes *Celsr1*$^{Crsh/Scy}$ exhibit craniorachischisis, a severe neural tube defect due to a failure to initiate neural tube closure in the cervical region (Curtin et al., 2003). A role for *Celsr1* in neural tube closure was recently confirmed by identification of six *Celsr1* mutations in human fetuses with craniorachischisis. In *in vitro* assays, those mutations all impair trafficking of Celsr1 protein, reducing its membrane localization (Robinson et al., 2011).

As *Celsr1*^{Crsh} and *Celsr1*^{Scy} homozygous mutants are embryonic lethal, a conditional allele (*Celsr1*^{f}) was generated to allow tissue-specific inactivation upon expression of Cre recombinase (Ravni, Qu, Goffinet, & Tissir, 2009), a null allele *Celsr1*^{ko} was obtained by crosses with germline-expressing Cre mice, and Western blot with an antibody against the N-terminal region of Celsr1 confirmed absence of Celsr1 protein in embryonic mouse brain extracts. Unlike heterozygous *Celsr1*^{Crsh} and *Celsr1*^{Scy}, heterozygous *Celsr1*$^{ko/+}$ mice have no perceptible phenotype. In contrast, homozygous *Celsr1*$^{ko/ko}$ mutants display abnormal behavior such as circling and hyperactivity. About 20% of them die at embryonic stages with various degrees of neural tube closure defects, and many have a looping tail. Some have striking skin hair patterning defects with whorls and crests instead of regular caudally and distally pointing hairs on the body and limbs (Fig. 7.2) (Ravni et al., 2009). When *Celsr1* is inactivated in crosses with *Emx1-Cre* mice, inactivation of the gene in the apical ectodermal ridge induces a whorl in distal hindlimb, showing that the action of Celsr1 is ectodermal cell autonomous. The mechanism of Celsr1 action in the hair bulb has been scrutinized by the Fuchs lab (Devenport & Fuchs, 2008). They found that, prior to hair growth, Celsr1 becomes asymmetrically localized along the anterior/posterior (A/P) axis in basal epidermal cells in hair follicles and in interfollicular epithelium. This asymmetric localization is essential for A/P orientation of skin hair. Hair follicles fail to adopt the A/P orientation in skin explants isolated from E13.5 embryos, before the polarization of Celsr1. In contrast, when explants are isolated from E14.5 embryos, when Celsr1 was fully polarized, epidermis keeps the Celsr1 polarization established *in vivo*, and the A/P polarity. Consistent with this, hair follicles are misaligned in E18.5 embryos with

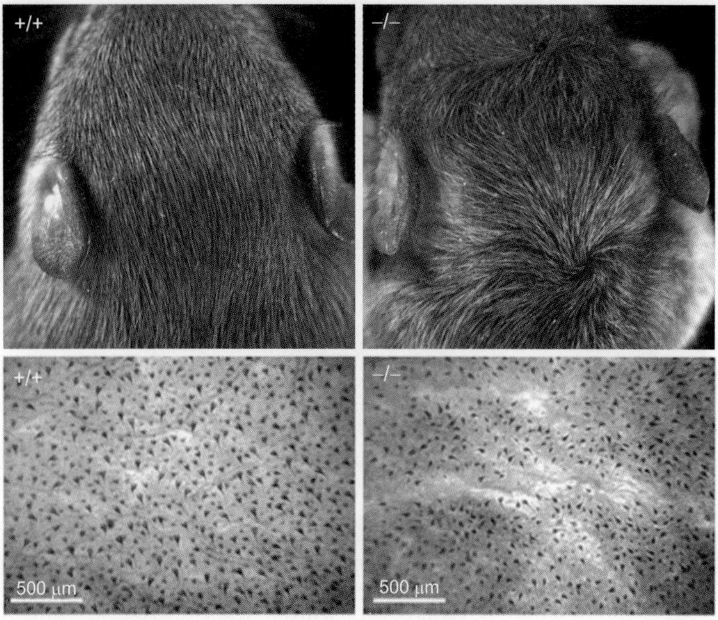

Figure 7.2 Hair patterning defects in *Celsr1* knockout mice. In normal mice (+/+, left), all body hairs point caudally and limb hairs point distally. Celsr1mutant mice (−/−, right) have fur whorls and crests variably located on the head, body, and on lower dorsal hindlimbs, where the trait is fully penetrant. Lower panels show whole mount preparations of skin at P1, stained with hematoxylin eosin to visualize disoriented hair follicles in the mutant (−/−, right) as compared to the wild type (+/+, right). (For color version of this figure, the reader is referred to the online version of this chapter.)

homozygous *Crsh* mutation. In mutant embryos, Celsr1 protein is produced as in the wild type, but no longer asymmetrically distributed. Moreover, the membrane recruitment of Fzd6 and the asymmetric localization of Vangl2 along the A/P axis are compromised, with some Vangl2 immunoreactivity forming intracellular puncta. Taken together, these results demonstrate that Celsr1 plays a critical role for PCP establishment in the developing skin and hair follicles (Devenport & Fuchs, 2008).

Importantly, PCP phenotypes in *Celsr1* mutants were also seen in mice with mutations in other core PCP genes (Table 7.1). The inner ear phenotype was reported in mice with mutated *Vangl2* (Montcouquiol et al., 2003, 2006), and *Fzd3* and *Fzd6* (Wang et al., 2006). Neural tube closure defects were observed in *Vangl2*, double *Dvl1* and *2*, and *Fzd3* and *6* (Kibar et al., 2001; Wang et al., 2006; Ybot-Gonzalez et al., 2007). Skin hair patterning defects were described in *Fzd6* and *Vangl2* (Devenport & Fuchs, 2008; Devenport,

Table 7.1 Summary of phenotypes in *Celsr1–3*, with comparison of some other PCP mutations

Gene	Allele	Phenotype
Celsr1	*Celsr1 Crsh* and *Scy*	Heterozygotes have inner ear PCP abnormalities, head shaking, looptail, and abnormal FBM neuron migration. Homozygotes are embryonic lethal with open neural tube.
	Celsr1 KO	Heterozygotes are normal and fertile.
		Homozygotes are viable but mostly sterile. About 20% die at embryonic stages with neural tube closure defects. Adults have abnormal behavioral traits, a looping tail, and abnormal skin hair patterning. The direction of FBM neuron migration is affected.
Celsr2	*Celsr2 Dgen*	Heterozygotes are normal and fertile.
		Homozygotes are fertile, except for some females with vaginal atresia. There is abnormal trajectory of FBM neuron migration and hydrocephalus due to defective ependymal ciliogenesis.
	Celsr2 Dgen	Heterozygotes are normal and fertile.
		Homozygotes are fertile and have abnormal FBM neuron migration.
Celsr3	*Celsr3 KO* and *Celsr3 F*	Heterozygotes are normal and fertile.
		Homozygotes die in few hours after birth and have severe axonal defects.
Fzd3	*Fzd3 KO*	Heterozygotes are normal and fertile.
		Homozygotes die in few hours after birth and have axonal defects similar to those in *Celsr3* mutants. Some have loop tail or open neural tube and PCP phenotype in the inner ear. The trajectory of FBM neuron migration is affected and mimics *Celsr2* and *3* double KO.
Fzd6	*Fzd6 KO*	Homozygous mutants survive and breed, and have abnormal skin hair patterning.
Vangl2	*Vangl2 Lp (Looptail)*	Some heterozygotes have kinked or looped tails. Homozygotes do not survive due to open neural tube, loop tail, inner ear PCP defects, FBM neuron migration defects, and other defects in cardiovascular, skeletal, and respiratory systems.
	Vangl2 KO and Vangl2 F	Heterozygotes are normal and fertile.
		Homozygote nulls have open neural tube and other PCP defects in the inner ear.
Vangl1	*Vangl1 gt*	Homozygotes are normal and fertile

Oristian, Heller, & Fuchs, 2011; Guo, Hawkins, & Nathans, 2004). These data provide strong evidence that Celsr1 is involved in classical PCP together with Fzd3 and 6, Dvl1 and 2, and Vangl2.

3. CELSR2 AND 3 IN CILIOGENESIS

Cilia are appendages that extend from the cell surface to the extracellular environment. Based on their structure and motility, cilia are classified into three types: motile, primary, and nodal cilia (Davenport & Yoder, 2005). Motile cilia garnish the apical surface of epithelial cell lining airways, reproductive tracts, and cerebral ventricles. They have a central pair of microtubules surrounded by nine doublet of microtubules, an organization known as 9+2, and are anchored to a basal body in the apical cortex. Motile cilia form tufts and their concerted beats generate directional flow at the surface of epithelial sheets. In the airways, they are crucial for clearing mucus and debris, and in genital tracts, they assist in the transit of sperm and eggs (Salathe, 2007; Voronina et al., 2009). In the mouse brain at late embryonic stage and during the first postnatal week, neuroepithelial cells that line the cerebral ventricles differentiate into a monolayer of ependymal cells. At birth, ependymal cells are not multiciliated yet (Fig. 7.3). Motile cilia develop progressively and reach their mature shape by the end of brain maturation (Fig. 7.3B). The basal body of each cilium has a lateral extension called "basal foot" that can be detected by transmission electron microscopy. Basal feet point in the effective beat direction and are used as a hallmark of cilia polarity (Gibbons, 1961; Hagiwara, Ohwada, Aoki, Suzuki, & Takata, 2008; Wallingford, 2010). To generate an efficient directional flow, cilia coordinate their beats within the same cell (each cell has dozens of motile cilia) and in all cells in the epithelial sheet. Thus, basal feet are aligned in the same orientation with respect to the tissue polarity axis, a process referred to as "rotational polarity" (Mirzadeh, Han, Soriano-Navarro, Garcia-Verdugo, & Alvarez-Buylla, 2010). Elegant studies in *Xenopus* ciliated epidermal explants isolated at different developmental stages showed that motile cilia acquire their rotational polarity in two steps. Early in development, genetic cues specify a rough planar axis that allows cells to produce a directional flow, thereby inducing a positive feedback loop that tunes basal feet polarity (Mitchell, Jacobs, Li, Chien, & Kintner, 2007). In mice, *Celsr1–3* are expressed in the developing ependymal layer. Mutation of *Celsr2* and *Celsr3* impairs ependymal ciliogenesis (Tissir et al., 2010). Although differentiation of ependymal cells occurs normally, their motile cilia never

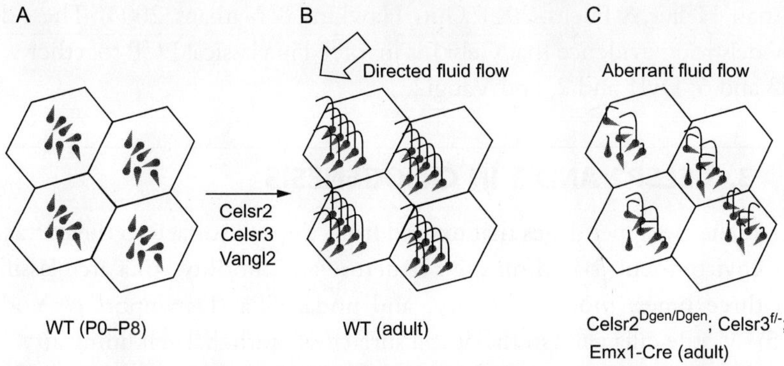

Figure 7.3 Celsr2 and Celsr3 regulate rotational polarity of ependymal cilia. Representation of motile cilia anchored to basal bodies (red dots) on the apical surface of ependymal cells. Cilia develop progressively according to ventral to dorsal, medial to lateral, and caudal to rostral gradients. During the first postnatal week (P0–P8), basal feet (green arrowheads) point in different directions (A). In normal animals, they rotate and adopt a more uniform orientation between P8 and P21, in response to planar cell polarity signals and hydrodynamic forces. This uniform orientation—also referred to as rotational polarity—enables cilia to beat in a coordinated manner and to generate a directed fluid flow (B). In Celsr2 and Celsr3 mutant ependyma (C), basal feet exhibit divergent orientation and fail to generate a directed fluid flow. (For interpretation of the references to color in this figure legend, the reader is referred to the online version of this chapter.)

develop in normal numbers and those that develop display abnormalities in position and planar organization. At the cellular level, basal feet are misoriented, a rotational polarity defect, and some basal bodies assemble deep in the cytoplasm. At the tissue level, ciliary tufts from neighboring cells point aberrantly in divergent directions, generating a translational polarity defect (Fig. 7.3C). Mutant cilia display the typical "9+2" structure. They are still able to beat with the same frequency as the controls. Yet, because of their abnormal orientation, they fail to generate a robust and directed flow (Fig. 7.3C). The membrane localization of Vangl2 and Fzd3 is disrupted in mutant ependymal cells, providing strong evidence that Celsr2 and Celsr3 regulate ciliogenesis via PCP signaling. In accord with this, downregulation of core PCP genes *Disheveled1–3*, and PCP effectors *Inturned* and *Fuzzy* affects the orientation of multicilia in *Xenopus* (Park, Haigo, & Wallingford, 2006; Park, Mitchell, Abitua, Kintner, & Wallingford, 2008). Furthermore, ependymal cells with the "*looptail*" mutation in *Vangl2* fail to align their motile cilia in response to hydrodynamic forces in *in vitro* assays (Guirao et al., 2010).

4. CELSR1–3 IN NEURONAL MIGRATION

The migration of facial branchiomotor (FBM) neurons in the developing rhombencephalon is an intriguing case that combines tangential and radial migration modes. FBM neurons, which innervate muscles responsible for facial expression (Chandrasekhar, 2004; Garel, Garcia-Dominguez, & Charnay, 2000; Guthrie, 2007), are generated in medial rhombomere4 (r4) at E10.5 and immediately extend their axons laterally toward muscle targets. At E11.5, their cell bodies initiate a tangential caudal migration from r4 to r6. They migrate in the subventricular region, pass medial to the nucleus abducens (nVI) in r5 (Song et al., 2006), and then move laterally and dorsally in r6. Finally, they undergo a radial migration in r6 to reach their subpial location where they form the motor nucleus of the facial (nVII) nerve (Fig. 7.4A) (Chandrasekhar, 2004). The caudal soma translocation of FBM neurons, with looping of their axons (so-called *genu* of facial nerve) is conserved from fish to mammals, with important species differences. For example, it is blunted in chick (Gilland & Baker, 2005). The first indication that *Celsr* genes are involved in FBM neurons migration came from an ENU mutagenesis screen in zebrafish that identified four point mutations in the *Celsr2/off road* locus (Wada et al., 2006). In these mutants, FBM neurons fail to migrate caudally to r6, moving instead prematurely into lateral r4–r5. Morpholino knockdown experiments showed that *Celsr1a* and *1b* have adjunct functions in FBM neuron migration. Whereas combined downregulation of *Celsr1a* and *Celsr1b* has little effect on its own, it worsens the *off road* mutant phenotype, with more cells stacked in r4 in *Celsr1a*; *Celsr1b*; *Celsr2* triple mutants than in *Celsr2* mutants. In mice, FBM neuron migration was studied using *in situ* hybridization for *Islet1* or *Tbx20*, two established markers. In wild-type embryos, FBM neurons form streams from medial r4 to lateral r6, with a sharp rostral edge. In *Celsr1*$^{ko/ko}$ mice, FBM neurons are still able to move out of r4, but a subset migrate rostrally into r3 and r2 rather than caudally, a phenotype never seen in any other vertebrate. This rostral migration phenotype is fully penetrant, although with variable expressivity. Caudally directed *Celsr1*$^{ko/ko}$ neurons move through r5, medially to the abducens nucleus (nVI), before moving laterally in r6 like their wild-type counterparts. Hence, a facial (nVII) nucleus forms in its normal location in lateral r6 by E13.5. In addition, rostrally migrating FBM neurons form an ectopic nucleus adjacent or

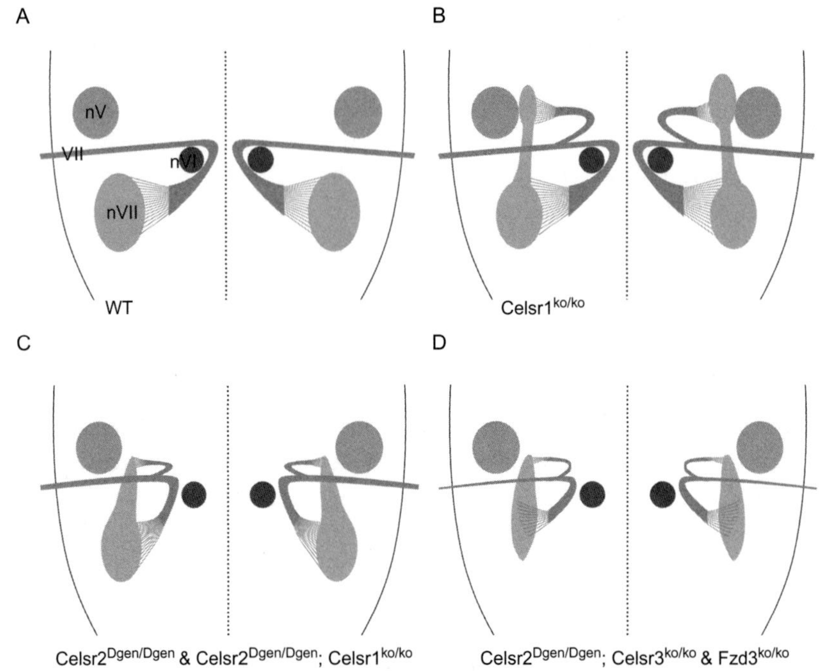

Figure 7.4 Migration of facial branchiomotor (FBM) neurons in normal and *Celsr1–3* knockout mice. Drawings summarize the phenotype at P0. In wild-type animals (A), FBM neurons form a single nucleus (nVII) in a region that derives from r6. Their axons loop around the abducens nucleus (nVI) before exiting the rhombencephalon dorsally, caudal to the trigeminal nucleus (nV). In *Celsr1*$^{ko/ko}$ mutants (B), in addition to the normal nVII, FBM neurons form another ectopic nucleus at the level of nV, but their axons leave the rhombencephalon at the right position. This is due to aberrant migration of FBM neurons in lateral r2 during embryogenesis. In *Celsr2*$^{Dgen/Dgen}$ and *Celsr2*$^{Dgen/Dgen}$; *Celsr1*$^{ko/ko}$ (C), the facial nerve *genu* is completely abnormal and axons do not loop around nVI because FBM neurons migrate prematurely in lateral r4–r5, forming lateral heterotopias. In *Celsr2*$^{Dgen/Dgen}$; *Celsr3*$^{ko/ko}$ and in *Fzd3*$^{ko/ko}$, in addition to the absence of the *genu* of the facial nerve, the size of nVII is reduced because of embryonic cell death (D). "r," rhombomere; nV, motor trigeminal nucleus; nVI, abducens nucleus; nVII, facial nerve nucleus. (See Color Insert.)

even rostral to the trigeminal (nV) nucleus but send their axons normally in the facial nerve (Fig. 7.4B). In the hindbrain, *Celsr1* is expressed in progenitors and in the floor plate, but not in postmitotic FBM neurons. Consistent with this, conditional inactivation of *Celsr1* in progenitors under *Nk6.2-Cre* recombination induces abnormal rostral migration, whereas its deletion in FBM neurons using *Isl1-Cre* does not.

In *Celsr2* mutant mice (*Celsr2*$^{Dgen/Dgen}$), like in *off road* mutant fishes, the caudal FBM neuron migration stream is severely truncated. Cells turn laterally, rostral to the abducens nucleus, so that axons do not loop around nVI and an elongated facial nucleus forms in r4 and r5 instead of r6 (Fig. 7.4C). When *Celsr2* is specifically inactivated in FBM neurons by crossing with *Isl1-Cre* mice, a similar phenotype is generated, suggesting that *Celsr2* is required cell autonomously in FBM neurons for their caudal migration (Qu et al., unpublished). This does not rule out the possibility that *Celsr2* may also be necessary in neuroepithelial cells along the migration path, as suggested in zebrafish (Wada et al., 2006). The aberrant migration of FBM neurons in *Celsr2* mutant mice is reminiscent of the situation in birds, which have no facial nerve *genu* and, intriguingly, lack a *Celsr2* gene (Formstone, 2010). Whether there is a causal link between the absence of *Celsr2* and blunted FBM neuron migration remains an open question. In zebrafish, *Celsr2*, *Celsr1a*, and *Celsr1b* redundantly regulate FBMN caudal migration. In mouse, double inactivation of *Celsr1* and *Celsr2* phenocopies single *Celsr2* mutation, with no obvious rostral migration of FBM neurons, suggesting that *Celsr2* is epistatic to *Celsr1* (Fig. 7.4C). It looks as if *Celsr2* deficiency hampers or slows down neuronal migration, thus masking the *Celsr1* mutant phenotype. Although *Celsr3* is required for tangential migration of calretinin-positive interneurons and radial migration of calbindin-positive interneurons in the forebrain (Ying et al., 2009), its deficiency does not affect the migration of FBM neurons. It does, however, exacerbate the *Celsr2* phenotype: the facial nucleus is greatly reduced in size in *Celsr2*$^{Dgen/Dgen}$; *Celsr3*$^{ko/ko}$ and *Celsr2*$^{Dgen/Dgen}$; *Celsr3*^{Isl1cko} as compared to *Celsr2*$^{Dgen/Dgen}$ mutant (Fig. 7.4C and D). Celsr3 deficiency increases apoptosis in a cell-autonomous manner. The phenotype of *Celsr2*$^{Dgen/Dgen}$; *Celsr3*$^{ko/ko}$ double mutants is similar to that of *Fzd3* mutants, and other PCP-related genes such as *Vangl2* and *Wnt5a* have been implicated in FBM neuron migration, suggesting that *Celsr1–3* regulate neuronal migration along the rostrocaudal axis by PCP-dependent mechanisms (Qu et al., 2010; Vivancos et al., 2009) (Table 7.1). In line with this, normal function of PCP genes such as *van gogh-like 2* (*vangl2*), *fzd3a*, *celsr2*, *prickle1a*, and *prickle1b* is required for caudal migration of FBM neurons in fish (Bingham, Higashijima, Okamoto, & Chandrasekhar, 2002; Carreira-Barbosa et al., 2003; Jessen et al., 2002; Mapp, Wanner, Rohrschneider, & Prince, 2010; Rohrschneider, Elsen, & Prince, 2007; Wada et al., 2006).

5. CELSR2 AND CELSR3 IN BRAIN WIRING

Functional neuronal networks are crucial for brain function. Network formation is finely orchestrated at the cellular and molecular levels by genetic programs and interactions with the environment. After reaching their location, postmitotic neurons extend axons that are guided to their targets by intrinsic programs, guidepost cells, and attractive and repulsive molecular cues. They ramify receptive dendritic fields according to tiling rules. In addition to their role in neuronal migration, *Celsrs* are implicated in dendrite development and axon guidance, from *Caenorhabditis elegans* and Drosophila to mammals (Berger-Muller & Suzuki, 2011; Gao, Kohwi, Brenman, Jan, & Jan, 2000; Matsubara, Horiuchi, Shimono, Usui, & Uemura, 2011).

In flies, sensory neurons extend dendrites dorsally toward the midline. Dendrites from homologous neurons in the two opposite hemisegments avoid each other, leading to a dendrite-free zone near the dorsal midline. This reciprocal inhibition of dendrite growth, akin to tiling, is defective in fmi/stan mutants, where dendrites do grow across the midline to occupy fields overlapping those of homologous neurons (Gao et al., 2000). A role for fmi/stan in the tiling of dendrites is further supported by the identification of fmi/stan mutants in genetic screens for aberrant dendritic extension of mushroom body neurons (Reuter et al., 2003). The function of fmi/stan in dendrite growth does not rely solely on adhesion mediated by the extracellular domain because overexpression of an N-terminally truncated fmi that lacks cadherin, EGF-like, and laminin G motifs rescues the dendritic phenotype partially (Kimura, Usui, Tsubouchi, & Uemura, 2006). Intriguingly, neither loss or gain of function of *frizzled* nor overexpression of *dsh* phenocopy the *fmi/stan* dendritic phenotype. Moreover, to our knowledge, no such phenotype has been described in other PCP mutants. In the mammalian nervous system, *Celsr2* is required for the maintenance of dendritic arbors: RNAi-induced downregulation of *Celsr2* in brain slices reduces the length of dendrites in cortical pyramidal neurons, and the complexity of dendritic trees of Purkinje cells (Shima, Kengaku, Hirano, Takeichi, & Uemura, 2004). By contrast, the silencing of *Celsr3* results in dendrite overextension. The opposite effects of Celsr2 and Celsr3 are related to a single amino acid change in the first intracellular loop: like fmi/stan, mammalian Celsr3 has a histidine at position 2413, and both repress growth and/or induce retraction of dendrites. That histidine residue is replaced with an arginine in Celsr2, which promotes dendrite growth and maintenance (Shima et al., 2007).

Early observations showed that *fmi/stan* is essential for the development of axonal tracts. Null mutations *fmi*^{E45} and *fmi*^{E59} are embryonic lethal, and in addition to PCP defects, mutant embryos display abnormalities in longitudinal axonal tracts. Rescue of axonal, but not PCP defects, by brain specific expression of the *fmi* cDNA restores viability (Usui et al., 1999). Subsequent elegant studies in the visual system showed that fmi/stan mediates axon–axon and axon–target interactions required for guidance of photoreceptor axons (Lee et al., 2003; Senti et al., 2003). The fmi-mediated choice of postsynaptic target is dose dependent and non-cell autonomous. It looks as if growth cones "compare" their fmi/stan levels with that of their neighbors to "decide" whether to establish contact or not (Chen & Clandinin, 2008). In mice, *Celsr3* is crucial for axon guidance. *Celsr3* mutant mice display marked defects in major tracts such as the anterior commissure, internal capsule, medial lemniscus, and corticospinal tract (Table 7.1). *Celsr3* is also essential for the anterior–posterior organization of monoaminergic axon in the brainstem (Fenstermaker et al., 2010) and for the rostral turning of commissural axons after midline crossing in the spinal cord (Price et al., 2006; Wang & Nathans, 2007). Celsr3 deficiency does not affect axonal growth *per se*, but rather guidance, resulting in stalling at intermediate targets or misrouting of axons (Tissir, Bar, Jossin, De Backer, & Goffinet, 2005). For instance, during development of corticothalamic reciprocal projections, mutant corticofugal axons travel normally in the subcortical layer but stall underneath the pallial–subpallial boundary. Reciprocally, thalamic axons never reach their cortical targets in layer 4; instead, they course ventrally along the hypothalamus and then turn externally toward the cortical marginal zone. Conditional removal of *Celsr3* in a stream that extends in the basal forebrain and ventral diencephalon precludes the entry of corticosubcortical and thalamocortical fibers in the "corridor" and results in defective internal capsule. This result demonstrates that *Celsr3* is required in intermediate targets to connect the cortex with subcortical structures. Target cells are molecularly defined by expression of Dlx5/6 and qualify as "guidepost cells" (Zhou et al., 2008). To test whether guidance of cortifugal axon is mediated by homophilic interactions of *Celsr3* in navigating growth cones and guidepost cells, *Celsr3* was specifically deleted in corticofugal axons by crossing the *Celsr3*^{f} allele with *Emx1-Cre*. Celsr3-defective corticothalamic axons develop normally, strongly suggesting that those fibers use guidance mechanisms independent of Celsr3–Celsr3 homophilic interactions. Consistent with this finding, a stall phenotype of axons at intermediate target cells was reported for fmi-defective sensory neurons in flies (Steinel & Whitington, 2009). These phenotypes can be rescued by fmi constructs that lack most of the extracellular domains, indicating that the advance of sensory axons in flies does not depend on fmi–fmi homophilic interactions (Steinel & Whitington, 2009). Contrary to corticothalamic axons, corticospinal axons do not develop in *Celsr3*^{Emx1cKO}, suggesting that *Celsr3* mediates homophilic interactions between corticospinal growth cones and guidepost cells. In support of this, in the fly visual system, fmi mediates interactions between the growth cones of photoreceptor axons and their targets in the medulla (Hakeda-Suzuki et al., 2011). Like in flies and mammals, FMI-1, the sole *C. elegans* fmi/stan ortholog, enables navigation of both pioneer and follower axons in the worm's ventral nerve cord. Rescue of FMI-1 mutant phenotype with different portions of FMI-1 revealed that the C- and N-termini are required for guidance of pioneer and follower axons, respectively (Steimel et al., 2010). Taken together, these data show that flamingo and its mammalian ortholog *Celsr3* are major players in axon guidance and that their mechanisms of action are context dependent and involve both homophilic and heterophilic interactions.

5.1. Motifs of Celsr important for their functions

As mentioned in the introduction, Celsr proteins contain a large extracellular N-terminus with nine cadherin repeats, EGF-like and laminin G-like motifs, a HRM, a G-proteolysis site, followed with seven transmembrane domains and a variable intracellular C-terminal tail. Although the functional relevance of these motifs remains poorly understood, some studies have revealed a few hints. In Drosophila S2 cells—which have no self aggregation properties—expression of full-length fmi induces cell aggregation, whereas expression of a form lacking most of the ectodomain does not, showing that the extracellular domain indeed promotes homophilic cell adhesion (Usui et al., 1999). Expression of mutant forms of mammalian Celsr2 in S2 cells assigned adhesive properties to the cadherin repeats (Shima et al., 2004).

Dendritic overgrowth and tiling defects of *fmi* mutants as well as fly viability can be rescued by expression of full-length fmi, although the rescue can be ascribed to two different functions of the protein. In transgenic rescue experiments, ΔNfmi—lacking all extracellular domains except the HRM—can rescue the dendrite overgrowth but not the tiling phenotype. Unlike full-length fmi, ΔNfmi does not promote S2 cell aggregation *in vitro*.

By contrast, C-terminally deleted fmi (ΔCFmi) retains the ability to mediate homophilic binding *in vitro* and rescues the tiling phenotype, but not the dendritic overgrowth or fly viability, pointing to a key role of the C-terminus. Thus, a dual molecular function of Fmi plays pivotal roles in dendrite morphogenesis. In the initial growing phase, Fmi might function as a receptor for an unidentified ligand, and this hypothetical heterophilic interaction would be responsible for limiting branch elongation. At a later stage, homophilic Fmi-binding at dendro-dendritic interfaces would elicit avoidance between dendritic terminals from opposing neurons (Kimura et al., 2006). Likewise, studies in worms, flies, and mice show that the function of fmi/Celsr in axon guidance does not rely only on its ability to mediate homophilic binding of cadherin repeats but also requires interactions of the C-terminus with unknown molecules (Steimel et al., 2010; Steinel & Whitington, 2009; Zhou et al., 2008).

The concept of distinct roles of the extra- and intracellular domains of fmi/Celsr is supported by mutation analysis in vertebrates. In rodents, Celsr2 and Celsr3 have opposing roles on dendrites in brain slice in culture, with Celsr2 promoting and Celsr3 restricting dendrite growth (Shima et al., 2004, 2007). Studies of chimeric constructs in which the ectodomains of Celsr2 and Celsr3 are swapped show that, whereas homophilic interactions (Celsr2–Celsr2 and Celsr3–Celsr3) are important, the transmembrane domains and C-terminus determine the dendrite enhancing or suppressing action. Celsr2 and Celsr3 have different effects on calcium release and activate, respectively, CamKII and Calcineurin signaling, and an amino acid change (R2413H) in the first intracellular loop is crucial to these distinct functions.

Another evidence for the importance of the C-terminus comes from a study in zebrafish (Carreira-Barbosa et al., 2009). Injection of a C-terminally truncated form of Celsr (ΔC-celsr; lacking 6 TM and tail) in wild-type embryos generates epiboly defects. This is accompanied by sequestration of ΔC-Celsr in the Golgi, where the protein behaves as a dominant negative, dimerizing with and precluding trafficking of the wild-type protein. Dimerization is thought to involve a conserved arginine-rich sequence N-terminal to the first cadherin repeat. Injection of the CT of Celsr2 fused to the membrane localization signal from the Lyn tyrosine kinase (Lyn-celsr) generates convergent extension defects without affecting epiboly. Lyn-celsr perturbs the Frizzled-induced membrane localization of Disheveled and thus PCP signaling. In mammals, knock-in of the mouse Celsr1 C-tail fused to membrane localization signal Lyn in the ubiquitous Rosa26 locus also perturbs the trafficking of Dvl2, and Vangl2 as a well as the subcellular of F-actin (Trichas et al., 2011). Interestingly, when a conserved serine acidic amino acid-rich domain (SE/D) is deleted from the fish Lyn-Celsr, the membrane localization of Disheveled is restored (Carreira-Barbosa et al., 2009). Whereas the extracellular and TM domains are important for the distribution of Frizzled–Disheveled complexes at the membrane, this process depends critically on the Cter SE/D domain. The ability of Celsr to regulate epiboly is closely associated with its ability to modulate cell cohesive property, whereas its ability to interact with the PCP pathway to regulate convergent extension may not require cell adhesion mediated by the cadherin repeats.

In the mouse skin, Celsr1 is internalized during epidermal basal progenitor division, a process which is crucial for balanced distribution of PCP proteins Fzd6 and Vangl2 in daughter cells and requires a cytoplasmic dileucine motif. Whereas E-Cadherin is normally not internalized, its fusion with the Celsr1 cytoplasmic domain induces internalization during mitosis and confers to the chimeric protein the ability to recruit Fzd6 and Vangl2. When the two leucines (2748–2749) are mutated to alanines, chimeric E-Cad-Celsr1, like mutated Celsr1, no longer translocates to endosomes (Devenport et al., 2011).

The Celsr1–3 CTs display much less similarity than their extracellular and TM regions. As Celsr1 is directly implicated in PCP, and Celsr2–3 in more distantly related processes, could differences in the C-terminus account for functional differences? In *Drosophila*, fmi, fz, and dsh depend on each other for their membrane localization (Das et al., 2002). Fmi interacts physically with fz via a region encompassing its HRM and TM domains, and selectively recruits fz and vang (strabismus) to opposing cell boundaries (Chen et al., 2008). In mice, the membrane localization of Fzd3 and Vangl2 in ependymal cells, Fzd6 and Vangl2 in skin epithelial cells, and Dvl2 and Vangl2 in the visceral endoderm cells depends on Celsr cadherins (Devenport & Fuchs, 2008; Tissir et al., 2010; Trichas et al., 2011). In *Celsr1*$^{Crsh/Crsh}$ mutant mice, a single amino acid substitution results in failure of the Celsr1 protein to reach the apical membrane (Devenport & Fuchs, 2008; Formstone et al., 2010; Ravni et al., 2009). Furthermore, all six Celsr1 mutations recently identified in human fetuses with craniorachischisis impair membrane trafficking of Celsr1 in *in vitro* assays (Robinson et al., 2011). Although not directly useful for definition of structure–function relationships, this exquisite sensitivity to minor sequence changes underscores the importance of fmi/Celsr conformation for their membrane insertion as well as for proper targeting of partner proteins. This raises the question of the role of putative proteins assisting in folding, like receptor-activating protein RAP does for the lipoprotein receptor family (Herz, 2006), or the beta-2 microglobulin for major histocompatibility complexes (Hansen, Connolly, Gould, & Fremont, 2010). Would Celsr, Fzd, and Vangl help each other's folding and complex formation in the endoplasmic reticulum and traffic as a complex to the Golgi and to the membrane, or are they assisted by other unidentified partners, and what would be the role of chaperonins in the process?

REFERENCES

Adler, P. N. (2002). Planar signaling and morphogenesis in Drosophila. *Developmental Cell, 2*, 525–535.

Bastock, R., Strutt, H., & Strutt, D. (2003). Strabismus is asymmetrically localised and binds to Prickle and Dishevelled during Drosophila planar polarity patterning. *Development, 130*, 3007–3014.

Beall, S. A., Boekelheide, K., & Johnson, K. J. (2005). Hybrid GPCR/cadherin (Celsr) proteins in rat testis are expressed with cell type specificity and exhibit differential Sertoli cell-germ cell adhesion activity. *Journal of Andrology, 26*, 529–538.

Bellaiche, Y., Gho, M., Kaltschmidt, J. A., Brand, A. H., & Schweisguth, F. (2001). Frizzled regulates localization of cell-fate determinants and mitotic spindle rotation during asymmetric cell division. *Nature Cell Biology, 3*, 50–57.

Berger-Muller, S., & Suzuki, T. (2011). Seven-pass transmembrane cadherins: Roles and emerging mechanisms in axonal and dendritic patterning. *Molecular Neurobiology, 44*, 313–320.

Bingham, S., Higashijima, S., Okamoto, H., & Chandrasekhar, A. (2002). The Zebrafish trilobite gene is essential for tangential migration of branchiomotor neurons. *Developmental Biology, 242*, 149–160.

Carreira-Barbosa, F., Concha, M. L., Takeuchi, M., Ueno, N., Wilson, S. W., & Tada, M. (2003). Prickle 1 regulates cell movements during gastrulation and neuronal migration in zebrafish. *Development, 130*, 4037–4046.

Carreira-Barbosa, F., Kajita, M., Morel, V., Wada, H., Okamoto, H., Martinez Arias, A., et al. (2009). Flamingo regulates epiboly and convergence/extension movements through cell cohesive and signalling functions during zebrafish gastrulation. *Development, 136*, 383–392.

Chae, J., Kim, M. J., Goo, J. H., Collier, S., Gubb, D., Charlton, J., et al. (1999). The Drosophila tissue polarity gene starry night encodes a member of the protocadherin family. *Development, 126*, 5421–5429.

Chandrasekhar, A. (2004). Turning heads: Development of vertebrate branchiomotor neurons. *Developmental Dynamics, 229*, 143–161.

Chen, W. S., Antic, D., Matis, M., Logan, C. Y., Povelones, M., Anderson, G. A., et al. (2008). Asymmetric homotypic interactions of the atypical cadherin flamingo mediate intercellular polarity signaling. *Cell, 133*, 1093–1105.

Chen, P. L., & Clandinin, T. R. (2008). The cadherin Flamingo mediates level-dependent interactions that guide photoreceptor target choice in Drosophila. *Neuron, 58*, 26–33.

Curtin, J. A., Quint, E., Tsipouri, V., Arkell, R. M., Cattanach, B., Copp, A. J., et al. (2003). Mutation of Celsr1 disrupts planar polarity of inner ear hair cells and causes severe neural tube defects in the mouse. *Current Biology, 13*, 1129–1133.

Das, G., Reynolds-Kenneally, J., & Mlodzik, M. (2002). The atypical cadherin Flamingo links Frizzled and Notch signaling in planar polarity establishment in the Drosophila eye. *Developmental Cell, 2*, 655–666.

Davenport, J. R., & Yoder, B. K. (2005). An incredible decade for the primary cilium: A look at a once-forgotten organelle. *American Journal of Physiology. Renal Physiology, 289*, F1159–F1169.

Deans, M. R., Antic, D., Suyama, K., Scott, M. P., Axelrod, J. D., & Goodrich, L. V. (2007). Asymmetric distribution of prickle-like 2 reveals an early underlying polarization of vestibular sensory epithelia in the inner ear. *The Journal of Neuroscience, 27*, 3139–3147.

Devenport, D., & Fuchs, E. (2008). Planar polarization in embryonic epidermis orchestrates global asymmetric morphogenesis of hair follicles. *Nature Cell Biology, 10*, 1257–1268.

Devenport, D., Oristian, D., Heller, E., & Fuchs, E. (2011). Mitotic internalization of planar cell polarity proteins preserves tissue polarity. *Nature Cell Biology, 13*, 893–902.

Feiguin, F., Hannus, M., Mlodzik, M., & Eaton, S. (2001). The ankyrin repeat protein Diego mediates Frizzled-dependent planar polarization. *Developmental Cell, 1*, 93–101.

Fenstermaker, A. G., Prasad, A. A., Bechara, A., Adolfs, Y., Tissir, F., Goffinet, A., et al. (2010). Wnt/planar cell polarity signaling controls the anterior-posterior organization of monoaminergic axons in the brainstem. *The Journal of Neuroscience, 30*, 16053–16064.

Formstone, C. J. (2010). 7TM-Cadherins: Developmental roles and future challenges. *Advances in Experimental Medicine and Biology, 706*, 14–36.

Formstone, C. J., & Little, P. F. (2001). The flamingo-related mouse Celsr family (Celsr1-3) genes exhibit distinct patterns of expression during embryonic development. *Mechanisms of Development, 109*, 91–94.

Formstone, C. J., Moxon, C., Murdoch, J., Little, P., & Mason, I. (2010). Basal enrichment within neuroepithelia suggests novel function(s) for Celsr1 protein. *Molecular and Cellular Neurosciences, 44*, 210–222.

Gao, F. B., Kohwi, M., Brenman, J. E., Jan, L. Y., & Jan, Y. N. (2000). Control of dendritic field formation in Drosophila: The roles of flamingo and competition between homologous neurons. *Neuron, 28*, 91–101.

Garel, S., Garcia-Dominguez, M., & Charnay, P. (2000). Control of the migratory pathway of facial branchiomotor neurones. *Development, 127*, 5297–5307.

Gibbons, I. R. (1961). The relationship between the fine structure and direction of beat in gill cilia of a lamellibranch mollusc. *The Journal of Biophysical and Biochemical Cytology, 11*, 179–205.

Gilland, E., & Baker, R. (2005). Evolutionary patterns of cranial nerve efferent nuclei in vertebrates. *Brain, Behavior and Evolution, 66*, 234–254.

Gubb, D., Green, C., Huen, D., Coulson, D., Johnson, G., Tree, D., et al. (1999). The balance between isoforms of the prickle LIM domain protein is critical for planar polarity in Drosophila imaginal discs. *Genes & Development, 13*, 2315–2327.

Guirao, B., Meunier, A., Mortaud, S., Aguilar, A., Corsi, J. M., Strehl, L., et al. (2010). Coupling between hydrodynamic forces and planar cell polarity orients mammalian motile cilia. *Nature Cell Biology, 12*, 341–350.

Guo, N., Hawkins, C., & Nathans, J. (2004). Frizzled6 controls hair patterning in mice. *Proceedings of the National Academy of Sciences of the United States of America, 101*, 9277–9281.

Guthrie, S. (2007). Patterning and axon guidance of cranial motor neurons. *Nature Reviews. Neuroscience, 8*, 859–871.

Hadjantonakis, A. K., Formstone, C. J., & Little, P. F. (1998). mCelsr1 is an evolutionarily conserved seven-pass transmembrane receptor and is expressed during mouse embryonic development. *Mechanisms of Development, 78*, 91–95.

Hadjantonakis, A. K., Sheward, W. J., Harmar, A. J., de Galan, L., Hoovers, J. M., & Little, P. F. (1997). Celsr1, a neural-specific gene encoding an unusual seven-pass transmembrane receptor, maps to mouse chromosome 15 and human chromosome 22qter. *Genomics, 45*, 97–104.

Hagiwara, H., Ohwada, N., Aoki, T., Suzuki, T., & Takata, K. (2008). The primary cilia of secretory cells in the human oviduct mucosa. *Medical Molecular Morphology*, *41*, 193–198.

Hakeda-Suzuki, S., Berger-Muller, S., Tomasi, T., Usui, T., Horiuchi, S. Y., Uemura, T., et al. (2011). Golden Goal collaborates with Flamingo in conferring synaptic-layer specificity in the visual system. *Nature Neuroscience*, *14*, 314–323.

Hansen, T. H., Connolly, J. M., Gould, K. G., & Fremont, D. H. (2010). Basic and translational applications of engineered MHC class I proteins. *Trends in Immunology*, *31*, 363–369.

Herz, J. (2006). The switch on the RAPper's necklace. *Molecular Cell*, *23*, 451–455.

Jenny, A., Darken, R. S., Wilson, P. A., & Mlodzik, M. (2003). Prickle and Strabismus form a functional complex to generate a correct axis during planar cell polarity signaling. *The EMBO Journal*, *22*, 4409–4420.

Jessen, J. R., Topczewski, J., Bingham, S., Sepich, D. S., Marlow, F., Chandrasekhar, A., et al. (2002). Zebrafish trilobite identifies new roles for Strabismus in gastrulation and neuronal movements. *Nature Cell Biology*, *4*, 610–615.

Johnson, K. J., Patel, S. R., & Boekelheide, K. (2000). Multiple cadherin superfamily members with unique expression profiles are produced in rat testis. *Endocrinology*, *141*, 675–683.

Kibar, Z., Vogan, K. J., Groulx, N., Justice, M. J., Underhill, D. A., & Gros, P. (2001). Ltap, a mammalian homolog of Drosophila Strabismus/Van Gogh, is altered in the mouse neural tube mutant Loop-tail. *Nature Genetics*, *28*, 251–255.

Kimura, H., Usui, T., Tsubouchi, A., & Uemura, T. (2006). Potential dual molecular interaction of the Drosophila 7-pass transmembrane cadherin Flamingo in dendritic morphogenesis. *Journal of Cell Science*, *119*, 1118–1129.

Krasnoperov, V., Lu, Y., Buryanovsky, L., Neubert, T. A., Ichtchenko, K., & Petrenko, A. G. (2002). Post-translational proteolytic processing of the calcium-independent receptor of alpha-latrotoxin (CIRL), a natural chimera of the cell adhesion protein and the G protein-coupled recent—Role of the G protein-coupled receptor proteolysis site (GPS) motif. *The Journal of Biological Chemistry*, *277*, 46518–46526.

Lee, R. C., Clandinin, T. R., Lee, C. H., Chen, P. L., Meinertzhagen, I. A., & Zipursky, S. L. (2003). The protocadherin Flamingo is required for axon target selection in the Drosophila visual system. *Nature Neuroscience*, *6*, 557–563.

Mapp, O. M., Wanner, S. J., Rohrschneider, M. R., & Prince, V. E. (2010). Prickle1b mediates interpretation of migratory cues during zebrafish facial branchiomotor neuron migration. *Developmental Dynamics*, *239*, 1596–1608.

Matsubara, D., Horiuchi, S. Y., Shimono, K., Usui, T., & Uemura, T. (2011). The seven-pass transmembrane cadherin Flamingo controls dendritic self-avoidance via its binding to a LIM domain protein, Espinas, in Drosophila sensory neurons. *Genes & Development*, *25*, 1982–1996.

Mirzadeh, Z., Han, Y. G., Soriano-Navarro, M., Garcia-Verdugo, J. M., & Alvarez-Buylla, A. (2010). Cilia organize ependymal planar polarity. *The Journal of Neuroscience*, *30*, 2600–2610.

Mitchell, B., Jacobs, R., Li, J., Chien, S., & Kintner, C. (2007). A positive feedback mechanism governs the polarity and motion of motile cilia. *Nature*, *447*, 97–101.

Mlodzik, M. (1999). Planar polarity in the Drosophila eye: A multifaceted view of signaling specificity and cross-talk. *The EMBO Journal*, *18*, 6873–6879.

Montcouquiol, M., Rachel, R. A., Lanford, P. J., Copeland, N. G., Jenkins, N. A., & Kelley, M. W. (2003). Identification of Vangl2 and Scrb1 as planar polarity genes in mammals. *Nature*, *423*, 173–177.

Montcouquiol, M., Sans, N., Huss, D., Kach, J., Dickman, J. D., Forge, A., et al. (2006). Asymmetric localization of Vangl2 and Fz3 indicate novel mechanisms for planar cell polarity in mammals. *The Journal of Neuroscience*, *26*, 5265–5275.

Park, T. J., Haigo, S. L., & Wallingford, J. B. (2006). Ciliogenesis defects in embryos lacking inturned or fuzzy function are associated with failure of planar cell polarity and Hedgehog signaling. *Nature Genetics*, *38*, 303–311.

Park, T. J., Mitchell, B. J., Abitua, P. B., Kintner, C., & Wallingford, J. B. (2008). Dishevelled controls apical docking and planar polarization of basal bodies in ciliated epithelial cells. *Nature Genetics*, *40*, 871–879.

Price, D. J., Kennedy, H., Dehay, C., Zhou, L. B., Mercier, M., Jossin, Y., et al. (2006). The development of cortical connections. *The European Journal of Neuroscience*, *23*, 910–920.

Qu, Y. B., Glasco, D. M., Zhou, L. B., Sawant, A., Ravni, A., Fritzsch, B., et al. (2010). Atypical cadherins Celsr1-3 differentially regulate migration of facial branchiomotor neurons in mice. *The Journal of Neuroscience*, *30*, 9392–9401.

Ravni, A., Qu, Y., Goffinet, A. M., & Tissir, F. (2009). Planar cell polarity cadherin Celsr1 regulates skin hair patterning in the mouse. *The Journal of Investigative Dermatology*, *129*, 2507–2509.

Reuter, J. E., Nardine, T. M., Penton, A., Billuart, P., Scott, E. K., Usui, T., et al. (2003). A mosaic genetic screen for genes necessary for Drosophila mushroom body neuronal morphogenesis. *Development*, *130*, 1203–1213.

Robinson, A., Escuin, S., Doudney, K., Vekemans, M., Stevenson, R. E., de Greene, N., et al. (2011). Mutations in the planar cell polarity genes CELSR1 and SCRIB are associated with the severe neural tube defect craniorachischisis. *Human Mutation*, *33*, 440–447.

Rohrschneider, M. R., Elsen, G. E., & Prince, V. E. (2007). Zebrafish Hoxb1a regulates multiple downstream genes including prickle1b. *Developmental Biology*, *309*, 358–372.

Salathe, M. (2007). Regulation of mammalian ciliary beating. *Annual Review of Physiology*, *69*, 401–422.

Segalen, M., & Bellaiche, Y. (2009). Cell division orientation and planar cell polarity pathways. *Seminars in Cell & Developmental Biology*, *20*, 972–977.

Senti, K. A., Usui, T., Boucke, K., Greber, U., Uemura, T., & Dickson, B. J. (2003). Flamingo regulates R8 axon-axon and axon-target interactions in the Drosophila visual system. *Current Biology*, *13*, 828–832.

Shima, Y., Copeland, N. G., Gilbert, D. J., Jenkins, N. A., Chisaka, O., Takeichi, M., et al. (2002). Differential expression of the seven-pass transmembrane cadherin genes Celsr1-3 and distribution of the Celsr2 protein during mouse development. *Developmental Dynamics*, *223*, 321–332.

Shima, Y., Kawaguchi, S. Y., Kosaka, K., Nakayama, M., Hoshino, M., Nabeshima, Y., et al. (2007). Opposing roles in neurite growth control by two seven-pass transmembrane cadherins. *Nature Neuroscience*, *10*, 963–969.

Shima, Y., Kengaku, M., Hirano, T., Takeichi, M., & Uemura, T. (2004). Regulation of dendritic maintenance and growth by a mammalian 7-pass transmembrane cadherin. *Developmental Cell*, *7*, 205–216.

Song, M. R., Shirasaki, R., Cai, C. L., Ruiz, E. C., Evans, S. M., Lee, S. K., et al. (2006). T-Box transcription factor Tbx20 regulates a genetic program for cranial motor neuron cell body migration. *Development*, *133*, 4945–4955.

Steimel, A., Wong, L., Najarro, E. H., Ackley, B. D., Garriga, G., & Hutter, H. (2010). The Flamingo ortholog FMI-1 controls pioneer-dependent navigation of follower axons in C. elegans. *Development*, *137*, 3663–3673.

Steinel, M. C., & Whitington, P. M. (2009). The atypical cadherin Flamingo is required for sensory axon advance beyond intermediate target cells. *Developmental Biology*, *327*, 447–457.

Strutt, D., Johnson, R., Cooper, K., & Bray, S. (2002). Asymmetric localization of frizzled and the determination of notch-dependent cell fate in the Drosophila eye. *Current Biology*, *12*, 813–824.

Taylor, J., Abramova, N., Charlton, J., & Adler, P. N. (1998). Van Gogh: A new Drosophila tissue polarity gene. *Genetics*, *150*, 199–210.

Tissir, F., Bar, I., Jossin, Y., De Backer, O., & Goffinet, A. M. (2005). Protocadherin Celsr3 is crucial in axonal tract development. *Nature Neuroscience*, *8*, 451–457.

Tissir, F., De Backer, O., Goffinet, A. M., & Lambert de Rouvroit, C. (2002). Developmental expression profiles of Celsr (Flamingo) genes in the mouse. *Mechanisms of Development*, *112*, 157–160.

Tissir, F., Qu, Y., Montcouquiol, M., Zhou, L., Komatsu, K., Shi, D., et al. (2010). Lack of cadherins Celsr2 and Celsr3 impairs ependymal ciliogenesis, leading to fatal hydrocephalus. *Nature Neuroscience*, *13*, 700–707.

Trichas, G., Joyce, B., Crompton, L. A., Wilkins, V., Clements, M., Tada, M., et al. (2011). Nodal dependent differential localisation of dishevelled-2 demarcates regions of differing cell behaviour in the visceral endoderm. *PLoS Biology*, *9*, e1001019.

Usui, T., Shima, Y., Shimada, Y., Hirano, S., Burgess, R. W., Schwarz, T. L., et al. (1999). Flamingo, a seven-pass transmembrane cadherin, regulates planar cell polarity under the control of Frizzled. *Cell*, *98*, 585–595.

Vivancos, V., Chen, P., Spassky, N., Qian, D., Dabdoub, A., Kelley, M., et al. (2009). Wnt activity guides facial branchiomotor neuron migration, and involves the PCP pathway and JNK and ROCK kinases. *Neural Development*, *4*, 7.

Volynski, K. E., Silva, J. P., Lelianova, V. G., Atiqur Rahman, M., Hopkins, C., & Ushkaryov, Y. A. (2004). Latrophilin fragments behave as independent proteins that associate and signal on binding of LTX(N4C). *The EMBO Journal*, *23*, 4423–4433.

Voronina, V. A., Takemaru, K. I., Treuting, P., Love, D., Grubb, B. R., Hajjar, A. M., et al. (2009). Inactivation of Chibby affects function of motile airway cilia. *The Journal of Cell Biology*, *185*, 225–233.

Wada, H., Tanaka, H., Nakayama, S., Iwasaki, M., & Okamoto, H. (2006). Frizzled3a and Celsr2 function in the neuroepithelium to regulate migration of facial motor neurons in the developing zebrafish hindbrain. *Development*, *133*, 4749–4759.

Wallingford, J. B. (2010). Planar cell polarity signaling, cilia and polarized ciliary beating. *Current Opinion in Cell Biology*, *22*, 597–604.

Wang, Y., Guo, N., & Nathans, J. (2006). The role of Frizzled3 and Frizzled6 in neural tube closure and in the planar polarity of inner-ear sensory hair cells. *The Journal of Neuroscience*, *26*, 2147–2156.

Wang, Y., & Nathans, J. (2007). Tissue/planar cell polarity in vertebrates: New insights and new questions. *Development*, *134*, 647–658.

Wolff, T., & Rubin, G. M. (1998). Strabismus, a novel gene that regulates tissue polarity and cell fate decisions in Drosophila. *Development*, *125*, 1149–1159.

Ybot-Gonzalez, P., Savery, D., Gerrelli, D., Signore, M., Mitchell, C. E., Faux, C. H., et al. (2007). Convergent extension, planar-cell-polarity signalling and initiation of mouse neural tube closure. *Development*, *134*, 789–799.

Ying, G., Wu, S., Hou, R., Huang, W., Capecchi, M. R., & Wu, Q. (2009). The protocadherin gene Celsr3 is required for interneuron migration in the mouse forebrain. *Molecular and Cellular Biology*, *29*, 3045–3061.

Zhou, L., Bar, I., Achouri, Y., Campbell, K., De Backer, O., Hebert, J. M., et al. (2008). Early forebrain wiring: Genetic dissection using conditional Celsr3 mutant mice. *Science*, *320*, 946–949.

CHAPTER EIGHT

The Kidney and Planar Cell Polarity

Thomas J. Carroll*[,1], Jing Yu[†,1]

*Department of Internal Medicine (Nephrology) and Molecular Biology, UT Southwestern Medical Center, Dallas, Texas, USA

[†]Department of Cell Biology, University of Virginia, School of Medicine, Charlottesville, Virginia, USA

[1]Corresponding authors: e-mail address: Thomas.Carroll@UTSouthwestern.edu; jy4m@virginia.edu

Contents

Abstract

Planar cell polarity (PCP) or tissue polarity describes a coordinated polarity at the plane of the tissue where most or all cells within a tissue are polarized in one direction. It is perpendicular to the apical–basal polarity of the cell. PCP is manifested readily in the Drosophila wing and cuticle bristles, Drosophila eye ommatidia, and mammalian hair and inner ear hair bundles, and less evidently, in cellular processes such as in the coordinated, directional cell movements, and oriented cell divisions that are important for tissue morphogenesis. Several distinct molecular and cellular processes have been implicated in the regulation of PCP. Here, we review potential roles for PCP during mouse kidney development and maintenance, including ureteric bud branching morphogenesis, renal medulla elongation, tubule diameter establishment/maintenance, glomerulogenesis, and response to injury. The potential mechanisms underlying these processes, including oriented cell division and coordinated cell migration/cell intercalation, are discussed. In addition, we discuss some unaddressed research topics related to PCP in the kidney that we hope will spur further discussion and investigation.

Current Topics in Developmental Biology, Volume 101
ISSN 0070-2153
http://dx.doi.org/10.1016/B978-0-12-394592-1.00011-9
© 2012 Elsevier Inc.
All rights reserved.

1. INTRODUCTION

Although it is well known that most cells show polarity along an apical/basal (A/B) axis, it has become increasingly evident that cells are also polarized along a second axis perpendicular to the A/B axis. This is easy to visualize for migrating cells as the leading and lagging edges. However, even epithelial cells can show this second axis of polarity, and when all or most of the cells within the tissue coordinate their polarity in one direction, it is referred to as tissue or planar cell polarity (PCP).

Although intensively studied in the fly for many years, PCP was largely ignored by vertebrate biologists. However, over the past 10 years, mutagenesis of vertebrate orthologs of genes necessary for PCP establishment in flies has revealed multiple requirements during normal development and tissue maintenance. In some tissues, the connection between the PCP genes and the actual PCP is clear, while in others, they are a complete mystery.

In this review, we characterize a number of processes that are thought to be regulated by PCP during kidney development and maintenance. We review what is known about the molecular regulators and the cellular processes controlled.

2. PLANAR CELL POLARITY

PCP describes polarity within the plane of a tissue, perpendicular to the apical–basal polarity of the cells. It can manifest itself externally, for example, in the directional alignment of Drosophila wing hairs and cuticular bristles (all point in the same direction), the orientation of mammalian body hair, or the orientation of stereociliary bundles of the inner ear. Less readily identifiable but essential examples are the organization of the ommatidia of the Drosophila eye or the uniform cellular orientation, coordinated directional cell migration, and oriented cell division that drive tissue morphogenesis. It is important to emphasize that mere asymmetry alone is not equivalent to PCP. Individual cells can show asymmetry, but if that asymmetry is not coordinated between all (or most) cells within a tissue, it is not PCP.

2.1. The Fat/Dachsous pathway

Mutagenesis screens in Drosophila have identified a number of genes necessary for establishing PCP in all or most tissues. These genes are broken up into distinct pathways based on where and when they act and also their relationship to one another. One such pathway is referred to as the Fat/Dachsous (Ds) pathway (Thomas & Strutt, 2012). Fat and Dachsous are atypical cadherins that interact heterotypically between adjacent cells (Matakatsu & Blair, 2004). Members of the Fat/Ds group show signs of subcellular localization (Ambegaonkar, Pan, Mani, Feng, & Irvine, 2012; Bosveld et al., 2012; Brittle, Thomas, & Strutt, 2012). The Fat and Dachsous proteins are enriched on one side or the other. Although the subcellular localization is not as strict as is seen for other PCP determinants (see below), its strength is reinforced by Fat/Ds intercellular dimerization as well as processes that regulate the activity of these two proteins (Ambegaonkar et al., 2012; Brittle et al., 2012).

Four-jointed (Fj) is a Golgi-associated kinase that phosphorylates the extracellular domain of Fat and Ds (Simon, Xu, Ishikawa, & Irvine, 2010). Phosphorylation of Fat by Fj increases its affinity for Ds, while phosphorylation of Ds decreases its affinity for Fat (Brittle, Repiso, Casal, Lawrence, & Strutt, 2010; Simon et al., 2010). In many tissues, Fj is expressed in a graded pattern, which results in graded activity of Fat and Ds (Simon, 2004). The net result of the action of Fj and the polarized distribution of Fat and Ds is that, although the proteins may be distributed throughout the cell, the active forms will segregate to opposite sides. In this way, the proteins establish PCP. Although there are some exceptions, neither does the polarization of active Fat and Ds require the activity of the core determinants (see below) nor does the Fat/Ds pathway affect core protein localization in individual cells (it does affect global polarization) (Casal, Lawrence, & Struhl, 2006). Thus, it appears that for the most part the Fat/Ds and core pathways act in parallel to each other rather than in a simple linear pathway (Lawrence, Struhl, & Casal, 2007).

2.2. The core pathway

A second major pathway identified in Drosophila is referred to as the "core" pathway (Gray, Roszko, & Solnica-Krezel, 2011; Maung & Jenny, 2011). It is composed of the seven-pass transmembrane receptor Frizzled (Fz), the cytoplasmic PDZ domain-containing protein Disheveled (Dsh), the atypical cadherin Flamingo/Starry night (Fmi/Stan), the tetramembrane-spanning protein Van gogh/strabismus (Vang/Stbm), the Lim and Pet domain-containing cytoplasmic protein Prickle (Pk), and the ankyrin domain-containing protein Diego (Dgo).

Although PCP affects polarity over an entire tissue, it is established at the level of individual cells. Within a cell, PCP results in the polarized

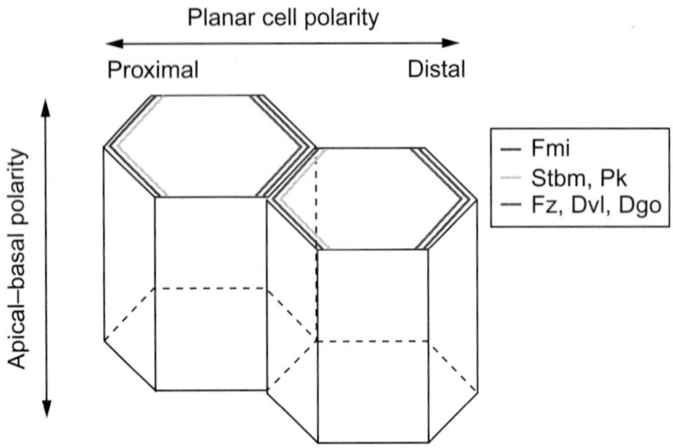

Figure 8.1 Schematic illustration of planar cell polarity (PCP). PCP is perpendicular to the apico-basal polarity of the epithelial tissue. PCP proteins are asymmetrically localized at the cell junctions between neighboring cells. This asymmetry defines the tissue polarity. (See Color Insert.)

localization of proteins, macromolecules, and organelles. Indeed, in some tissues, many of the PCP determinants are themselves localized within the plane of the cell and are required for the subcellular localization of other pathway determinants (Fig. 8.1). For instance, during PCP establishment in the fly wing, Fz and Dsh are localized on the distal side of the cell, while Vang and Pk are localized on the proximal side (Axelrod, 2001; Bastock, Strutt, & Strutt, 2003; Das, Jenny, Klein, Eaton, & Mlodzik, 2004; Jenny, Darken, Wilson, & Mlodzik, 2003; Shimada, Usui, Yanagawa, Takeichi, & Uemura, 2001; Tree, Shulman, et al., 2002). The precise mechanisms regulating the planar-polarized positions of each protein are still not perfectly clear although they appear to be reinforced by positive and negative interactions with other core proteins. It is still not clear if the polarized localization of the core or Fat/Ds determinants is maintained in all vertebrate tissues manifesting PCP. Ultimately, the Fz/PCP and Ft/Ds pathways regulate the polarization of the cytoskeleton (Tree, Ma, & Axelrod, 2002).

2.3. The Wnt pathway

Wnt ligands play a crucial role in establishing PCP in vertebrates. Wnts signal through an ever-expanding group of receptors and coreceptors (Buechling & Boutros, 2011). Nearly all Wnt signaling depends on interaction of the ligand with a Fz receptor. Note that this is the same Fz molecule that is a component of the core PCP group. Dsh also appears to play a dual role as a core PCP determinant and a component of the Wnt pathway. These observations led many to speculate that perhaps Wnts established PCP. However, there is no evidence that Wnt ligands have a role in PCP in flies.

Depending on the receptor complex and intracellular environment within the receiving cell, the Wnt signal is transduced down one of several different pathways (Cadigan & Liu, 2006). If the ligand binds a Fz and a low-density lipoprotein receptor, it activates the "canonical pathway" (Wehrli et al., 2000). Canonical activity results in stabilization of β-catenin and the formation of a β-catenin/Lef/Tcf transcriptional complex. If the Wnt interacts with a Fz and Ror1/2, then a noncanonical/β-catenin-independent cascade is activated (Green, Inoue, & Sternberg, 2008; Mikels & Nusse, 2006; Oishi et al., 2003). The noncanonical transduction cascade is still not well understood, although, in some instances, it may involve activation of Rho GTPases and Jnk (Boutros, Paricio, Strutt, & Mlodzik, 1998; Habas, Dawid, & He, 2003; Kohn & Moon, 2005; Oishi et al., 2003). Ultimately, activation of the noncanonical pathway results in polarization of the cytoskeleton and PCP. The two pathways appear to antagonize each other so that in one particular cell at one particular moment, only one pathway may be active (Green et al., 2008; Liao et al., 2006; Mikels & Nusse, 2006).

Although it would be attractive to think that gradients of Wnt proteins play a role in establishing PCP, there is little evidence that this is the case (Strutt, 2009). In some instances, it appears that Wnts play permissive roles in PCP while in others they are instructive (Gao et al., 2011; Goldstein, Takeshita, Mizumoto, & Sawa, 2006; Gong, Mo, & Fraser, 2004; Green et al., 2008; Heisenberg et al., 2000; Matsui et al., 2005; Prasad & Clark, 2006; Schlesinger, Shelton, Maloof, Meneghini, & Bowerman, 1999; Ulrich et al., 2003; Witze, Litman, Argast, Moon, & Ahn, 2008). How the activation of the Wnt pathway affects planar polarity is still not clear.

2.4. The primary cilium

The primary cilium is a small projection present on the apical side of most cells, which consists of membrane surrounding a cytoskeletal structure known as the axoneme. The axoneme is anchored within the basal bodies that further act as anchoring or nucleation sites for other cytoskeletal components and various cytosolic proteins as well as forming a part of the centrioles during cell division (Pan, Wang, & Snell, 2005). In some tissues, the cilia show planar-polarized localization within the apical membrane

(Bayly & Axelrod, 2011; Ganner et al., 2009; Guirao et al., 2010; Park, Mitchell, Abitua, Kintner, & Wallingford, 2008).

Several recent studies have suggested that the cilium regulates Wnt pathway usage and PCP. This is an extremely complicated issue that cannot be dealt with appropriately here (for a more detailed discussion, see Wallingford & Mitchell, 2011). However, we briefly discuss this issue here, as it is an important topic.

Mice carrying mutations in genes that are necessary for ciliogenesis show pathologies that have been linked to PCP defects including cystic kidney tubules (Hildebrandt, Attanasio, & Otto, 2009; Lin et al., 2003; Patel, Chowdhury, & Igarashi, 2009; Sattar & Gleeson, 2011; Sharma, Berbari, & Yoder, 2008; Waters & Beales, 2011). Although there is strong evidence that PCP signaling is necessary for the planar-polarized location of the cilia and the polarized beating of the cilia in some tissues, neither of these situations appears to be relevant to the mammalian kidney. Kidney cilia are nonmotile and do not show obvious planar polarization.

Ciliary signaling has been linked to oriented cell divisions, a form of PCP, and activation of the noncanonical Wnt pathway. However, at this point, the mechanisms connecting the cilia, Wnt signaling, and PCP signaling (if there is such a connection) are still unclear. It is important to point out that the mere presence of a defect in PCP in a mutant background does not necessarily indicate that the mutated gene or structure plays a role in PCP. It may simply be necessary for the cell type-specific effect of PCP. Further, characterization will be required to gain a greater understanding of the cilia to Wnt signaling and PCP.

2.5. Junctional remodeling

One of the processes affected by PCP are directed cell movements that occur during morphogenesis (Aigouy et al., 2010; Carreira-Barbosa et al., 2009; Phillips, Murdoch, Chaudhry, Copp, & Henderson, 2005; Shnitsar & Borchers, 2008; Skoglund & Keller, 2010; Takeuchi et al., 2003; Ulrich et al., 2003; Veeman, Slusarski, Kaykas, Louie, & Moon, 2003; Wallingford & Harland, 2002). Although such movements take place during axis elongation in Drosophila, they do not appear to be dependent on any of the "canonical" PCP determinants discussed above. Instead, these cell movements appear to be regulated by the spatial expression of factors along the anterior/posterior axis of the embryo (Blankenship, Backovic, Sanny, Weitz, & Zallen, 2006; Zallen & Blankenship, 2008).

A/P patterning regulates cell adhesion and junctional remodeling that leads to changes in cell geometry, resulting in the formation of multicellular rosette structures. The rosettes resolve themselves along the A/P axis, resulting in tissue thinning and elongation, a process that can broadly be described as convergent extension (CE) (Blankenship et al., 2006).

A distinct set of factors that are localized to the adherens junctions such as the Par3, Scribbled, and nonmuscle myosin homologs are required for CE during Drosophila germband extension (Blankenship et al., 2006; Djiane, Yogev, & Mlodzik, 2005; Gray et al., 2011; Montcouquiol et al., 2003; Zallen & Blankenship, 2008). Interestingly, although these factors show planar-polarized expression within cells similar to what is observed for the core PCP proteins, mutations in Fz and Dsh do not affect these morphogenetic movements. These studies emphasize that not all PCP is equivalent. Multiple independent pathways most likely regulate PCP in tissue-specific manners. Therefore, defects in seemingly unrelated cellular processes can lead to similar effects on the entire tissue. This is an important point, especially when one begins to consider all the pathologies that have been attributed to PCP defects in the kidney.

3. KIDNEY DEVELOPMENT

The formation of the mammalian permanent kidney, or metanephros, is initiated when an epithelial bud (the ureteric bud) forms from the Wolffian duct and invades the surrounding mesenchyme (known as the metanephric mesenchyme, MM). The MM consists of at least two cell types. Cells that lie closest to and surround the ureteric bud are collectively referred to as the cap mesenchyme. A second population of cells lies distal to the ureteric bud and adjacent to the cap mesenchyme and is referred to as the cortical interstitium. Following its invasion into the mesenchyme, the ureteric bud epithelium undergoes branching morphogenesis forming a highly complex, tree-like system of tubes that make up the renal collecting duct system of the mature kidney.

The cap mesenchyme contains a self-renewing, multipotent progenitor cell population that proliferates with the branching ureteric buds so that each new bud tip is covered by cap mesenchyme (Boyle et al., 2008; Kobayashi et al., 2008). After each round of branching, a subpopulation of the cap mesenchyme will undergo a mesenchymal-to-epithelial transition just below the ureteric bud tips forming a pretubular aggregate and then a primitive epithelial structure known as the renal vesicle. The renal vesicle will undergo extensive morphogenesis to form the nonvascular portion of the nephron.

The cortical interstitium also consists of a stem/progenitor cell that gives rise to renal interstitial cells including the pericytes, smooth muscle, and the mesangial cells within and about the renal corpuscles (Humphreys et al., 2010). Signals provided by both the cap mesenchyme and the interstitium regulate the branching morphogenesis of the ureteric bud and potentially the patterning of the nephron and collecting ducts (Costantini & Kopan, 2010; Yang et al., 2002).

The epithelial components of the kidney undergo extensive growth during the embryonic period and continue postnatally. Morphogenesis establishes the ultimate diameter, length, and shape of the kidney tubules, which are absolutely essential for normal organ function. The cellular and molecular processes regulating this morphogenesis are still poorly understood although studies over the past several years have provided some insights.

Karner et al. (2009) found that, during the embryonic period, orientation of the mitotic spindle of most of the epithelial cells appears to be random relative to the proximal–distal axis of the tubule. Without some sort of compensatory mechanism, one would then predict that tubule diameter (or at least the number of cells within the cross-sectional circumference of the tubule) would increase during the embryonic period. It does not. In fact, it was found that the number of cells within the circumferential wall decreased as the tubule elongated (Karner et al., 2009). How can this be? One possibility is that a significant numbers of epithelial cells are culled by apoptosis. However, several groups have reported that the rate of epithelial cell apoptosis is negligible during the embryonic period. A second possibility is that some sort of cellular rearrangement is occurring that causes the tubules to thin and contributes to their elongation. Indeed, cellular rearrangements of this type have been observed during gastrulation and neurulation in multiple species (Axelrod & McNeill, 2002; Concha & Adams, 1998; Darken et al., 2002; Goto & Keller, 2002; Heisenberg et al., 2000; Honda, Nagai, & Tanemura, 2008; Nikolaidou & Barrett, 2005; Torban, Kor, & Gros, 2004; Wallingford, Fraser, & Harland, 2002; Wallingford et al., 2000; Wang et al., 2006; Winklbauer, 2009). Although the mechanism regulating tissue elongation and thinning varies, the phenotypic outcome is essentially the same and these processes are collectively referred to as CE.

Several distinct types of cell behavior have now been described that lead to CE movements including mediolateral intercalation and multicellular rosette formation and resolution (Fig. 8.2). Each of these processes has distinct, defining characteristics that can indicate whether they are occurring. For example, cells undergoing intercalation frequently show mediolateral

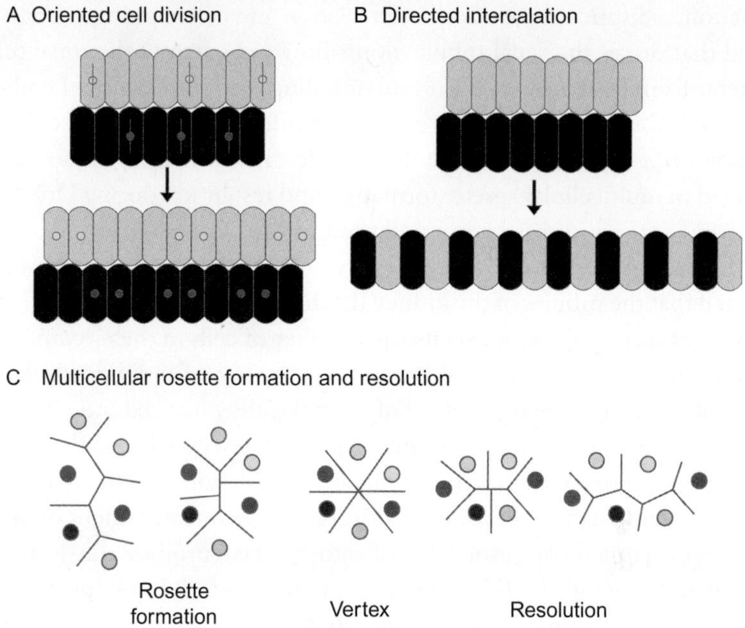

Figure 8.2 Processes regulating tubule elongation. (A) Oriented cell division leads to lengthening of the tubule, while the diameter stays the same. (B) Directed intercalation leads to the tubule elongating while it also becomes thinner. (C) Multicellular rosette formation and resolution (adapted from Zallen & Blankenship, 2008) describe the changes in cell morphology and adhesiveness that result in a transient rosette with a vertex forming. The rosette resolves resulting in the tubule becoming longer and thinner. Only oriented cell division has been demonstrated in the kidney. Panels (B) and (C) require live imaging to confirm. (See Color Insert.)

elongation perpendicular to the axis of extension (a form of PCP) (Bertet, Sulak, & Lecuit, 2004; Nikolaidou & Barrett, 2005; Shih & Keller, 1992; Skoglund & Keller, 2010; Wallingford et al., 2000; Wang et al., 2005; Zallen & Blankenship, 2008). Kidney epithelial cells show such morphology. However, kidney epithelia also form multicellular rosettes (C.M. Karner & T.C. Carroll, unpublished observations). Unfortunately, the issue of which (if any) type of movement is taking place during tubule formation can only be conclusively demonstrated using live imaging, which has not been accomplished in a vertebrate as of yet. However, CE-like movements would explain how a proliferating epithelium with randomly oriented cell division does not increase its diameter.

Cell intercalation/rearrangement appears to be a conserved mechanism in regulating renal tubule size. During Drosophila Malpighian tubule formation, a distinct phase of cell intercalation and cell rearrangement is observed that drives the renal tubule elongation and narrows the renal tubule diameter from 6–10 cells to 2 cells surrounding the lumen (Jung, Denholm, Skaer, & Affolter, 2005 and references therein). Though the molecular machinery is not well characterized, nonmuscle myosin II heavy chain, a gene involved in multicellular rosette formation and resolution during Drosophila germband extension, has been implicated in this process.

After birth and for the first several weeks of life, Fischer et al. (2006) reported that the tubules of the kidney (both the nephrons and the collecting ducts) continue to elongate, while the number of cells in the circumference remains fairly constant. By following the progeny of individually labeled postnatal tubule cells over a period of 5 weeks, they noticed little divergent cell migration or nonclonal cell intercalation. They concluded that, at least in a normal postnatal tubule, CE/cellular intercalation is unlikely to be involved in maintenance of tube diameter. However, measurement of mitotic angles in outer medullary Aqp2+ (collecting ducts), uromodulin+ (ascending limb of loops-of-Henle), or Lotus tetragonolobus (LTL)+ (proximal tubules) tubules identified rigorously oriented cell division (Fig. 8.2) along the proximal/distal axis of tubules such that 95% of the mitotic angles were within 34° and the average is 11°. This finding suggests that oriented cell division drives tubule elongation and diameter establishment during postnatal tubular development. Yu et al. (2009) found a similar process was involved in the growth of the renal papilla. Thus, oriented cell divisions appear to be essential for much of the growth that drives the ultimate shape of the kidney.

Interestingly, Luyten et al. (2010) found that the average number of cells in the circumference of the collecting duct continues to decrease postnatally, suggesting that some intercalation continues after birth. The apparent discrepancy in the involvement of cell intercalation in postnatal collecting ducts in regard with the findings of Fischer et al. and Luyten et al. remains to be resolved. Although the observations of Fischer et al. could be explained if cell intercalation in collecting ducts is clonal, this explanation cannot explain a decrease in circumference. *Pkd1* mutant cells displayed reduced rates in scattered cell migration in the *in vitro* wound healing assay, suggesting that defective cell intercalation may contribute to cyst formation in *Pkd1* mutants and PC1 plays a role in regulating cell intercalation (Luyten et al., 2010). Thus, it appears that the ultimate length and diameter of the kidney tubules depend on a combination of directed cell movements and oriented cell divisions.

4. PCP IN KIDNEY DEVELOPMENT

Many of the processes proposed to regulate kidney development such as directed cell migration during UB branching and potentially nephron formation, CE movements during tubule diameter establishment, and oriented cell division during elongation depend on PCP. Several orthologs of the PCP determinants are expressed in the developing kidney. Characterization of mouse kidneys mutated for these factors has uncovered a number of defects, some expected and some unexpected. In many cases, the defects have not been attributed to any specific deficit in PCP. On the other hand, several mutants have recently been found to have defects in some aspects of PCP. In many of these cases, it is not clear how the genes affected relate to the establishment of PCP. In this section, we describe our current understanding of the role of all types of PCP during kidney development.

4.1. The Fat/Ds group

4.1.1 Branching morphogenesis

There are four mammalian paralogs of *Fat* (*Fat1-4*) and two orthologs of *Ds* (*Dchs1* and *Dchs2*). Although *Fat4* has the highest overall degree of similarity to Drosophila Fat, the other three paralogs show conservation in important functional regions (Matakatsu & Blair, 2012).

Kidneys lacking *Dchs1* show mild defects in early UB branching morphogenesis, resulting in kidneys that are reduced in size (Mao et al., 2011). *Fat4* mutants show a very similar phenotype, suggesting that the interaction between Fat and Dachsous proteins is conserved in mice (Mao et al., 2011). Indeed, compound heterozygotes show phenotypes nearly identical to the individual homozygotes, further supporting an interaction (Mao et al., 2011).

The precise causes of the defects leading to reduced branching in these mutants are unclear. Both show increased rates of apoptosis within the ureteric bud and decreased rates of proliferation within the nephron progenitors and the UB. These sorts of changes on cell number are not typically associated with defects in PCP. Indeed, no specific defects in PCP were identified in mutants during branching morphogenesis although there are clear defects during tubule elongation (see below). Interestingly, *Fat4* and *Dchs1* proteins are present at high levels in the interstitium with lower levels in the nephron progenitors. Levels are very low to undetectable in the ureteric bud (Mao et al., 2011). Thus, it is possible that *Fat4* and *Dchs1* regulate an unknown PCP in the interstitium or nephron progenitors and the branching defects seen in mutants are indirect effects caused by disrupted communication between one of these cell types and the UB. However, definitive conclusions will require analysis of kidneys with tissue-specific ablation of these genes.

4.1.2 Tubule diameter establishment/maintenance

Dchs1 and *Fat4* mutants show dilated distal nephron segments during embryogenesis (Mao et al., 2011). *Fat4* mutants show defects in the orientation of cell division in postnatal collecting ducts (Saburi et al., 2008). Whether this is the cause of tubular dilation is not clear although orientation of cell division has been implicated in maintaining tubule diameter in other mouse models (Fischer et al., 2006).

Although *Fat4* and *Dchs1* mutants both show kidney defects, no defects are observed in kidneys lacking the only known Fj ortholog, *Fjx1* (Saburi, Hester, Goodrich, & McNeill, 2012). Although this finding in itself is not all that surprising (mutation of *Fj* alone has a relatively mild phenotype in the fly wing), *Fjx1* does not seem to genetically interact with *Fat4* in the developing kidney (Saburi et al., 2012). There are a number of possible explanations for this finding. It is possible that there is another functional homolog of *Fj* in the mouse. A second possibility is that *Fat4* and *Dchs1* are not regulating PCP during kidney tubule morphogenesis. (*Fat4* and *Fjx1* do show strong genetic interaction in other tissues.) The final possibility is that the interaction between *Fjx1* and *Fat4* (and *Dchs1*?) is not conserved in mouse kidneys. Consistent with this last possibility, unlike Drosophila, in the mouse kidney, *Fjx1* does not appear to be expressed in the same cells as *Fat4* or *Dchs1* (Carroll & Das, 2011; Mao et al., 2011). Further analysis will need to be performed to determine which, if any, of these scenarios is correct.

4.2. The core pathway

4.2.1 Branching morphogenesis

4.2.1.1 *Vangl2*

Although many of the core determinants are expressed in the developing kidney, there is relatively little known of their function. Yates et al. (2010) recently described the phenotype of the *Loop-tail* (*Lp*) mice that bear a point mutation in the *Vangl2* gene (ortholog of the Drosophila core PCP determinant *Vang/Stbm*) that produces a potential dominant-negative mutation. Reduced ureteric bud branching morphogenesis was observed in mutant kidneys from E13.5. The mutant kidneys are most significantly reduced in size along their anterior–posterior axis.

Neither were significant changes noted in the expression of other core determinants, nor were the rates of proliferation or apoptosis affected. There was also no description of defects in any aspect of PCP in these mutants. During ureteric bud branching morphogenesis, dilated ampullae are resolved into two stalks with narrower diameter. This process may involve PCP-dependent cell rearrangements. Cell movements have been observed in ureteric bud tips during branching morphogenesis. It will be interesting to examine whether directional cell movements occur during branching morphogenesis and, if so, whether Wnt/PCP signaling affects this process cell autonomously in the ureteric bud epithelium. However, for now, the mechanism underlying this phenotype is unclear.

4.2.1.2 *Fz4/8*

Fz4 and *Fz8* are expressed specifically in the ureteric bud epithelium during early kidney development and mice lacking both genes show mildly disrupted branching morphogenesis (Ye, Wang, Rattner, & Nathans, 2011). Fz receptors can act within the core PCP pathway, the canonical Wnt pathway, or the noncanonical Wnt pathway. In *Fz4/8* mutants, it is not clear which aspect of Fz signaling is disrupted. The mutants show some similarities to *Wnt11* mutants including reduced branching and decreased levels of the *Ret* receptor tyrosine kinase in the UB and its ligand *Gdnf* in the mesenchyme (Majumdar, Vainio, Kispert, McMahon, & McMahon, 2003). Indeed, *Wnt11* is able to signal through *Fz4* and *Fz8 in vitro* (Ye et al., 2011). However, *Wnt11* can activate both canonical (Tao et al., 2005) and noncanonical (Garriock, D'Agostino, Pilcher, & Krieg, 2005; Marlow, Topczewski, Sepich, & Solnica-Krezel, 2002; Matsui et al., 2005; Tada & Smith, 2000; Takeuchi et al., 2003; Ye et al., 2011) pathways providing little insight into pathway usage in the kidney.

It is important to note that the *Fz4/8* mutant UBs also show some similarity to kidneys that have had β-catenin removed from the UB (reduction of Ret/Gdnf expression and activity, reduced branching, and reduced proliferation rates) (Marose, Merkel, McMahon, & Carroll, 2008). Although the authors did not examine the expression of β-catenin targets such as *Axin2* in *Fz4/8* mutants, these data may support a canonical role for *Fz4/8* in UB branching. However, at this point, it is impossible to rule out noncanonical or PCP roles for the Frizzleds.

4.2.1.3 Inversin

Inversin, encoded by the *Nphp2* gene, was identified by positional cloning of the Inv mouse that was generated in a mouse mutagenesis screen (Morgan et al., 1998; Otto et al., 2003; Simons et al., 2005). Along with defects in left/right asymmetry, Inv mutants formed cystic kidneys. Sequencing of the Nphp2 gene revealed that it shared high homology with fly Dgo. Indeed, Invs and Dgo can substitute for each other functionally (Simons et al., 2005).

Inversin physically and functionally interacts with Dvl (Simons et al., 2005). *Inversin* has been proposed as a Wnt pathway switch, supporting noncanonical signaling while repressing canonical signaling. Whether this function is active during kidney development and contributes to the defects seen in mutant kidneys is still under debate. Analysis of the expression of a Lef/Tcf/β-catenin reporter in *Inv* mutants did not show any changes (Sugiyama, Tsukiyama, Yamaguchi, & Yokoyama, 2011). However, whether these reporters truly report all Wnt activity is still not clear. A more detailed analysis of cellular phenotypes or genetic studies would certainly shed light onto this question.

4.2.2 Tubule elongation

4.2.2.1 *Vangl2*

A possible direct role for PCP signaling in renal tubule morphogenesis comes from examination of the *Lp* mice. Renal medulla formation is defective in both heterozygous and homozygous mutants (Yates et al., 2010; Q. Ren and J. Yu unpublished observations).

In the developing renal medulla, *Vangl2* is expressed at modest levels in both the ureteric bud epithelium and the interstitium and weakly in the loops-of-Henle (J. Yu, unpublished observations). The nature of the *Lp* mutation means that Vangl2 activity is compromised in all these cell types in *Lp* mutants. Thus, it is still unknown whether Wnt/PCP signaling is required cell autonomously within the renal tubules or within the interstitium. This question of where the PCP pathway is active and functioning needs to be addressed with cell type-specific ablation of *Vangl2* or other Wnt/PCP signaling components. Interestingly, contrary to what is observed in flies, Vangl2 protein does not show obvious planar-polarized localization in any cell types in the kidney (T. J. Carroll & J. Yu, unpublished observations).

Interestingly, the presence of a single *Lp* allele on the *Fat4* homozygous mutant background leads to an enhancement of tubular dilation (Saburi et al., 2012). Interpretation of this result is difficult. The simplest explanation is that both genes are regulating PCP but they function in separate, parallel pathways. Although reduction of Vangl2 activity alone does not lead to significant tubular dilation or cyst formation, its function is accentuated in the context of *Fat4* loss. Once again, it will be interesting to assess PCP including cellular orientation in both individual and compound mutants as well as assessing the specific cell types in which these two gene products function.

4.2.2.2 *Fz3*

Fz3 has been implicated in PCP in other mouse organ systems. *Fz3* expression levels were increased in cysts from human ADPKD kidneys as well as *Pkd1* mutant kidneys, suggesting it may be involved in cyst progression (Luyten et al., 2010). However, no single or compound mutation of Frizzleds has been reported to result in cyst formation. Further, Fz3 levels were unchanged in precystic *Pkd1* mutant tubules (Luyten et al., 2010), suggesting that, though it may be linked to cyst expansion, it is unlikely to cause the disease.

4.2.3 Glomerulogenesis

4.2.3.1 *Vangl2*

Planar polarity phenomenon has not been defined in renal corpuscles, but it is reported that its development was disrupted in *Lp* homozygotes by the criteria of the number of capillary lumens per glomerular tuft and glomerular tuft diameter at E18.5 (Yates et al., 2010). Only half of the *Lp* homozygous mutant renal corpuscles contain two capillary lumens, and a small fraction of glomeruli contain distorted capillary tufts, which are never observed in wild-type kidneys. The crenellated pattern of actin filaments in podocytes is less prominent in Lp mutants (Yates et al., 2010). Knockdown of *Vangl2* in podocyte cells also leads to decreased number of cell projections, stress fibers and cell motility (Babayeva et al., 2011). This suggests that tissue polarity may be at play in some aspects of renal corpuscle maturation.

4.3. The Wnt pathway

4.3.1 Wnt7b

Wnt7b is expressed in the nonbranching/stalk region of the collecting ducts during the embryonic period (Yu et al., 2009). Ablation of *Wnt7b* results in a failure to form a renal medulla (Yu et al., 2009). Prospective medullary collecting ducts fail to elongate and instead become dilated. Orientation of cell division in this segment of a wild-type ureteric epithelium is biased toward longitudinal axis of the ducts, though the trend is not as tight as in postnatal collecting ducts. In mutants, the orientation of cell division tends to be perpendicular to the longitudinal axis of the duct (Fig. 8.3). This suggests that *Wnt7b* regulates oriented cell division and that this process is involved in medullary collecting duct elongation and renal medulla formation.

As mentioned, oriented cell division depends on PCP. *Wnt7b* has been shown to be capable of activating both canonical and noncanonical signaling in different cellular contexts (Tu et al., 2007). During kidney development, *Wnt7b* activates a β-catenin-dependent transcriptional program within the medullary interstitium. Interestingly, Wnt7b/β-catenin targets include

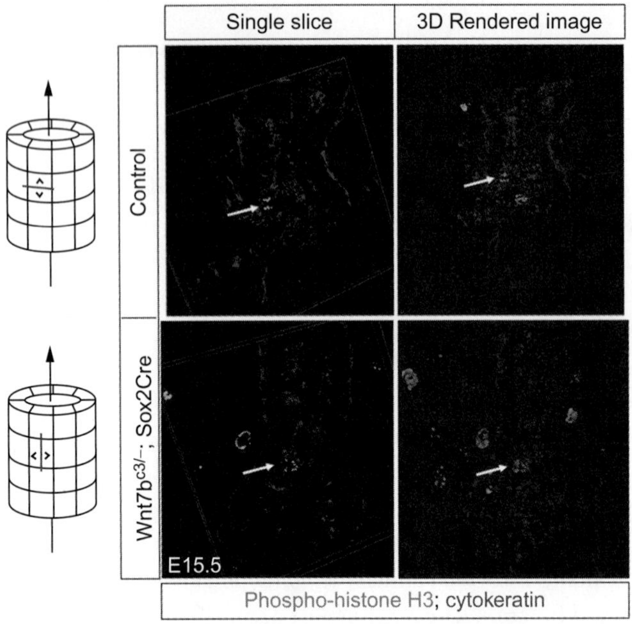

Figure 8.3 Orientation of cell division in the prospective medullary collecting duct cells is disrupted in Wnt7b mutants. The cell division plane (red line in the schematic drawing) tends to bias toward perpendicular to the longitudinal axis of the prospective medullary collecting ducts (the arrow in the schematic drawing) in wild-type kidney, which favors elongation of the collecting ducts. In contrast, the plane of cell division tends to be parallel to the longitudinal axis of the prospective medullary collecting ducts in Wnt7b mutant kidneys, which favors an increase in the diameter of collecting ducts. (For interpretation of the references to color in this figure legend, the reader is referred to the online version of this chapter.)

Wnt11, *Wnt4*, and to a lesser extent, *Wnt5a* mRNAs (Yu et al., 2009). All three of these Wnts are thought to primarily signal through the noncanonical pathway. Thus, it appears that the regulation of PCP by *Wnt7b* within the context of the kidney collecting ducts may be indirect. A reasonable model based on the data is that *Wnt7b*, signaling through β-catenin, activates the expression of noncanonical ligands within the interstitium that signal back to the collecting ducts to regulate PCP and oriented cell division, thus driving medullary collecting duct extension.

4.3.2 *Wnt9b*

Like *Wnt7b*, *Wnt9b* is also expressed in the embryonic collecting ducts although Wnt9b expression extends more toward the branching tips. Mice carrying a germline null mutation of *Wnt9b* do not form renal vesicles or maintain their cap mesenchyme (Carroll, Park, Hayashi, Majumdar, &

McMahon, 2005). However, mice that have had *Wnt9b* deleted in the nonbranching/stalk portion of the collecting ducts or mice carrying a hypomorphic allele of *Wnt9b* develop cysts (Karner et al., 2009). Interestingly, even though *Wnt9b* is expressed in the collecting ducts, cysts form in all nephron segments. In fact, cysts are first apparent in the proximal tubules.

From the earliest stages of development examined (E13.5 onward), *Wnt9b* mutants show a significant increase in the number of cells making up the cross-sectional circumference of the collecting ducts and proximal tubules compared to wild-type littermates (Karner et al., 2009). As mentioned, at these early stages, cell division is random and *Wnt9b* mutants have no significant change in the orientation of their mitotic spindles, suggesting that the cause of the tubular dilation in these mutants is defective cell movement/CE. Indeed, although the epithelial cells of mutants are elongated similar to what is seen in wild-type tubules, their orientation relative to the proximal–distal axis becomes randomized (Karner et al., 2009) (Fig. 8.4). This sort of defect in cell orientation would be predicted to lead to defects in mediolateral intercalation (if such a process is occurring), which could contribute to tubule dilation.

Around the time of birth, cell division becomes tightly oriented throughout the collecting duct system (excluding the still branching UB tips). Interestingly, when noncystic collecting ducts were examined in postnatal *Wnt9b* mutants, it was found that the orientation of cell division was randomized (Karner et al., 2009). If the orientation of cell division is controlled in part by cellular orientation, this makes sense and the same cellular defect could be affecting both CE movements and oriented cell division in *Wnt9b* mutants. However, it is possible that *Wnt9b* regulates multiple aspects of PCP that regulate distinct cellular processes.

Although it has been shown that *Wnt9b* signals through β-catenin to the mesenchyme during the initial stages of mesenchymal-to-epithelial transition (Carroll et al., 2005; Karner et al., 2011), the target cell and signal transduction cascade used during tubule morphogenesis are still not clear. *Wnt9b* could be directly affecting PCP within the epithelia or it could be acting through the interstitium as Wnt7b does. Although Karner et al. (2009) found no defects in the expression of β-catenin targets in *Wnt9b* mutants, most of these are expressed in the epithelia. As the target cell of *Wnt9b* is unknown, the utility of epithelial targets is unknown. The authors did find decreased levels of GTP-bound (active) Rho and phosphorylated Jnk (targets of noncanonical Wnt signaling) in mutants, although these effects could also be indirect. Clearly, more work needs to be done in this field.

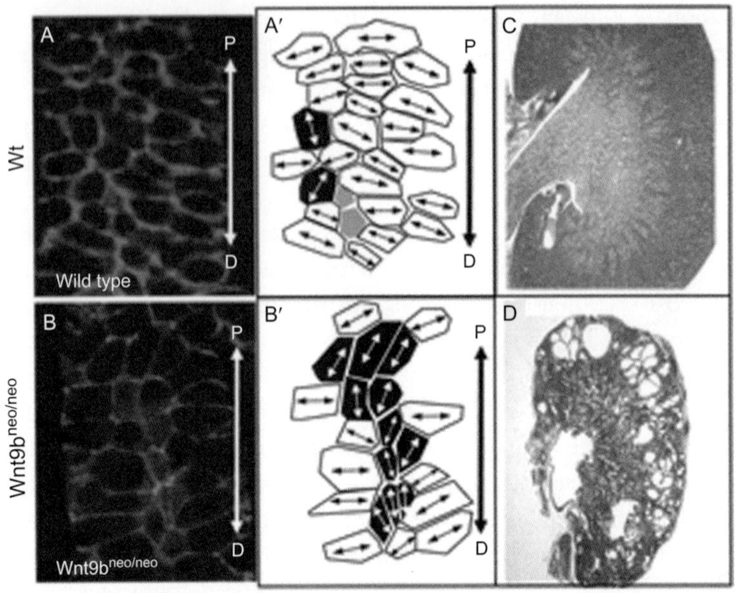

Figure 8.4 Wnt9b mutants show defects in PCP. Wild-type (A and C) and Wnt9b$^{neo/neo}$ kidneys from E15.5 (A and B) and P30 kidneys (C and D). (A and B) Outlines of cells viewed from the apical side. The majority of wild-type cells are elongated, and their long axis lies between 45° and 90° relative to the proximal/distal axis of the tubule. This orientation becomes randomized in Wnt9b mutants. (A′ and B′) Outlines of the cells from (A) and (B) with their long axes indicated by arrows. Cells lying between 0° and 45° are black, while 45–90° are in white. (C and D) H and E sections of postnatal day 30 kidneys. Note that Wnt9b mutants are cystic. (For color version of this figure, the reader is referred to the online version of this chapter.)

4.3.3 Wnt11

Wnt11 is traditionally thought of as a "noncanonical" Wnt although it clearly can activate the canonical pathway in some contexts (indeed, *Wnt11* seems to be "the canonical Wnt" during Xenopus axis induction) (Tao et al., 2005). Although *Wnt11* is expressed at the tips of the ureteric bud and within the medullary interstitium, it is not capable of inducing tubule formation in isolated mesenchyme, a process thought to be primarily mediated by β-catenin (Kispert, Vainio, & McMahon, 1998). This could indicate that *Wnt11* signals noncanonically in the kidney or it does not signal to the cap mesenchyme. In cell culture, *Wnt11* can signal through *Fz4/8* receptors, which are normally expressed in the ureteric bud epithelium, and activate both canonical and noncanonical Wnt pathways (Ye et al., 2011), suggesting that *Wnt11* acts cell autonomously on the ureteric bud epithelium. It remains to be determined which Wnt pathway *Wnt11* activates in the UB cells.

Wnt11 mutants have mild branching defects similar to what is observed in *Fat4*, *Dachs1*, and *Vangl2* mutants. Thus, it is tempting to speculate that *Wnt11* works in the same pathway as one or more of these factors to regulate PCP. However, it is not clear how similar the defects viewed in these mutants are. Further, it is not clear how noncanonical Wnt signaling relates to the Fat/Ds or core pathways. In fact, the three pathways may function in parallel. However, Gao et al. recently showed that *Wnt5a*, signaling through the noncanonical receptor *Ror2*, mediated the phosphorylation and activation of Vangl2, thus demonstrating a direct connection between noncanonical Wnt and "core" PCP signaling (Gao et al., 2011). The authors suggested that the Vangl2 activity gradient was the result of a Wnt5a gradient and this process contributed to PCP. *Ror2* is expressed at high levels in the developing kidney although kidney defects have not been described in *Ror2* mutants (Al-Shawi, Ashton, Underwood, & Simons, 2001; Matsuda et al., 2001). It would be interesting to determine if a Wnt/Ror2/Vangl2 pathway is active in the kidney.

4.4. The primary cilium

Polycystic kidney disease (PKD) is a common genetic disorder characterized by cyst formation and overgrowth of the kidneys. Mutations in numerous genes can cause PKD. One commonality that connects several of the gene products is that they appear to be necessary for the formation and/or function of the primary cilia. This has led to PKD frequently being referred to as a ciliopathy (Hildebrandt et al., 2009; Patel et al., 2009; Sattar & Gleeson, 2011; Sharma et al., 2008; Waters & Beales, 2011).

Over the past several years, several reports have indicated that defects in PCP may contribute to PKD. Specifically, randomized orientation of cell division was observed prior to overt tubular dilation in ciliary mutants (Fischer et al., 2006). These data strongly suggest that the orientation in which a cell divides contributes to the formation of a cyst. However, misoriented cell division alone cannot be causal as most of the cell division that occurs in the embryo is randomly oriented and genetic lesions have been identified that lead to randomized cell division in tubules without cyst formation (Karner et al., 2009; Nishio et al., 2010). Indeed, if mechanisms for regulated cell movement and oriented cell division are both present in a growing tubule, it would seem that defects in both processes would have to occur to increase diameter.

As mentioned above, there is some evidence that the cilia can control PCP perhaps through regulating Wnt pathway usage. Thus, a simple model would be that loss of the cilia or ciliary signaling contributes to PKD in part by disrupting PCP, which then leads to defects in oriented cell division and CE/directed cell movements. However, other than analyzing the orientation of cell division and Wnt pathway activation, no other examination of PCP has been reported in ciliary mutants. It will be of great interest to see how ciliary mutants regulate cell elongation and orientation and otherwise compare to mice lacking PCP determinants. It will also be interesting to determine how defects in the core and Fat/Ds pathways affect ciliary signaling.

5. PCP AND TUBULAR REPAIR

Much of the discussion to this point has concerned the role of PCP in kidney development. A question arises as to the relevance of PCP to kidney diseases, such as autosomal dominant PKD, that do not appear to be congenital disorders. Recent studies suggest that tubular injury may be required for cysts to form (which may explain the high level of pathological heterogeneity in family members carrying the same mutation) (Happe et al., 2009; Patel et al., 2008; Takakura et al., 2009; Weimbs, 2006). After a cell dies within an epithelium, it must quickly be replaced without compromising the epithelial barrier function. This appears to be taken care of by a combination of oriented cell division and directed cell movement and/or junctional remodeling. How do cells know in which direction to move or divide in response to injury? In flies, members of both the core and the Fat/Ds PCP pathways are required (Li, Han, & Xi, 2010; Li, Zepeda-Orozco, et al., 2009). Growing evidence suggests that these processes may also be involved in proper response to injury in the kidney.

Patel et al. (2008) showed that injury of adult kidneys lacking cilia resulted in cyst formation. Further, they showed that, although cell division was oriented in wild-type tubules after injury, it was randomized in the ciliary mutants prior to overt signs of dilation or cystogenesis. Similar results have been reported in other mutant backgrounds (Bonnet et al., 2009; Fischer et al., 2006; Happe et al., 2009; Li, Kale, & Baker, 2009; Saburi et al., 2008; Simons & Walz, 2006; Sugiyama et al., 2011). However, Nishio et al. (2010) have found that defects in oriented cell division are neither necessary nor sufficient to cause cysts supporting the model that defective cell movement may also be involved.

Are the known PCP determinants involved in recovery from kidney injury? Li, Kale, et al. (2009) found that subcellular localization and levels of expression of Fz3, a Fz shown to mediate Wnt/PCP signaling, are altered in tubules with misoriented cell division following urinary tract obstruction, supporting an involvement of this pathway in epithelial repair. However, conclusive evidence of the role of PCP in repair will require assaying injured kidneys that have had PCP determinants mutated after development is complete.

6. CONCLUSION

PCP is currently a hot topic in the field of kidney research. However, as frequently happens in hot fields, some of the less exciting, descriptive work is forsaken in the attempt to be the first to publish. There are still numerous, basic questions about PCP in the kidney that have not been (but in our opinion should be) answered. Although there is increasing evidence that PCP plays an essential role in kidney development and disease, we still do not know what cellular processes it is controlling. Is it controlling cell elongation, orientation, movement, division, or all of the above? Are different PCP pathways affecting distinct cellular processes? Where does the signal that establishes the direction of PCP come from? When is it established? Does it depend on A/P patterning of the tubules? Given the expression of numerous PCP determinants in the interstitial fibroblasts, what is the role of this cell type in PCP? How do all the factors that affect PCP (especially, the cystogenic factors) relate to one another and the cellular processes regulated during PCP? Polarized, subcellular localization of many PCP determinants appears to be essential for their function and for PCP in flies and other vertebrate organ systems, but such a localization pattern has so far not been observed for any PCP proteins examined in renal tubules (Luyten et al., 2010; T. J. Carroll & J. Yu, unpublished observations). Is this aspect of PCP conserved, and if not, how is PCP established in the kidney? We speak of CE or directed cell movements regulating diameter, but there is actually no evidence that these movements occur and, if they do, what they look like. Although the process of directed intercalation/CE is perfectly plausible for a flat tissue, it does not work in a tubule as intercalation of cells toward the top of a tubule would lead to convergence and extension on the top side but divergence and retraction on the bottom. Clearly, a more complicated process must be involved.

There are still many questions to be answered in this exciting, growing field. Kidney researchers can take some solace in the fact that fly geneticists have been studying this question for decades and there are still many unanswered questions. Over the next several years, it is certain that the number of processes regulated by PCP will continue to grow and this area of investigation will expand our knowledge of kidney development and maintenance.

REFERENCES

Aigouy, B., Farhadifar, R., Staple, D. B., Sagner, A., Roper, J. C., Julicher, F., et al. (2010). Cell flow reorients the axis of planar polarity in the wing epithelium of Drosophila. *Cell, 142*, 773–786.

Al-Shawi, R., Ashton, S. V., Underwood, C., & Simons, J. P. (2001). Expression of the Ror1 and Ror2 receptor tyrosine kinase genes during mouse development. *Development Genes and Evolution, 211*, 161–171.

Ambegaonkar, A. A., Pan, G., Mani, M., Feng, Y., & Irvine, K. D. (2012). Propagation of dachsous-fat planar cell polarity. *Current Biology, 22*, 1302–1308.

Axelrod, J. D. (2001). Unipolar membrane association of Dishevelled mediates Frizzled planar cell polarity signaling. *Genes & Development, 15*, 1182–1187.

Axelrod, J. D., & McNeill, H. (2002). Coupling planar cell polarity signaling to morphogenesis. *The Scientific World Journal, 2*, 434–454.

Babayeva, S., Zilber, Y., & Torban, E. (2011). Planar cell polarity pathway regulates actin rearrangement, cell shape, motility and nephrin distribution in podocytes. *Am. J. Physiol. Renal Physiol, 300*, F549–560.

Bastock, R., Strutt, H., & Strutt, D. (2003). Strabismus is asymmetrically localised and binds to Prickle and Dishevelled during Drosophila planar polarity patterning. *Development, 130*, 3007–3014.

Bayly, R., & Axelrod, J. D. (2011). Pointing in the right direction: New developments in the field of planar cell polarity. *Nature Reviews. Genetics, 12*, 385–391.

Bertet, C., Sulak, L., & Lecuit, T. (2004). Myosin-dependent junction remodelling controls planar cell intercalation and axis elongation. *Nature, 429*, 667–671.

Blankenship, J. T., Backovic, S. T., Sanny, J. S., Weitz, O., & Zallen, J. A. (2006). Multicellular rosette formation links planar cell polarity to tissue morphogenesis. *Developmental Cell, 11*, 459–470.

Bonnet, C. S., Aldred, M., von Ruhland, C., Harris, R., Sandford, R., & Cheadle, J. P. (2009). Defects in cell polarity underlie TSC and ADPKD-associated cystogenesis. *Human Molecular Genetics, 18*, 2166–2176.

Bosveld, F., Bonnet, I., Guirao, B., Tlili, S., Wang, Z., Petitalot, A., et al. (2012). Mechanical control of morphogenesis by fat/dachsous/four-jointed planar cell polarity pathway. *Science, 336*, 724–727.

Boutros, M., Paricio, N., Strutt, D. I., & Mlodzik, M. (1998). Dishevelled activates JNK and discriminates between JNK pathways in planar polarity and wingless signaling. *Cell, 94*, 109–118.

Boyle, S., Misfeldt, A., Chandler, K. J., Deal, K. K., Southard-Smith, E. M., Mortlock, D. P., et al. (2008). Fate mapping using Cited1-CreERT2 mice demonstrates that the cap mesenchyme contains self-renewing progenitor cells and gives rise exclusively to nephronic epithelia. *Developmental Biology, 313*, 234–245.

Brittle, A. L., Repiso, A., Casal, J., Lawrence, P. A., & Strutt, D. (2010). Four-jointed modulates growth and planar polarity by reducing the affinity of dachsous for fat. *Current Biology*, *20*, 803–810.

Brittle, A., Thomas, C., & Strutt, D. (2012). Planar polarity specification through asymmetric subcellular localization of Fat and Dachsous. *Current Biology*, *22*, 907–914.

Buechling, T., & Boutros, M. (2011). Wnt signaling signaling at and above the receptor level. *Current Topics in Developmental Biology*, *97*, 21–53.

Cadigan, K. M., & Liu, Y. I. (2006). Wnt signaling: Complexity at the surface. *Journal of Cell Science*, *119*, 395–402.

Carreira-Barbosa, F., Kajita, M., Morel, V., Wada, H., Okamoto, H., Martinez Arias, A., et al. (2009). Flamingo regulates epiboly and convergence/extension movements through cell cohesive and signalling functions during zebrafish gastrulation. *Development*, *136*, 383–392.

Carroll, T. J., & Das, A. (2011). Planar cell polarity in kidney development and disease. *Organogenesis*, *7*, 180–190.

Carroll, T. J., Park, J. S., Hayashi, S., Majumdar, A., & McMahon, A. P. (2005). Wnt9b plays a central role in the regulation of mesenchymal to epithelial transitions underlying organogenesis of the mammalian urogenital system. *Developmental Cell*, *9*, 283–292.

Casal, J., Lawrence, P. A., & Struhl, G. (2006). Two separate molecular systems, Dachsous/Fat and Starry night/Frizzled, act independently to confer planar cell polarity. *Development*, *133*, 4561–4572.

Concha, M. L., & Adams, R. J. (1998). Oriented cell divisions and cellular morphogenesis in the zebrafish gastrula and neurula: A time-lapse analysis. *Development*, *125*, 983–994.

Costantini, F., & Kopan, R. (2010). Patterning a complex organ: Branching morphogenesis and nephron segmentation in kidney development. *Developmental Cell*, *18*, 698–712.

Darken, R. S., Scola, A. M., Rakeman, A. S., Das, G., Mlodzik, M., & Wilson, P. A. (2002). The planar polarity gene strabismus regulates convergent extension movements in Xenopus. *The EMBO Journal*, *21*, 976–985.

Das, G., Jenny, A., Klein, T. J., Eaton, S., & Mlodzik, M. (2004). Diego interacts with Prickle and Strabismus/Van Gogh to localize planar cell polarity complexes. *Development*, *131*, 4467–4476.

Djiane, A., Yogev, S., & Mlodzik, M. (2005). The apical determinants aPKC and dPatj regulate Frizzled-dependent planar cell polarity in the Drosophila eye. *Cell*, *121*, 621–631.

Fischer, E., Legue, E., Doyen, A., Nato, F., Nicolas, J. F., Torres, V., et al. (2006). Defective planar cell polarity in polycystic kidney disease. *Nature Genetics*, *38*, 21–23.

Ganner, A., Lienkamp, S., Schafer, T., Romaker, D., Wegierski, T., Park, T. J., et al. (2009). Regulation of ciliary polarity by the APC/C. *Proceedings of the National Academy of Sciences of the United States of America*, *106*, 17799–17804.

Gao, B., Song, H., Bishop, K., Elliot, G., Garrett, L., English, M. A., et al. (2011). Wnt signaling gradients establish planar cell polarity by inducing Vangl2 phosphorylation through Ror2. *Developmental Cell*, *20*, 163–176.

Garriock, R. J., D'Agostino, S. L., Pilcher, K. C., & Krieg, P. A. (2005). Wnt11-R, a protein closely related to mammalian Wnt11, is required for heart morphogenesis in Xenopus. *Developmental Biology*, *279*, 179–192.

Goldstein, B., Takeshita, H., Mizumoto, K., & Sawa, H. (2006). Wnt signals can function as positional cues in establishing cell polarity. *Developmental Cell*, *10*, 391–396.

Gong, Y., Mo, C., & Fraser, S. E. (2004). Planar cell polarity signalling controls cell division orientation during zebrafish gastrulation. *Nature*, *430*, 689–693.

Goto, T., & Keller, R. (2002). The planar cell polarity gene strabismus regulates convergence and extension and neural fold closure in Xenopus. *Developmental Biology*, *247*, 165–181.

Gray, R. S., Roszko, I., & Solnica-Krezel, L. (2011). Planar cell polarity: Coordinating morphogenetic cell behaviors with embryonic polarity. *Developmental Cell, 21*, 120–133.

Green, J. L., Inoue, T., & Sternberg, P. W. (2008). Opposing Wnt pathways orient cell polarity during organogenesis. *Cell, 134*, 646–656.

Guirao, B., Meunier, A., Mortaud, S., Aguilar, A., Corsi, J. M., Strehl, L., et al. (2010). Coupling between hydrodynamic forces and planar cell polarity orients mammalian motile cilia. *Nature Cell Biology, 12*, 341–350.

Habas, R., Dawid, I. B., & He, X. (2003). Coactivation of Rac and Rho by Wnt/Frizzled signaling is required for vertebrate gastrulation. *Genes & Development, 17*, 295–309.

Happe, H., Leonhard, W. N., Van der Wal, A., Van de Water, B., Lantinga-van Leeuwen, I. S., Breuning, M. H., et al. (2009). Toxic tubular injury in kidneys from Pkd1-deletion mice accelerates cystogenesis accompanied by dysregulated planar cell polarity and canonical Wnt signaling pathways. *Human Molecular Genetics, 18*, 2532–2542.

Heisenberg, C. P., Tada, M., Rauch, G. J., Saude, L., Concha, M. L., Geisler, R., et al. (2000). Silberblick/Wnt11 mediates convergent extension movements during zebrafish gastrulation. *Nature, 405*, 76–81.

Hildebrandt, F., Attanasio, M., & Otto, E. (2009). Nephronophthisis: Disease mechanisms of a ciliopathy. *Journal of the American Society of Nephrology, 20*, 23–35.

Honda, H., Nagai, T., & Tanemura, M. (2008). Two different mechanisms of planar cell intercalation leading to tissue elongation. *Developmental Dynamics, 237*, 1826–1836.

Humphreys, B. D., Lin, S. L., Kobayashi, A., Hudson, T. E., Nowlin, B. T., Bonventre, J. V., et al. (2010). Fate tracing reveals the pericyte and not epithelial origin of myofibroblasts in kidney fibrosis. *The American Journal of Pathology, 176*, 85–97.

Jenny, A., Darken, R. S., Wilson, P. A., & Mlodzik, M. (2003). Prickle and Strabismus form a functional complex to generate a correct axis during planar cell polarity signaling. *The EMBO Journal, 22*, 4409–4420.

Jung, A. C., Denholm, B., Skaer, H., & Affolter, M. (2005). Renal tubule development in Drosophila: A closer look at the cellular level. *Journal of the American Society of Nephrology, 16*, 322–328.

Karner, C. M., Chirumamilla, R., Aoki, S., Igarashi, P., Wallingford, J. B., & Carroll, T. J. (2009). Wnt9b signaling regulates planar cell polarity and kidney tubule morphogenesis. *Nature Genetics, 41*, 793–799.

Karner, C. M., Das, A., Ma, Z., Self, M., Chen, C., Lum, L., et al. (2011). Canonical Wnt9b signaling balances progenitor cell expansion and differentiation during kidney development. *Development, 138*, 1247–1257.

Kispert, A., Vainio, S., & McMahon, A. P. (1998). Wnt-4 is a mesenchymal signal for epithelial transformation of metanephric mesenchyme in the developing kidney. *Development, 125*, 4225–4234.

Kobayashi, A., Valerius, M. T., Mugford, J. W., Carroll, T. J., Self, M., Oliver, G., et al. (2008). Six2 defines and regulates a multipotent self-renewing nephron progenitor population throughout mammalian kidney development. *Cell Stem Cell, 3*, 169–181.

Kohn, A. D., & Moon, R. T. (2005). Wnt and calcium signaling: Beta-catenin-independent pathways. *Cell Calcium, 38*, 439–446.

Lawrence, P. A., Struhl, G., & Casal, J. (2007). Planar cell polarity: One or two pathways? *Nature Reviews. Genetics, 8*, 555–563.

Li, X., Han, Y., & Xi, R. (2010). Polycomb group genes Psc and Su(z)2 restrict follicle stem cell self-renewal and extrusion by controlling canonical and noncanonical Wnt signaling. *Genes & Development, 24*, 933–946.

Li, W., Kale, A., & Baker, N. E. (2009). Oriented cell division as a response to cell death and cell competition. *Current Biology, 19*, 1821–1826.

Li, L., Zepeda-Orozco, D., Patel, V., Truong, P., Karner, C. M., Carroll, T. J., et al. (2009). Aberrant planar cell polarity induced by urinary tract obstruction. *American Journal of Physiology. Renal Physiology*, *297*, F1526–F1533.

Liao, G., Tao, Q., Kofron, M., Chen, J. S., Schloemer, A., Davis, R. J., et al. (2006). Jun NH2-terminal kinase (JNK) prevents nuclear beta-catenin accumulation and regulates axis formation in Xenopus embryos. *Proceedings of the National Academy of Sciences of the United States of America*, *103*, 16313–16318.

Lin, F., Hiesberger, T., Cordes, K., Sinclair, A. M., Goldstein, L. S., Somlo, S., et al. (2003). Kidney-specific inactivation of the KIF3A subunit of kinesin-II inhibits renal ciliogenesis and produces polycystic kidney disease. *Proceedings of the National Academy of Sciences of the United States of America*, *100*, 5286–5291.

Luyten, A., Su, X., Gondela, S., Chen, Y., Rompani, S., Takakura, A., et al. (2010). Aberrant regulation of planar cell polarity in polycystic kidney disease. *Journal of the American Society of Nephrology*, *21*, 1521–1532.

Majumdar, A., Vainio, S., Kispert, A., McMahon, J., & McMahon, A. P. (2003). Wnt11 and Ret/Gdnf pathways cooperate in regulating ureteric branching during metanephric kidney development. *Development*, *130*, 3175–3185.

Mao, Y., Mulvaney, J., Zakaria, S., Yu, T., Morgan, K. M., Allen, S., et al. (2011). Characterization of a Dchs1 mutant mouse reveals requirements for Dchs1-Fat4 signaling during mammalian development. *Development*, *138*, 947–957.

Marlow, F., Topczewski, J., Sepich, D., & Solnica-Krezel, L. (2002). Zebrafish Rho kinase 2 acts downstream of Wnt11 to mediate cell polarity and effective convergence and extension movements. *Current Biology*, *12*, 876–884.

Marose, T. D., Merkel, C. E., McMahon, A. P., & Carroll, T. J. (2008). Beta-catenin is necessary to keep cells of ureteric bud/Wolffian duct epithelium in a precursor state. *Developmental Biology*, *314*, 112–126.

Matakatsu, H., & Blair, S. S. (2004). Interactions between Fat and Dachsous and the regulation of planar cell polarity in the Drosophila wing. *Development*, *131*, 3785–3794.

Matakatsu, H., & Blair, S. S. (2012). Separating planar cell polarity and Hippo pathway activities of the protocadherins Fat and Dachsous. *Development*, *139*, 1498–1508.

Matsuda, T., Nomi, M., Ikeya, M., Kani, S., Oishi, I., Terashima, T., et al. (2001). Expression of the receptor tyrosine kinase genes, Ror1 and Ror2, during mouse development. *Mechanisms of Development*, *105*, 153–156.

Matsui, T., Raya, A., Kawakami, Y., Callol-Massot, C., Capdevila, J., Rodriguez-Esteban, C., et al. (2005). Noncanonical Wnt signaling regulates midline convergence of organ primordia during zebrafish development. *Genes & Development*, *19*, 164–175.

Maung, S. M., & Jenny, A. (2011). Planar cell polarity in Drosophila. *Organogenesis*, *7*, 165–179.

Mikels, A. J., & Nusse, R. (2006). Purified Wnt5a protein activates or inhibits beta-catenin-TCF signaling depending on receptor context. *PLoS Biology*, *4*, e115.

Montcouquiol, M., Rachel, R. A., Lanford, P. J., Copeland, N. G., Jenkins, N. A., & Kelley, M. W. (2003). Identification of Vangl2 and Scrb1 as planar polarity genes in mammals. *Nature*, *423*, 173–177.

Morgan, D., Turnpenny, L., Goodship, J., Dai, W., Majumder, K., Matthews, L., et al. (1998). Inversin, a novel gene in the vertebrate left-right axis pathway, is partially deleted in the inv mouse. *Nature Genetics*, *20*, 149–156.

Nikolaidou, K. K., & Barrett, K. (2005). Getting to know your neighbours; a new mechanism for cell intercalation. *Trends in Genetics*, *21*, 70–73.

Nishio, S., Tian, X., Gallagher, A. R., Yu, Z., Patel, V., Igarashi, P., et al. (2010). Loss of oriented cell division does not initiate cyst formation. *Journal of the American Society of Nephrology*, *21*, 295–302.

Oishi, I., Suzuki, H., Onishi, N., Takada, R., Kani, S., Ohkawara, B., et al. (2003). The receptor tyrosine kinase Ror2 is involved in non-canonical Wnt5a/JNK signalling pathway. *Genes to Cells*, *8*, 645–654.

Otto, E. A., Schermer, B., Obara, T., O'Toole, J. F., Hiller, K. S., Mueller, A. M., et al. (2003). Mutations in INVS encoding inversin cause nephronophthisis type 2, linking renal cystic disease to the function of primary cilia and left-right axis determination. *Nature Genetics*, *34*, 413–420.

Pan, J., Wang, Q., & Snell, W. J. (2005). Cilium-generated signaling and cilia-related disorders. *Laboratory Investigation*, *85*, 452–463.

Park, T. J., Mitchell, B. J., Abitua, P. B., Kintner, C., & Wallingford, J. B. (2008). Dishevelled controls apical docking and planar polarization of basal bodies in ciliated epithelial cells. *Nature Genetics*, *40*, 871–879.

Patel, V., Chowdhury, R., & Igarashi, P. (2009). Advances in the pathogenesis and treatment of polycystic kidney disease. *Current Opinion in Nephrology and Hypertension*, *18*, 99–106.

Patel, V., Li, L., Cobo-Stark, P., Shao, X., Somlo, S., Lin, F., et al. (2008). Acute kidney injury and aberrant planar cell polarity induce cyst formation in mice lacking renal cilia. *Human Molecular Genetics*, *17*, 1578–1590.

Phillips, H. M., Murdoch, J. N., Chaudhry, B., Copp, A. J., & Henderson, D. J. (2005). Vangl2 acts via RhoA signaling to regulate polarized cell movements during development of the proximal outflow tract. *Circulation Research*, *96*, 292–299.

Prasad, B. C., & Clark, S. G. (2006). Wnt signaling establishes anteroposterior neuronal polarity and requires retromer in C. elegans. *Development*, *133*, 1757–1766.

Saburi, S., Hester, I., Fischer, E., Pontoglio, M., Eremina, V., Gessler, M., et al. (2008). Loss of Fat4 disrupts PCP signaling and oriented cell division and leads to cystic kidney disease. *Nature Genetics*, *40*, 1010–1015.

Saburi, S., Hester, I., Goodrich, L., & McNeill, H. (2012). Functional interactions between Fat family cadherins in tissue morphogenesis and planar polarity. *Development*, *139*, 1806–1820.

Sattar, S., & Gleeson, J. G. (2011). The ciliopathies in neuronal development: A clinical approach to investigation of Joubert syndrome and Joubert syndrome-related disorders. *Developmental Medicine and Child Neurology*, *53*, 793–798.

Schlesinger, A., Shelton, C. A., Maloof, J. N., Meneghini, M., & Bowerman, B. (1999). Wnt pathway components orient a mitotic spindle in the early Caenorhabditis elegans embryo without requiring gene transcription in the responding cell. *Genes & Development*, *13*, 2028–2038.

Sharma, N., Berbari, N. F., & Yoder, B. K. (2008). Ciliary dysfunction in developmental abnormalities and diseases. *Current Topics in Developmental Biology*, *85*, 371–427.

Shih, J., & Keller, R. (1992). Cell motility driving mediolateral intercalation in explants of Xenopus laevis. *Development*, *116*, 901–914.

Shimada, Y., Usui, T., Yanagawa, S., Takeichi, M., & Uemura, T. (2001). Asymmetric colocalization of Flamingo, a seven-pass transmembrane cadherin, and Dishevelled in planar cell polarization. *Current Biology*, *11*, 859–863.

Shnitsar, I., & Borchers, A. (2008). PTK7 recruits dsh to regulate neural crest migration. *Development*, *135*, 4015–4024.

Simon, M. A. (2004). Planar cell polarity in the Drosophila eye is directed by graded Four-jointed and Dachsous expression. *Development*, *131*, 6175–6184.

Simon, M. A., Xu, A., Ishikawa, H. O., & Irvine, K. D. (2010). Modulation of fat: Dachsous binding by the cadherin domain kinase four-jointed. *Current Biology*, *20*, 811–817.

Simons, M., Gloy, J., Ganner, A., Bullerkotte, A., Bashkurov, M., Kronig, C., et al. (2005). Inversin, the gene product mutated in nephronophthisis type II, functions as a molecular switch between Wnt signaling pathways. *Nature Genetics*, *37*, 537–543.

Simons, M., & Walz, G. (2006). Polycystic kidney disease: Cell division without a c(l)ue? *Kidney International*, *70*, 854–864.

Skoglund, P., & Keller, R. (2010). Integration of planar cell polarity and ECM signaling in elongation of the vertebrate body plan. *Current Opinion in Cell Biology*, *22*, 589–596.

Strutt, D. (2009). Gradients and the specification of planar polarity in the insect cuticle. *Cold Spring Harbor Perspectives in Biology*, *1*, a000489.

Sugiyama, N., Tsukiyama, T., Yamaguchi, T. P., & Yokoyama, T. (2011). The canonical Wnt signaling pathway is not involved in renal cyst development in the kidneys of inv mutant mice. *Kidney International*, *79*, 957–965.

Tada, M., & Smith, J. C. (2000). Xwnt11 is a target of Xenopus Brachyury: Regulation of gastrulation movements via Dishevelled, but not through the canonical Wnt pathway. *Development*, *127*, 2227–2238.

Takakura, A., Contrino, L., Zhou, X., Bonventre, J. V., Sun, Y., Humphreys, B. D., et al. (2009). Renal injury is a third hit promoting rapid development of adult polycystic kidney disease. *Human Molecular Genetics*, *18*, 2523–2531.

Takeuchi, M., Nakabayashi, J., Sakaguchi, T., Yamamoto, T. S., Takahashi, H., Takeda, H., et al. (2003). The prickle-related gene in vertebrates is essential for gastrulation cell movements. *Current Biology*, *13*, 674–679.

Tao, Q., Yokota, C., Puck, H., Kofron, M., Birsoy, B., Yan, D., et al. (2005). Maternal wnt11 activates the canonical wnt signaling pathway required for axis formation in Xenopus embryos. *Cell*, *120*, 857–871.

Thomas, C., & Strutt, D. (2012). The roles of the cadherins Fat and Dachsous in planar polarity specification in Drosophila. *Developmental Dynamics*, *241*, 27–39.

Torban, E., Kor, C., & Gros, P. (2004). Van Gogh-like2 (Strabismus) and its role in planar cell polarity and convergent extension in vertebrates. *Trends in Genetics*, *20*, 570–577.

Tree, D. R., Ma, D., & Axelrod, J. D. (2002). A three-tiered mechanism for regulation of planar cell polarity. *Seminars in Cell & Developmental Biology*, *13*, 217–224.

Tree, D. R., Shulman, J. M., Rousset, R., Scott, M. P., Gubb, D., & Axelrod, J. D. (2002). Prickle mediates feedback amplification to generate asymmetric planar cell polarity signaling. *Cell*, *109*, 371–381.

Tu, X., Joeng, K. S., Nakayama, K. I., Nakayama, K., Rajagopal, J., Carroll, T. J., et al. (2007). Noncanonical Wnt signaling through G protein-linked PKCdelta activation promotes bone formation. *Developmental Cell*, *12*, 113–127.

Ulrich, F., Concha, M. L., Heid, P. J., Voss, E., Witzel, S., Roehl, H., et al. (2003). Slb/Wnt11 controls hypoblast cell migration and morphogenesis at the onset of zebrafish gastrulation. *Development*, *130*, 5375–5384.

Veeman, M. T., Slusarski, D. C., Kaykas, A., Louie, S. H., & Moon, R. T. (2003). Zebrafish prickle, a modulator of noncanonical Wnt/Fz signaling, regulates gastrulation movements. *Current Biology*, *13*, 680–685.

Wallingford, J. B., Fraser, S. E., & Harland, R. M. (2002). Convergent extension: The molecular control of polarized cell movement during embryonic development. *Developmental Cell*, *2*, 695–706.

Wallingford, J. B., & Harland, R. M. (2002). Neural tube closure requires Dishevelled-dependent convergent extension of the midline. *Development*, *129*, 5815–5825.

Wallingford, J. B., & Mitchell, B. (2011). Strange as it may seem: The many links between Wnt signaling, planar cell polarity, and cilia. *Genes & Development*, *25*, 201–213.

Wallingford, J. B., Rowning, B. A., Vogeli, K. M., Rothbacher, U., Fraser, S. E., & Harland, R. M. (2000). Dishevelled controls cell polarity during Xenopus gastrulation. *Nature*, *405*, 81–85.

Wang, J., Hamblet, N. S., Mark, S., Dickinson, M. E., Brinkman, B. C., Segil, N., et al. (2006). Dishevelled genes mediate a conserved mammalian PCP pathway to regulate convergent extension during neurulation. *Development*, *133*, 1767–1778.

Wang, J., Mark, S., Zhang, X., Qian, D., Yoo, S. J., Radde-Gallwitz, K., et al. (2005). Regulation of polarized extension and planar cell polarity in the cochlea by the vertebrate PCP pathway. *Nature Genetics*, *37*, 980–985.

Waters, A. M., & Beales, P. L. (2011). Ciliopathies: An expanding disease spectrum. *Pediatric Nephrology*, *26*, 1039–1056.

Wehrli, M., Dougan, S. T., Caldwell, K., O'Keefe, L., Schwartz, S., Vaizel-Ohayon, D., et al. (2000). Arrow encodes an LDL-receptor-related protein essential for Wingless signalling. *Nature*, *407*, 527–530.

Weimbs, T. (2006). Regulation of mTOR by polycystin-1: Is polycystic kidney disease a case of futile repair? *Cell Cycle*, *5*, 2425–2429.

Winklbauer, R. (2009). Cell adhesion in amphibian gastrulation. *International Review of Cell and Molecular Biology*, *278*, 215–275.

Witze, E. S., Litman, E. S., Argast, G. M., Moon, R. T., & Ahn, N. G. (2008). Wnt5a control of cell polarity and directional movement by polarized redistribution of adhesion receptors. *Science*, *320*, 365–369.

Yang, J., Blum, A., Novak, T., Levinson, R., Lai, E., & Barasch, J. (2002). An epithelial precursor is regulated by the ureteric bud and by the renal stroma. *Developmental Biology*, *246*, 296–310.

Yates, L. L., Papakrivopoulou, J., Long, D. A., Goggolidou, P., Connolly, J. O., Woolf, A. S., et al. (2010). The planar cell polarity gene Vangl2 is required for mammalian kidney-branching morphogenesis and glomerular maturation. *Human Molecular Genetics*, *19*, 4663–4676.

Ye, X., Wang, Y., Rattner, A., & Nathans, J. (2011). Genetic mosaic analysis reveals a major role for frizzled 4 and frizzled 8 in controlling ureteric growth in the developing kidney. *Development*, *138*, 1161–1172.

Yu, J., Carroll, T. J., Rajagopal, J., Kobayashi, A., Ren, Q., & McMahon, A. P. (2009). A Wnt7b-dependent pathway regulates the orientation of epithelial cell division and establishes the cortico-medullary axis of the mammalian kidney. *Development*, *136*, 161–171.

Zallen, J. A., & Blankenship, J. T. (2008). Multicellular dynamics during epithelial elongation. *Seminars in Cell & Developmental Biology*, *19*, 263–270.

CHAPTER NINE

Dishevelled: *In Vivo* Roles of a Multifunctional Gene Family During Development

Anthony Wynshaw-Boris[1]
UCSF School of Medicine, San Francisco, California, USA
[1]Corresponding author: e-mail address: wynshawborist@peds.ucsf.edu

Contents

Abstract

Dishevelled proteins are highly conserved throughout evolution and participate in canonical and noncanonical Wnt pathways. In particular, Dvls participate in the planar cell polarity (PCP) pathway that is important for a variety of developmental processes. There are three *Dvl* genes in mammals. These three genes are highly conserved and broadly

Current Topics in Developmental Biology, Volume 101
ISSN 0070-2153
http://dx.doi.org/10.1016/B978-0-12-394592-1.00007-7

© 2012 Elsevier Inc.
All rights reserved.

expressed throughout development and in the adult. Studies of mice with disruption of a single *Dvl* gene have revealed phenotypes that range from defective social behavior to developmental defects that include abnormal heart, skeletal, and cochlear morphogenesis, suggesting that each of the *Dvl* genes had unique functions. However, analysis of double or triple *Dvl* knock-outs revealed novel phenotypes or more severe phenotypes displayed by single knock-outs suggesting overlapping or redundant functions of these *Dvl* genes. Most if not all of the phenotypes displayed by the *Dvl* mutants appear to be the result of PCP pathway functions, not of the canonical Wnt pathway. This suggests that the PCP pathway is sensitive to reduction of *Dvls*, but only one *Dvl* allele of six may be required for Wnt pathway.

The well-conserved canonical and noncanonical Wnt pathways are important for all aspects of mammalian development (Wang, Etheridge, & Wynshaw-Boris, 2006), and various human diseases, including developmental diseases and cancer, are caused by abnormal Wnt signaling. Signals from extracellular Wnt ligands are received by three classes of coreceptors. These signals are interpreted and transduced to the nucleus and/or cytoskeleton via a number of intracellular proteins, including Dishevelleds (Dvls). Depending on whether the pathways result in β-catenin-mediated gene transcription, we can divide the multiple Wnt pathways into canonical (β-catenin-dependent) and noncanonical (β-catenin-independent) Wnt pathways. Activation of the canonical Wnt pathway results in the stabilization of β-catenin to alter transcriptional activity in the nucleus, while activation of the Wnt/planar cell polarity (Wnt/PCP) pathway, a noncanonical Wnt pathway, results in changes in epithelial polarity and tissue reorganization by modulating cytoskeletal organization and adhesion. Understanding how the various Wnt pathways that regulate development are integrated *in vivo* is an important outstanding question. Dvls are outstanding candidates to address this question, as these conserved proteins are required in all eukaryotes for both canonical and noncanonical Wnt pathways.

Over the past 15 years, our laboratory has produced null alleles of each of the three *Dvl* genes in mice, and we have used these mice to uncover partially unique but predominantly redundant functions among the three *Dvl* genes. In support of unique functions for each of the Dvls, single mutants for *Dvl1* display novel social behavior abnormalities (Lijam et al., 1997), while both *Dvl2* (Hamblet et al., 2002) and *Dvl3* (Etheridge et al., 2008) mutants die at birth conotruncal heart defects and display cochlear abnormalities. *Dvl1;Dvl2* (Wang, Hamblet, et al., 2006; Wang et al., 2005) and *Dvl2;Dvl3* double mutants (Etheridge et al., 2008) display severe neural tube defects (craniorachischisis) and severe cochlear defects, while *Dvl1;*

Dvl3 double mutants die during midgestation (Etheridge et al., 2008). Loss of five of six Dvl alleles results in defects in nodal cilia posterior localization and left–right asymmetry defects (Hashimoto et al., 2010). In further support of redundancy, *Dvl1;Dvl2;Dvl3* triple homozygous mutants are unable to undergo gastrulation and do not form mesoderm (unpublished results from our lab and the lab of Hiroshi Hamada). We have begun to dissect the *in vivo* pathways that *Dvls* regulate normal development and are disrupted in the *Dvl* mutants to produce these phenotypes. Taking advantage of studies in *Drosophila* and *Xenopus* that identified domains in Dvl proteins required for either canonical Wnt or noncanonical Wnt/PCP pathway function, we produced *in vivo* alleles in mice that can distinguish these two pathways. We used these alleles to provide definitive evidence that the craniorachischisis phenotype displayed by *Dvl1;Dvl2* double mutants resulted from disruption of convergent extension movements via the Wnt/PCP pathway.

In this review, I highlight these *in vivo* studies of *Dvl* function in mice. I briefly discuss the major Wnt pathways. Then, as most if not all of the identified phenotypes appear to be the result of defective noncanonical Wnt/PCP signaling, and this volume focuses on the PCP pathway, I focus on these functions of *Dvl* in mammals.

1. CANONICAL Wnt PATHWAY

In the absence of Wnt signal, glycogen synthase kinase-3 (Gsk3) is active and phosphorylates β-catenin. This results in the ubiquitination and degradation of β-catenin through the proteosome pathway. Transcription factors of the TCF/LEF family are associated with corepressors in the nucleus to repress transcription of genes containing response elements for these transcription factors. In the presence of Wnt, a secreted protein associated with the extracellular matrix (ECM), the Wnt ligand binds to its receptor Fz and coreceptors Lrp5/6 and activates Dishevelled (Dvl in mice, Dsh in flies). Dvl/Dsh activation leads to the inactivation of Gsk3 as part of a large multiprotein complex including axin, β-catenin and APC, the product of the gene mutated in the human disorder adenomatous polyposis coli. As a result, β-catenin is stabilized and binds to TCF/LEF. β-Catenin binding dismisses the associated corepressors from TCF and LEFs and activates transcription of genes containing response elements for these transcription factors. Thus, the result of Wnt signaling is to convert extracellular signals into changes in transcriptional activity. Many of the murine homologues of this pathway are expressed throughout development, and the canonical Wnt pathway is critical for the development of all organs (reviewed in Wang, Etheridge, et al., 2006). As an example, I will discuss the role of this pathway during gastrulation.

During gastrulation, the one-layered epiblast gives rise to the three germ layers (ectoderm, mesoderm, and endoderm) with defined anterior–posterior (A–P) and dorsal–ventral (D–V) axes. Using a Wnt reporter line where a *LacZ* transgene is placed behind multiple TCF/LEF binding sites and a minimal promoter, Wnt signaling (β-gal) activity was detected shortly before the onset of gastrulation and primitive streak formation at embryonic day 6.0 (E6.0) in an asymmetric pattern of epiblast cells near the embryonic-extraembryonic boundary. As gastrulation proceeds, β-gal staining was observed in the posterior part of the embryo and progresses distally, marking the forming primitive streak (Mohamed, Clarke, & Dufort, 2004).

The canonical Wnt pathway plays a critical role in primitive streak formation, based on the phenotype of several mouse mutants. The primitive streak fails to form in *Lrp5*$^{-/-}$; *Lrp6*$^{-/-}$ double homozygotes, as well as a classical mouse mutant *mesd* (*mesoderm development*) that deletes the gene encoding Mesd, a specific chaperone for Lrp5 and Lrp6 trafficking to the cell membrane from the endoplasmic reticulum (Hsieh et al., 2003; Kelly, Pinson, & Skarnes, 2004). Consequently, gastrulation is completely disrupted and mesoderm and definitive endoderm does not form. A similar defect is also observed in the *Wnt3* mutants (Liu et al., 1999) and *β-catenin* null mutants (Huelsken & Birchmeier, 2001; Huelsken et al., 2000; Morkel et al., 2003), demonstrating that Wnt/β-catenin signaling is an essential driving force of gastrulation in mammals. *β-Catenin* null mutants also display defective A–P axis specification as evidenced by the failure to orient distal visceral endoderm (DVE) anteriorly to form the anterior visceral endoderm (AVE). The A–P patterning defect appears to be unique to *β-catenin* mutants, as the AVE is correctly translocated and specified in *Wnt3*, *Lrp*5/6, and *mesd* mutants despite the absence of primitive streak (Hsieh et al., 2003; Kelly et al., 2004; Liu et al., 1999). The expression of *Cripto*, the gene encoding a coreceptor for nodal signaling that is essential for anterior translocation of the DVE, is missing in the epiblast of *β-catenin* mutants (Ding et al., 1998; Morkel et al., 2003), suggesting that *Cripto* may be a direct target of the LEF/TCF/β-catenin complex (Morkel et al., 2003). The Wnt protein responsible for the induction of *β-catenin* at this early time is unknown. The *β-catenin* knock-out phenotype is more severe than the *Wnt3* knock-out, which does undergo axis rotation, but does not undergo mesoderm induction in the primitive streak. In contrast, in *mesd* mutants, *Cripto* is expressed properly. Mesoderm induction occurs when β-catenin directly induces the transcription of *Brachyury* and other targets. Therefore, *β-catenin* has two distinct effects during gastrulation: an early effect on D–V to A–P axis rotation via *Cripto* and an unknown Wnt and a later effect on mesoderm induction via *Brachyury* and other genes due to *Wnt3* (Morkel et al., 2003). It is important to note that in addition to mediating Wnt signaling, β-catenin also has an essential function in the formation of the cadherins junction (Nelson & Nusse, 2004). Although removal of β-catenin does not appear to alter intercellular adhesion, presumably due to the compensatory substitution by plakoglobin (Huelsken et al., 2000; Lickert et al., 2002), it might interfere with morphogenetic movements that require β-catenin-mediated dynamic regulation of cadherins junction formation and may play some role in PCP signaling.

2. PCP PATHWAY, A NONCANONICAL Wnt PATHWAY

Exposure to certain Wnts (such as Wnt5a) did not induce transformation of certain cultured mammalian cells or stabilization of β-catenin (Wong, Gavin, & McMahon, 1994), which is a typical response to treatment of Wnt1 or Wnt3a. Instead, a number of distinct pathways, including JNK signaling, small GTPase Rho/Rac/Cdc42 signaling, and Ca^{2+} signaling, have been proposed as the downstream targets of these "noncanonical" or Wnt5a class Wnts (Veeman, Axelrod, & Moon, 2003; Wallingford & Habas, 2005). As these targets are also subjected to regulation by numerous other signals, we do not have a definitive biochemical assay to measure the activation in response to each noncanonical Wnt. Our current understanding of the noncanonical Wnt pathway is based on studies in other model organisms including *Drosophila*, *Xenopus*, and zebrafish (Goodrich, 2008; Wang & Nathans, 2007; Zallen, 2007). The noncanonical Wnt/PCP pathway shares two critical components with the canonical Wnt pathway: the Fz receptor and a multifunctional cytoplasmic protein Dishevelled (Dsh in flies, XDsh in frogs, and Dvl in mice) (Wallingford & Habas, 2005). However, all the other members of this pathway are distinct from those of the canonical Wnt pathway and no Wnt ligand appears to be involved (Chen et al., 2008; Povelones, Howes, Fish, & Nusse, 2005). The main function of this pathway is to define PCP, the polarity of a cell within the plane of the epithelium, perpendicular to the apical–basal (AB) axis of the cell. Manifestations of this polarity include uniformly oriented trichomes on the wing epithelial cells and bristles on the thorax, abdomen, and leg and precisely coordinated orientation of ommatidial units of the compound eye.

Studies over the past few years have identified a large number of novel cytoplasmic or membrane proteins, including a "core" PCP group consisting of Flamingo, Diego, Prickle, and Strabismus/Van Gogh, in addition to Frizzled and Dishevelled. In *Xenopus* and zebrafish, almost all the homologues of the fly "core" PCP proteins have been found to control convergent extension, a coordinated extension of the A–P axis with concomitant narrowing of the medial–lateral axis (Carreira-Barbosa et al., 2003; Darken et al., 2002; Goto & Keller, 2002; Kinoshita, Iioka, Miyakoshi, & Ueno, 2003; Park & Moon, 2002; Takeuchi et al., 2003; Veeman, Slusarski, Kaykas, Louie, & Moon, 2003; Wallingford, Fraser, & Harland, 2002; Wallingford et al., 2000). In contrast to the PCP pathway in fly, both Wnt5a and Wnt11 have been implicated in regulating convergent extension in vertebrates (Heisenberg et al., 2000; Yamanaka et al., 2002). During convergent extension, several PCP proteins have been shown to control the polarity of lamellipodial protrusions that drive polarized cell intercalation (Jessen et al., 2002; Wallingford et al., 2000). In zebrafish, the PCP pathway appears to regulate convergent extension, in part, through determining the orientation of cell division (Gong, Mo, & Fraser, 2004).

3. CONSERVED MAMMALIAN PCP PATHWAY

Mouse genetics has provided compelling evidence that a homologous PCP pathway also exists in mammals to regulate a diverse array of morphogenetic processes.

3.1. PCP pathway in inner ear development

In *Loop-tail* (*Lp*) mutant mice, which harbor a loss-of-function point mutation in *Vangl2*, one of the mammalian homologues of the fly PCP gene *Strabismus/Van Gogh* (*Stbm/Vang*), the uniform orientation of stereocilia on the sensory hairs cells of the cochlea is disrupted (Kibar et al., 2001; Montcouquiol et al., 2003; Murdoch, Doudney, Paternotte, Copp, & Stanier, 2001). Orientation of these stereocilia, reminiscent of the trichome on epithelial cells of the fly wing, is a likely manifestation of PCP in mammals (Lewis & Davies, 2002). In addition to *Lp*, mutations of several other PCP homologues in the mouse result in misorientation of stereociliary bundles. These include *Crsh* and *Scy* (*Crash* and *Spin*) *cycle*, two distinct mutations of *Celsr1*, a homologue of the PCP pathway member *flamingo* (Curtin et al., 2003), $Fz3^{-/-};Fz6^{-/-}$ (Wang, Guo, & Nathans, 2006), and $Dvl1^{-/-};Dvl2^{-/-}$ (Wang et al., 2005) mutants (see below). Vangl2 binds to Dvl1–3 while the missense point mutation found in *Lp* perturbs Vangl2 interaction with all three Dvls (Torban, Wang, Groulx, & Gros, 2004). Consistent with the findings in fly where most PCP proteins show an asymmetric plasma membrane distribution, we found that a transgenic Dvl2–EGFP fusion protein, capable of rescuing the stereocilia orientation defects in $Dvl1^{-/-};Dvl2^{-/-}$ mutants, displays apically restricted asymmetric plasma membrane localization to the abneural side of the hair cell. The uniform asymmetric localization of Dvl2–EGFP is disrupted in *Lp/Lp* embryos (Wang et al., 2005). *Lp*-dependent asymmetric localization of Fz3 and Fz6 in the sensory hair cells, utricles, and cristae was reported (Wang, Guo, et al., 2006), suggesting that in mammals a conserved PCP pathway serves as the underlying mechanism in coordinating stereocilia orientation. Of note, $Fz6^{-/-}$ mice display misorientation of hairs throughout the body, implying that the mammalian PCP pathway also controls macroscopic hair patterning (Guo, Hawkins, & Nathans, 2004).

3.2. PCP pathway in neurulation

In addition to the inner ear polarity defects, *Lp*, *Crsh*, *Scy*, $Fz3^{-/-};Fz6^{-/-}$, and $Dvl1^{-/-};Dvl2^{-/-}$ mutants all result in a unique neural tube closure defect where the entire neural tube from midbrain to tail fails to close, a severe congenital neural tube defect termed craniorachischisis in humans (Curtin et al., 2003; Hamblet et al., 2002; Kibar et al., 2001; Murdoch et al., 2001; Murdoch et al. 2003; Wang, Guo, & Nathans, 2006). Based on experiments in *Xenopus* in which overexpression of XDsh and Stbm mutants that blocks convergent extension usually results in similar neural tube closure defects, the cause of craniorachischisis in these mutants appears to be due to a failure of convergent extension (Copp, Greene, & Murdoch, 2003; Darken et al., 2002; Goto & Keller, 2002; Ueno & Greene, 2003; Wallingford & Harland, 2002). In support of this hypothesis, we found that during neurulation the mouse neural plate undergoes concomitant lengthening and narrowing, a morphogenetic process that resembles convergent extension in other vertebrates. Both $Dvl1^{-/-};Dvl2^{-/-}$ and *Lp/Lp* mutants disrupt the coordinated lengthening and narrowing of the neural plate, as measured by the increase of the length-to-width ratio. Significantly, the reduction of length-to-width ratio in *Dvl1*$^{-/-}$; *Dvl2*$^{-/-}$ and *Lp/Lp* mutants is observed from 4-somite stage, the earliest time point when length-to-width ratio can be determined and several hours before neural tube closure occurs in control embryos (Wang, Hamblet, et al., 2006). Thus, a Dvl/Vangl-mediated homologous PCP pathway is important for a convergent extension-like morphogenetic process during neurulation to reduce the width of the neural plate, which may facilitate neural tube closure by shortening the distance between the opposing neural folds (Wallingford & Harland, 2002). Consistent with this idea, *Dvl2*$^{-/-}$; *Lp/+* mutants also show significantly reduced length-to-width ratio as *Dvl1*$^{-/-}$; *Dvl2*$^{-/-}$ and *Lp/Lp* and fail to closure their neural tubes, while *Lp/+* mutants display moderate reduction of the length-to-width ratio and delayed neural tube closure (Wang, Hamblet, et al., 2006).

3.3. PCP pathway in axonal tract development

In addition to its documented roles during neurulation and inner ear development, accumulating evidence has linked the noncanonical Wnt pathway to axonal tract development. Mutation of *Celsr3*, another *flamingo/starry night* mouse homolog, leads to the absence of the internal capsule, anterior commissure, and many of the longitudinal axonal bundles, disconnecting the cortex from subcortical structures (Tissir, Bar, Jossin, De Backer, & Goffinet, 2005). *Fz3* null mutants (Wang, Thekdi, Smallwood, Macke, & Nathans, 2002) or embryos with knock-down of *Dvls* (Zhang et al., 2007) display almost identical abnormalities, strongly implying that the noncanonical Wnt pathway/PCP is involved in axon tract development in the central nervous system.

3.4. PCP pathway during gastrulation: *mouse prickle-1* (*mpk1*) is essential for epiblast AB polarity

The canonical Wnt pathway is not the only Wnt-related pathway important for gastrulation. Recently, Tao et al. (2009) found that the deletion of *mpk1* gene resulted in early embryonic lethality, between E5.5 and E6.5, associated with failure of DVE migration and primitive streak formation. Mutant epiblast tissue was disorganized, and epiblast cells displayed abnormal cell shapes, mislocalized ECM proteins, and disrupted orientation of mitotic spindles with loss of AB polarity of epiblast cells. They demonstrated that *mpk1* genetically interacts with another core PCP gene *Vangl2/stbm* in epiblast formation, suggesting that PCP components are commonly required for the establishment and/or the maintenance of epiblast AB polarity. Thus, *mpk1* appears to play an important role in AB epiblast polarity formation as part of the PCP pathway.

4. *DVL* GENES: CROSSROADS OF WNT AND PCP PATHWAYS

As noted above, *Fz* and *Dishevelled* from *Drosophila melanogaster* participate in Wnt/Wg signaling and the PCP pathway. The PCP pathway in *Drosophila* and convergent extension in vertebrates are regulated by the same genetic pathway (reviewed in Wang & Nathans, 2007; Zallen, 2007). In *Drosophila* and *Xenopus* (Axelrod, Miller, Shulman, Moon, & Perrimon, 1998; Wallingford et al., 2000), it is known that distinct domains of Dvl proteins (DIX, DEP, and PDZ domains) are important for distinguishing between Wnt/Wg and PCP signaling. The DIX domain is mainly involved in the Wnt pathway: deletion of the DIX (ΔDIX) domain abolishes Wnt activity, but PCP activity is maintained. In contrast, deletion of the PDZ or C-terminal part of the DEP domain (ΔPDZ and ΔDEP, respectively) affects only the PCP but not the Wnt pathway in *Drosophila*. Thus, *in vivo* analysis of Dvl proteins can be used to define the pathways mediated by Dvls to produce phenotypic effects, including in mice. We have shown that similar mutational analysis can define *Dvl* functions *in vivo* in mammals (see below).

5. SINGLE MUTANTS FOR *DVL* GENES IN THE MOUSE: UNIQUE PHENOTYPES (FIG. 9.1)

We have been studying the *Dvl* family of genes in mice to gain insights into Dvl function and to investigate the function of canonical Wnt and non-canonical PCP pathways *in vivo*. The three murine *Dishevelled* genes, *Dvl1*, *Dvl2*, and *Dvl3*, are all widely expressed in embryonic and adult tissues, suggesting that there may be functional overlap. To test this, we have made mouse knock-outs for each of the *Dvl* genes by conventional gene targeting. Surprisingly, analysis of these single mutants uncovered unique phenotypes for each of the *Dvl* genes.

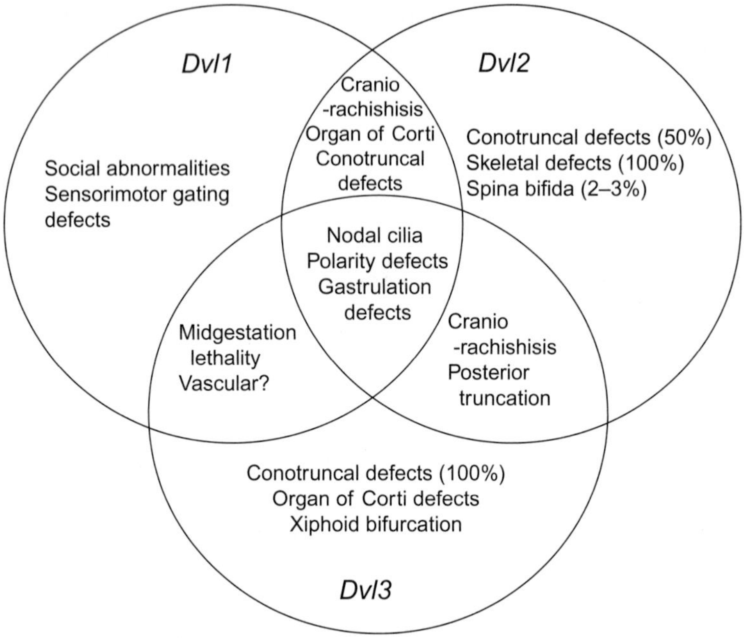

Figure 9.1 Summary of single, double, and triple *Dvl* mutant phenotypes in the form of a Venn diagram. The single gene knock-out phenotypes are illustrated by the large circles. The major double- and triple-mutant phenotypes are shown in the intersection of two or three circles, respectively. See text for details.

6. *DVL1* NULL MICE: SOCIAL INTERACTION DEFECTS

Mice completely deficient for *Dvl1* are viable, fertile, and structurally normal. Surprisingly, these mice exhibited reduced social interaction and were the first mutant organism of any species that displayed social behavior deficits (Lijam et al., 1997). When examining offspring housed as uniform genotypes, we noticed that most (50–75%) wild-type 129SvEv mice were completely devoid of whiskers and had trimmed facial hair. Whisker-trimming or barbering is a social behavior characteristic of several caged wild-type mouse strains, and 129SvEv mice (Taconic Farms) displayed this severe whisker-trimming behavior. In contrast, all 129SvEv *Dvl1*-deficient mice (100%) had full sets of whiskers and facial hair, while $Dvl1^{+/-}$ mice had trimmed whiskers. By video recording wild-type and mutant mice, we confirmed that whisker-trimming was a social behavior that was defective in $Dvl1^{-/-}$ mice.

To examine whether differences in social interaction were responsible for differences in whisker-trimming behavior, we combined one wild-type mouse, devoid of whiskers and facial hair, and one $Dvl1^{-/-}$ mouse, with full facial whiskers and hair, in 11 cages, and observed the changes in whisker pattern after mixing. In all 11 cages, wild-type mice regrew full sets of whiskers and facial hair within 2–4 weeks, while in five of the cages, *Dvl1*$^{-/-}$ mice lost all whiskers and facial hair. The wild-type mice with whiskers were then returned to their home cage with their original wild-type littermates. Within 2 weeks, the whiskers of the returned mice were now trimmed. In addition, whiskerless *Dvl1*$^{-/-}$ mutant mice regrew whiskers when returned to their home cage. Thus, the genotype of the cagemate determined the whisker pattern, demonstrating that this is in fact a social behavior, and *Dvl1*$^{-/-}$ mice displayed defective social interaction. Since barbering is often performed by dominant animals within a cage, a social dominance tube test was performed, pairing one wild-type against one *Dvl1*$^{-/-}$ mouse. In the majority of cases, the wild-type mouse won the tube test trial. Thus, wild-type mice are more dominant than the mutant mice, consistent with the hypothesis that the aberrant whisker-trimming behavior of the *Dvl1*-deficient mice is related to low levels of social interaction.

Dvl1-deficient mice also displayed less huddling contact during home-cage sleeping and displayed deficits in nest-building. Wild-type 129SvEv mice generally slept huddled together. If commercial nestlet wafers were provided to the mice, they built fluffy, well-formed nests within 45 min. In contrast, *Dvl1*-deficient mice tended to sleep in scattered, random patterns and did not build full nests. In fact, mutant mice tended to sleep on top of intact nestlet material. Taken together, the abnormal whisker-trimming behavior, the lack of huddling during sleep, and the poor nest-building displayed by *Dvl1*-deficient mice demonstrate an essential role for *Dvl1* in normal home-cage social behavior.

Other tests that measure motor, sensory, balance, and learning functions were normal in the *Dvl1* mutants, suggesting that the social deficits observed in these mice were not due to generalized neurological dysfunction. The only other test found to be abnormal in *Dvl1* knock-out mice was prepulse inhibition (PPI) of startle. PPI is thought to measure sensorimotor gating, a process where inhibitory neural pathways filter the multitude of stimuli that bombard the senses at any time, allowing attention to be focused on one stimulus. Sensorimotor gating has no known correlate in social behavior, but as it measures filtering inhibitory pathways, it would not be unreasonable to propose that the PPI and social interaction deficits are related. We replicated these findings of abnormal social behavior, but the PPI abnormalities were variable (Long et al., 2004). Thus, *Dvl1* mutant mice have specific defects in social behavior by several different measures, and *Dvl1* is an entry point to genetic pathways important for social interaction. Further studies will be needed to determine the role of *Dvls* in sensorimotor gating.

It is likely that the phenotypes displayed by *Dvl1*$^{-/-}$ mice are mediated by functions of Dvls in dendritic development (Rosso, Sussman, Wynshaw-Boris, & Salinas, 2005) and synapse formation (Ahmad-Annuar et al., 2006). Overexpression of Wnt7b or Dvl increases dendritic branching in cultured hippocampal neurons (Rosso et al., 2005), and hippocampal neurons from *Dvl1*-deficient mice display reduced dendritic arborization, an effect mediated by noncanonical Wnt signaling via activation of Rac and JNK, but not canonical Wnt signaling. Additionally, cerebellar slices from *Wnt7a/Dvl1* double-mutant mice display defects in neurotransmitter release at mossy fiber–granule cell synapses (Ahmad-Annuar et al., 2006). *Dvl1* loss decreased synaptic vesicle recycling in mossy fibers, while overexpression of Dvl increases the number of Bassoon clusters. Importantly, Dvls localize to synaptic sites. Thus, Dvl proteins appear to mediate critical behaviors such as social behavior and PPI via PCP pathway effects on dendritic development and synapse formation.

7. *DVL2* MUTANT MICE: CARDIAC, NEURULATION, AND SKELETAL DEFECTS

Mice homozygous for null alleles of *Dvl2* in either inbred 129S6 or mixed backgrounds can survive to adulthood and are fertile, but they are born in reduced numbers from heterozygous crosses. *Dvl2* homozygotes survive to term and are born, but 50% die at birth (Hamblet et al., 2002). We have observed patterning defects in *Dvl2* mutants: 50% display cardiac defects (the probable cause of lethality) in the outflow tract, including double outlet right ventricle and persistent truncus arteriosus. These abnormalities in conotruncal development are due to a defect in cardiac neural crest development during outflow tract formation, demonstrated by marker analysis in embryos (Hamblet et al., 2002; Kioussi et al., 2002). In addition to these cardiac defects, *Dvl2*$^{-/-}$ mice display two other malformations (Hamblet et al., 2002). Approximately 90% of *Dvl2*$^{-/-}$ mice have vertebral and rib malformations that affect the proximal as well as the distal parts of the ribs from defects in somite segmentation. Finally, 2–3% of *Dvl2*$^{-/-}$ embryos displayed thoracic spina bifida. Thus, *Dvl2* is essential for normal cardiac morphogenesis, somite segmentation, and neural tube closure.

The cardiac defects in *Dvl2* mutants may result from a failure in activating *Pitx2* by the Wnt signaling in the *Dvl2* mice in the neural crest (Kioussi et al., 2002). We have obtained genetic and biochemical evidence directly linking components of the Dvl–Pitx2 pathway in a dosage-dependent fashion to cardiac outflow tract development. Dvl2 acts upstream of Pitx2, and β-catenin directly regulates *Pitx2* gene transcription in pituitary and muscle. *Pitx2* can activate targets including cyclin D2, cyclin D1, and c-Myc and is required to regulate cell-specific proliferation events. β-Catenin plays a nuclear integrating function in Wnt-dependent proliferation events in specific cell types. With Pitx2 and other transcription factors, β-catenin converts transcriptional repressors to activators by dismissing HDACs and mediating sequential recruitment of specific, required coactivator complexes that include the CBP/p300 complex, the NLI/Ldb/CLIM coactivators of LIM homeodomain factors, and the TRAP/DRIP/ARC complexes.

8. *DVL3* MUTANT MICE

Mice homozygous for null alleles of *Dvl3* can rarely survive to adulthood and are fertile, but, similar to *Dvl2* mutants, 80% of *Dvl3* homozygotes in a mixed genetic background and 100% of *Dvl3* mutants in a 129/SvEv inbred background die at birth due to conotruncal heart defects (Etheridge et al., 2008) and display cochlear defects with abnormal orientation of hair cells. The *Dvl3* mutants are the only single *Dvl* mutants with cochlear defects, most likely the result of PCP pathway defects. We do not know the pathway(s) disrupted in conotruncal development in *Dvl3* mutants. However, further examination of heart morphogenesis led us to uncover a mechanism, whereby β-catenin activates a Wnt11-mediated PCP pathway during heart morphogenesis (Zhou et al., 2007). These studies suggest that *Dvl2* and/or *Dvl3* may participate in both canonical and noncanonical pathways during conotruncal development.

9. *DVL* DOUBLE MUTANTS: REDUNDANCY AMONG THE *DVL* GENES (FIG. 9.1)

Although the analysis of *Dvl* single mutants uncovered unique phenotypes for each of the *Dvl* genes, *Dvl* double-mutant phenotypes demonstrated that there is substantial overlapping and redundant functions among the three *Dvl* genes.

10. *DVL1/2* DOUBLE MUTANTS

All *Dvl1/Dvl2* double homozygotes display completely open neural tubes and exencephaly, demonstrating an essential role for *Dvl* genes in neural tube closure (Hamblet et al., 2002; Wang, Hamblet, et al., 2006). These double mutants display a novel phenotype: shortening of cochlear ducts with abnormal orientation of the hair cells of the inner ear (Wang, Hamblet, et al., 2006; Wang et al., 2005), a phenotype similar to *Dvl3* single mutants. Additionally, the skeletal and cardiac abnormalities seen in *Dvl2* mutants were also more severe in the *Dvl1/2* double mutants than in *Dvl2* single mutants. The neural tube and cochlear elongation defects result from convergent extension defects, while the inner ear hair cell defects are due to defects in polarity of these cells. All of these defects in *Dvl1/2* mutants result from defective Wnt/PCP pathway (Wang, Hamblet, et al., 2006, see below). We have not determined which of these pathways causes the somite defects that result in skeletal abnormalities in the *Dvl2* and *Dvl1/2* mutants, and there is evidence for both Wnt and PCP pathway defects to explain the conotruncal abnormalities (Kioussi et al., 2002; Zhou et al., 2007, see above).

The mammalian auditory sensory organ, the organ of Corti, consists of precise arrays of sensory hair cells with uniformly oriented stereocilia on the apical surfaces that display a distinct PCP parallel to the sensory epithelium. We demonstrated that the mature organ of Corti is formed from a thicker and shorter postmitotic primordium through unidirectional extension, characteristic of convergent extension (Wang et al., 2005). Mutations in the PCP pathway, including *Lp* and *Dvl1/2* mutants, interfered with this extension, resulting in the formation of a shortened and widened cochlear duct and its sensory organ, as well as misorientation of stereocilia. Analogous to the homologous pathway in *Drosophila*, Dvl2 displays PCP-dependent polarized subcellular localization across the organ of Corti. This study suggests a conserved underlying molecular mechanism for PCP pathway regulation of polarized extension and PCP of the organ of Corti in invertebrates and vertebrates and indicates that the mammalian PCP pathway might directly couple cellular intercalation movements to the establishment of PCP in the cochlea.

11. *DVL1/3* DOUBLE MUTANTS

Normal development was observed in *Dvl1*$^{-/-}$;*Dvl3*$^{-/-}$ embryos until E12.5 (Etheridge et al., 2008), but these mutants died between E13.5 and E15.5 of unknown causes. The timing of lethality suggests defective vascular or cardiac function, but further studies are required to determine the precise cause of death and the pathway(s) responsible.

12. *DVL2/3* DOUBLE MUTANTS

Dvl2$^{+/-}$;*Dvl3*$^{-/-}$ double mutants at E18.5 appear shorter along the A–P axis and show an abnormal head shape, forebrain defects, and a truncated snout, as well as a shortened and kinked tail. Craniorachischisis was observed in two *Dvl2*$^{+/-}$;*Dvl3*$^{-/-}$ mutants, and one of these also showed other severe phenotypes, including gastroschisis and absence of the tail. *Dvl2*$^{-/-}$;*Dvl3*$^{+/-}$ mutants display craniorachischisis, pericardial effusion, and abnormal looping of the heart, as well as severe posterior truncation, and do not survive beyond E9.5 (Etheridge et al., 2008). No *Dv2*$^{-/-}$; *Dvl3*$^{-/-}$ embryos have been recovered from litters collected from E8.5 onward, due to severe gastrulation defects and lethality earlier in development (see below). These findings are consistent with the interpretation that there is functional overlap among the *Dvl* genes.

13. *DVL1,2,3* TRIPLE MUTANTS DISPLAY DEFECTS IN THE POSTERIOR PLACEMENT OF NODAL CILIA (FIG. 9.1)

As noted for the *Dvl* double mutants, the phenotypes of *Dvl1/2/3* triple mutants further demonstrated that there is substantial overlapping and redundant functions among the three *Dvl* genes. In collaboration with Hiroshi Hamada (Osaka University), we found that the three *Dvl* genes participate in the planar polarization of node cells to determine the rotational axis of nodal cilia (Hashimoto et al., 2010). Rotational movement of the node cilia generates a leftward fluid flow in the mouse embryo because the cilia are posteriorly tilted (Nonaka et al., 2005; Okada, Takeda, Tanaka, Belmonte, & Hirokawa, 2005), but it was not known how A–P information is translated into the posterior tilt of the node cilia. We found that the basal body of node cilia is initially positioned centrally but then gradually shifts toward the posterior side of the node cells, a shift that was impaired along with malpositioning of the basal body and unidirectional nodal flow in compound mutant mice lacking *Dvl* genes. The basal body was normally positioned in the node cells of *Wnt3a*$^{-/-}$ embryos (coding for a canonical Wnt involved in gastrulation movements), but inhibition of Rac1, a component of the noncanonical Wnt signaling pathway, impaired the polarized localization of the basal body in wild-type embryos. Dvl2–EGFP and Dvl3–EYFP proteins were localized to the apical side of the node cells, and their location was polarized to the posterior side of the cells before the posterior positioning of the basal body. Thus, posterior positioning of the basal body to provide posterior tilt to node cilia is determined by planar polarization mediated by noncanonical Wnt/PCP signaling.

In a continuation of these studies (unpublished results from our lab and the lab of Hiroshi Hamada), we discovered the complete loss of all the three mouse *Dvl* genes in *Dvl1,2,3* triple null mutants die soon after implantation, without evidence of gastrulation. In addition, a slightly less severe phenotype was observed in *Dvl1*$^{+/+}$;*Dvl2*$^{-/-}$;*Dvl3*$^{-/-}$ and *Dvl1*$^{+/-}$;*Dvl2*$^{-/-}$; *Dvl3*$^{-/-}$ embryos, where gastrulation is often initiated. Based on the discussion above, these severe gastrulation defects may be due to the disruption of either the canonical Wnt pathway and or the noncanonical PCP pathway. Further studies will be required to define these phenotypes and the pathway(s) important for these severe defects.

14. *DVL* CONDITIONAL TAGGED ALLELES

To further investigate redundancy, we made fluorescently tagged transgenic alleles for each of the Dvl genes by BAC recombineering and transgenesis. The advantages of using the BAC transgenic approach are that the BAC transgene will likely recapitulate the endogenous gene expression pattern, as it contains large flanking genomic region where regulatory elements reside, and because usually only one to two copies will be integrated into the genome. In addition, each of the *Dvl* BAC transgenic alleles can be used to see if additional copies of any *Dvl* gene can substitute for other *Dvls*.

First, we produced two BAC transgenic alleles for *Dvl2* (Wang, Hamblet, et al., 2006). Both alleles had in-frame EGFP tags to allow us to examine intracellular distribution of Dvl2 protein, as well as a loxP site in intron 2 to differentiate this allele from the wild-type and the *Dvl2* knock-out allele. The second BAC allele also contains a second loxP site in the adjacent *Acadvl* gene. The latter allele acts as a conditional allele *in vivo*. Both *Dvl2* BAC transgenes were able to fully rescue *Dvl1*$^{-/-}$; *Dvl2*$^{-/-}$ and *Dvl2*$^{-/-}$; *Lp/+* mutants to fertile adults, demonstrating that the encoded Dvl2–EGFP is appropriately expressed for full phenotypic rescue. The *Dvl2–EGFP* BAC allele is also expressed at levels comparable to endogenous *Dvl2* allele (Wang, Hamblet, et al., 2006; Wang et al., 2005).

Similar to the Dvl2–EGFP BAC transgenic allele, we produced *Dvl* fluorescently tagged BAC alleles of *Dvl1* and *Dvl3*. These two BAC alleles,

Dvl1–ECFP and *Dvl3–EYFP* (Etheridge et al., 2008), are also expressed at levels comparable to endogenous alleles and completely rescue single- and double-mutant phenotypes. *Dvl1–ECFP* and *Dvl3–EYFP*, like the *Dvl2–EGFP* allele, are also conditional alleles *in vivo*.

15. *DVL* TRANSGENES DEMONSTRATE FUNCTIONAL REDUNDANCY

The generation of fully functional *Dvl* transgenes allowed us to use a genetic approach to further determine redundancy of function between the Dvl proteins during development (Etheridge et al., 2008). We determined whether an extra copy of either *Dvl1* or *Dvl2*, in the form of *Dvl1–ECFP* and *Dvl2–EGFP* BAC transgenes (which we will refer to here as *Dvl1TG* and *Dvl2TG*, respectively), was able to rescue lethal *Dvl3*$^{-/-}$ phenotype.

As noted above, *Dvl3*$^{-/-}$ mutants (which still have two copies of the *Dvl1* allele and two copies of the *Dvl2* allele) cannot survive. Surprisingly, addition of the *Dvl1TG* rescued the *Dvl3*$^{-/-}$ phenotype such that *Dvl3*$^{-/-}$; *Dvl1TG* mutants (now *Dvl3*$^{-/-}$ with three copies of *Dvl1* and two copies of *Dvl2*) survived in expected numbers. In a similar cross using the *Dvl2TG*, we found that an extra copy of the *Dvl2*, as the *Dvl2TG*, also rescued the *Dvl3*$^{-/-}$ lethal phenotype to at least 90% efficiency. As 50% of *Dvl2*$^{-/-}$ mutants die perinatally, we used a similar strategy to determine whether additional copies of *Dvl1* or *Dvl3* could rescue *Dvl2*$^{-/-}$ lethality. There was an approximately 96% rescue of the *Dvl2*$^{-/-}$ phenotype with an extra copy of *Dvl3*. However, the *Dvl1TG* was not able to rescue *Dvl2*$^{-/-}$ mutant phenotype, in contrast to the rescue of the *Dvl2*$^{-/-}$ phenotype by the *Dvl3TG* and the *Dvl3*$^{-/-}$ phenotype by the *Dvl2TG* or *Dvl1TG*. These results, along with the double-mutant phenotypes, strongly support the notion that the three *Dvl* genes have redundant functions during development.

16. *DVL2* ALLELIC SERIES TO DISTINGUISH WNT AND PCP PATHWAY PHENOTYPES *IN VIVO*

We wanted tools that would allow us to determine whether any of the Dvl phenotypes we observed were due to defects in Wnt and/or PCP pathway signaling. We took advantage of the studies in *Xenopus* and *Drosophila* (described above) that defined protein domains in Dvls required for canonical Wnt and PCP signaling. Therefore, we made a *Dvl2* allelic series (*ΔDIX*, *ΔDEP*, and a *dsh1* point mutant) to determine whether the canonical Wnt or noncanonical Wnt/PCP pathways were responsible for defects seen in *Dvl1/2* double mutants (Wang, Hamblet, et al., 2006). As noted above, two of the conserved domains in Dsh/Dvl, DIX and DEP, play different roles in the Wnt pathway and the PCP/convergent extension (see Wallingford & Habas, 2005). In *Xenopus*, mediolateral intercalation in the mesoderm requires the C-terminal DEP domain but not the N-terminal DIX domain of Xdsh. To confirm that mammalian neurulation requires the same domain of Dvl2, we generated mutant *Dvl2* BAC transgenes identical to the ΔDEP mutant in *Xenopus* (*ΔDEP–EGFP*). When crossed into *Dvl1*$^{-/-}$; *Dvl2*$^{-/-}$ background, *ΔDEP–EGFP* completely failed to rescue the neurulation defects. In contrast, a *ΔDIX–EGFP* transgene, deleting part of the DIX domain, fully rescue the neurulation defect in *Dvl1*$^{-/-}$; *Dvl2*$^{-/-}$ mutants. As the *ΔDIX–EGFP* transgene removed a VKEEIS motif in the N-terminal region essential for Dvl2 function in the canonical Wnt pathway (Capelluto et al., 2002), this result also confirmed that the neural tube closure defect in *Dvl1*$^{-/-}$; *Dvl2*$^{-/-}$ mutants was not due to the loss of function of Wnt signaling. When we investigated the distribution of the mutant Dvl2 protein during neurulation, we found that while ΔDIX–EGFP was still primarily membrane localized, indistinguishable from wild-type Dvl2–EGFP, ΔDEP–EGFP was evenly distributed in the cytoplasm.

Thus, the DEP domain was required for targeting Dvl2 to the plasma membrane during mammalian neurulation. However, there were differences in Dvl2–EGFP localization in the neuroepithelium and in the organ of Corti, where Dvl2–EGFP showed an asymmetric membrane distribution that was disrupted in *Lp/Lp* mutants, reminiscent of fly PCP establishment (see Wang et al., 2005). We tested whether the point mutation *dsh1* identified in fly, which specifically abolished the PCP (Axelrod et al., 1998), might affect convergent extension in mammals. The *dsh1* mutation results in a K to M missense mutation in the C-terminal DEP domain (Axelrod et al., 1998). Using BAC recombineering, we introduced an identical mutation into *Dvl2–EGFP* BAC and produced transgenic mice. All *Dvl1*$^{-/-}$; *Dvl2*$^{-/-}$; *dsh1–EGFP* embryos recovered at E9.5 or later displayed craniorachischisis, suggesting that the *dsh1* mutation completely disrupted the ability of Dvl2 to function during convergent extension. Also consistent with the notion that convergent extension underlies cochlea elongation (Wang et al., 2005), we observed a shorter and wider cochlea in *Dvl1*$^{-/-}$; *Dvl2*$^{-/-}$; *dsh1–EGFP* embryos (Wang, Hamblet, et al., 2006). However, in either wild-type or *Dvl1*$^{-/-}$; *Dvl2*$^{-/-}$ background, this mutation did not appear to appreciably affect the membrane localization of Dvl2–EGFP in neuroepithelium during neurulation (data not shown).

17. CONCLUSIONS AND FUTURE STUDIES

The phenotypes of single *Dvl* mutants (Fig. 9.1) reveal unique functions of each of the *Dvl* genes, but the phenotypes of double and triple *Dvl* mutants (Fig. 9.1) also reveal a high level of functional redundancy among this gene family. The level of redundancy is not surprising, given that all three *Dvl* genes are highly conserved and broadly expressed during development and in the adult. Why are there unique phenotypes? There may be some situations where one of the *Dvls* has predominant expression pattern. For example, *Dvl1* may play some unique role in the brain, based on the singular behavioral phenotype displayed by the *Dvl1* mutants. *Dvl2* and *Dvl3* may have higher levels of expression in the developing heart, which could explain the conotruncal defects displayed by these mutants.

Most if not all of the phenotypes displayed by the *Dvl* mutants appear to be the result of PCP pathway functions, not of the canonical Wnt pathway. Clearly, the neurulation, cochlear elongation, cochlear hair cell polarity, nodal cilia polarity, and axonal guidance phenotypes are the result of PCP pathway functions. There is evidence that the brain defects that could explain the *Dvl1* behavioral deficits are due to PCP pathway functions. The pathway dependence of the skeletal and severe gastrulation defects is not currently known. We will attempt to employ the *Dvl2* allelic series to test these phenotypes *in vivo*, based on the functional redundancy of the *Dvls in vivo*. However, this may not be successful, and allelic series for *Dvl1* and/or *Dvl3* may have to be produced and tested. It will also be helpful if downstream targets of the PCP pathway can be identified.

What is perhaps most interesting and surprising is that there are no clear canonical Wnt pathway defects yet identified in the *Dvl* mutants. This suggests that the PCP pathway is sensitive to reduction of *Dvls*, but only one *Dvl* allele of six may be required for Wnt pathway. Why? There may be unique as well as redundant *Dvl* functions in the PCP pathway, but the function of each of the Dvls is only redundant in the Wnt pathway, and a single *Dvl* allele may be sufficient for Wnt signaling necessary for grossly normal development.

Several of these and other unanswered questions may be approachable by the use of conditional *Dvl* alleles. Each of the fluorescently tagged BAC transgenic alleles is also conditional alleles and can be used for studying the redundancy and pathway dependence of *Dvl* function, and to examine later phenotypes that may be revealed when earlier embryonic lethal phenotypes are circumvented.

REFERENCES

Ahmad-Annuar, A., Ciani, L., Simeonidis, I., Herreros, J., Fredj, N. B., Rosso, S. B., et al. (2006). Signaling across the synapse: A role for Wnt and Dishevelled in presynaptic assembly and neurotransmitter release. *The Journal of Cell Biology*, *174*, 127–139.

Axelrod, J. D., Miller, J. R., Shulman, J. M., Moon, R. T., & Perrimon, N. (1998). Differential recruitment of Dishevelled provides signaling specificity in the planar cell polarity and Wingless signaling pathways. *Genes & Development*, *12*, 2610–2622.

Capelluto, D. G., Kutateladze, T. G., Habas, R., Finkielstein, C. V., He, X., & Overduin, M. (2002). The DIX domain targets dishevelled to actin stress fibres and vesicular membranes. *Nature*, *419*, 726–729.

Carreira-Barbosa, F., Concha, M. L., Takeuchi, M., Ueno, N., Wilson, S. W., & Tada, M. (2003). Prickle 1 regulates cell movements during gastrulation and neuronal migration in zebrafish. *Development*, *130*, 4037–4046.

Chen, W. S., Antic, D., Matis, M., Logan, C. Y., Povelones, M., Anderson, G. A., et al. (2008). Asymmetric homotypic interactions of the atypical cadherin flamingo mediate intercellular polarity signaling. *Cell*, *133*, 1093–1105.

Copp, A. J., Greene, N. D., & Murdoch, J. N. (2003). The genetic basis of mammalian neurulation. *Nature Reviews. Genetics*, *4*, 784–793.

Curtin, J. A., Quint, E., Tsipouri, V., Arkell, R. M., Cattanach, B., Copp, A. J., et al. (2003). Mutation of Celsr1 disrupts planar polarity of inner ear hair cells and causes severe neural tube defects in the mouse. *Current Biology*, *13*, 1129–1133.

Darken, R. S., Scola, A. M., Rakeman, A. S., Das, G., Mlodzik, M., & Wilson, P. A. (2002). The planar polarity gene strabismus regulates convergent extension movements in Xenopus. *The EMBO Journal*, *21*, 976–985.

Ding, J., Yang, L., Yan, Y. T., Chen, A., Desai, N., Wynshaw-Boris, A., et al. (1998). Cripto is required for correct orientation of the anterior-posterior axis in the mouse embryo. *Nature*, *395*, 702–707.

Etheridge, S. L., Ray, S., Li, S., Hamblet, N. S., Lijam, N., Tsang, M., et al. (2008). Murine Dishevelled 3 functions in redundant pathways with Dishevelled 1 and 2 in normal cardiac outflow tract, cochlea and neural tube development. *PLoS Genetics*, *4*, e1000259.

Gong, Y., Mo, C., & Fraser, S. E. (2004). Planar cell polarity signalling controls cell division orientation during zebrafish gastrulation. *Nature*, *430*, 689–693.

Goodrich, L. V. (2008). The plane facts of PCP in the CNS. *Neuron*, *60*, 9–16.

Goto, T., & Keller, R. (2002). The planar cell polarity gene strabismus regulates convergence and extension and neural fold closure in Xenopus. *Developmental Biology*, *247*, 165–181.

Guo, N., Hawkins, C., & Nathans, J. (2004). Frizzled6 controls hair patterning in mice. *Proceedings of the National Academy of Sciences of the United States of America*, *101*, 9277–9281.

Hamblet, N. S., Lijam, N., Ruiz-Lozano, P., Wang, J., Yang, Y., Luo, Z., et al. (2002). Dishevelled 2 is essential for cardiac outflow tract development, somite segmentation and neural tube closure. *Development*, *129*, 5827–5838.

Hashimoto, M., Shinohara, K., Wang, J., Ikeuchi, S., Yoshiba, S., Meno, C., et al. (2010). Planar polarization of node cells determines the rotational axis of the node cilia. *Nature Cell Biology*, *12*, 170–176.

Heisenberg, C. P., Tada, M., Rauch, G. J., Saude, L., Concha, M. L., Geisler, R., et al. (2000). Silberblick/Wnt11 mediates convergent extension movements during zebrafish gastrulation. *Nature*, *405*, 76–81.

Hsieh, J. C., Lee, L., Zhang, L., Wefer, S., Brown, K., DeRossi, C., et al. (2003). Mesd encodes an LRP5/6 chaperone essential for specification of mouse embryonic polarity. *Cell*, *112*, 355–367.

Huelsken, J., & Birchmeier, W. (2001). New aspects of Wnt signaling pathways in higher vertebrates. *Current Opinion in Genetics & Development, 11*, 547–553.

Huelsken, J., Vogel, R., Brinkmann, V., Erdmann, B., Birchmeier, C., & Birchmeier, W. (2000). Requirement for beta-catenin in anterior-posterior axis formation in mice. *The Journal of Cell Biology, 148*, 567–578.

Jessen, J. R., Topczewski, J., Bingham, S., Sepich, D. S., Marlow, F., Chandrasekhar, A., et al. (2002). Zebrafish trilobite identifies new roles for Strabismus in gastrulation and neuronal movements. *Nature Cell Biology, 4*, 610–615.

Kelly, O. G., Pinson, K. I., & Skarnes, W. C. (2004). The Wnt co-receptors Lrp5 and Lrp6 are essential for gastrulation in mice. *Development, 131*, 2803–2815.

Kibar, Z., Vogan, K. J., Groulx, N., Justice, M. J., Underhill, D. A., & Gros, P. (2001). Ltap, a mammalian homolog of Drosophila Strabismus/Van Gogh, is altered in the mouse neural tube mutant Loop-tail. *Nature Genetics, 28*, 251–255.

Kinoshita, N., Iioka, H., Miyakoshi, A., & Ueno, N. (2003). PKC delta is essential for Dishevelled function in a noncanonical Wnt pathway that regulates Xenopus convergent extension movements. *Genes & Development, 17*, 1663–1676.

Kioussi, C., Briata, P., Baek, S. H., Rose, D., Hamblet, N. S., Herman, T., et al. (2002). Identification of a Wnt/Dvl/b-catenin->Pitx2 pathway mediating cell type-specific proliferation during development. *Cell, 111*, 673–685.

Lewis, J., & Davies, A. (2002). Planar cell polarity in the inner ear: How do hair cells acquire their oriented structure? *Journal of Neurobiology, 53*, 190–201.

Lickert, H., Kutsch, S., Kanzler, B., Tamai, Y., Taketo, M. M., & Kemler, R. (2002). Formation of multiple hearts in mice following deletion of beta-catenin in the embryonic endoderm. *Developmental Cell, 3*, 171–181.

Lijam, N., Paylor, R., McDonald, M. P., Crawley, J. N., Deng, C. -X., Herrup, K., et al. (1997). Social interaction and sensorimotor gating abnormalities in mice lacking *Dvl-1*. *Cell, 90*, 895–905.

Liu, P., Wakamiya, M., Shea, M. J., Albrecht, U., Behringer, R. R., & Bradley, A. (1999). Requirement for Wnt3 in vertebrate axis formation. *Nature Genetics, 22*, 361–365.

Mohamed, O. A., Clarke, H. J., & Dufort, D. (2004). Beta-catenin signaling marks the prospective site of primitive streak formation in the mouse embryo. *Developmental Dynamics, 231*, 416–424.

Montcouquiol, M., Rachel, R. A., Lanford, P. J., Copeland, N. G., Jenkins, N. A., & Kelley, M. W. (2003). Identification of Vangl2 and Scrb1 as planar polarity genes in mammals. *Nature, 423*, 173–177.

Morkel, M., Huelsken, J., Wakamiya, M., Ding, J., van de Wetering, M., Clevers, H., et al. (2003). Beta-catenin regulates Cripto- and Wnt3-dependent gene expression programs in mouse axis and mesoderm formation. *Development, 130*, 6283–6294.

Murdoch, J. N., Doudney, K., Paternotte, C., Copp, A. J., & Stanier, P. (2001). Severe neural tube defects in the loop-tail mouse result from mutation of Lpp 1, a novel gene involved in floor plate specification. *Human Molecular Genetics, 10*, 2593–2601.

Murdoch, J. N., Henderson, D. J., Doudney, K., Gaston-Massuet, C., Phillips, H. M., Paternotte, C., et al. (2003). Disruption of scribble (Scrb1) causes severe neural tube defects in the circletail mouse. *Human Molecular Genetics, 12*, 87–98.

Nelson, W. J., & Nusse, R. (2004). Convergence of Wnt, beta-catenin, and cadherin pathways. *Science, 303*, 1483–1487.

Nonaka, S., Yoshiba, S., Watanabe, D., Ikeuchi, S., Goto, T., Marshall, W. F., et al. (2005). De novo formation of left-right asymmetry by posterior tilt of nodal cilia. *PLoS Biology, 3*, e268.

Okada, Y., Takeda, S., Tanaka, Y., Belmonte, J. C., & Hirokawa, N. (2005). Mechanism of nodal flow: A conserved symmetry breaking event in left-right axis determination. *Cell, 121*, 633–644.

Park, M., & Moon, R. T. (2002). The planar cell-polarity gene stbm regulates cell behaviour and cell fate in vertebrate embryos. *Nature Cell Biology*, *4*, 20–25.

Povelones, M., Howes, R., Fish, M., & Nusse, R. (2005). Genetic evidence that Drosophila frizzled controls planar cell polarity and Armadillo signaling by a common mechanism. *Genetics*, *171*, 1643–1654.

Rosso, S. B., Sussman, D., Wynshaw-Boris, A., & Salinas, P. C. (2005). Wnt signaling through Dishevelled, Rac and JNK regulates dendritic development. *Nature Neuroscience*, *8*, 34–42.

Takeuchi, M., Nakabayashi, J., Sakaguchi, T., Yamamoto, T. S., Takahashi, H., Takeda, H., et al. (2003). The prickle-related gene in vertebrates is essential for gastrulation cell movements. *Current Biology*, *13*, 674–679.

Tissir, F., Bar, I., Jossin, Y., De Backer, O., & Goffinet, A. M. (2005). Protocadherin Celsr3 is crucial in axonal tract development. *Nature Neuroscience*, *8*, 451–457.

Torban, E., Wang, H. J., Groulx, N., & Gros, P. (2004). Independent mutations in mouse vangl2 that cause neural tube defects in looptail mice impair interaction with members of the dishevelled family. *The Journal of Biological Chemistry*, *279*, 52703–52713.

Ueno, N., & Greene, N. D. (2003). Planar cell polarity genes and neural tube closure. *Birth Defects Research. Part C, Embryo Today : Reviews*, *69*, 318–324.

Veeman, M. T., Axelrod, J. D., & Moon, R. T. (2003). A second canon. Functions and mechanisms of beta-catenin-independent Wnt signaling. *Developmental Cell*, *5*, 367–377.

Veeman, M. T., Slusarski, D. C., Kaykas, A., Louie, S. H., & Moon, R. T. (2003). Zebrafish prickle, a modulator of noncanonical Wnt/Fz signaling, regulates gastrulation movements. *Current Biology*, *13*, 680–685.

Wallingford, J. B., Fraser, S. E., & Harland, R. M. (2002). Convergent extension: The molecular control of polarized cell movement during embryonic development. *Developmental Cell*, *2*, 695–706.

Wallingford, J. B., & Habas, R. (2005). The developmental biology of Dishevelled: An enigmatic protein governing cell fate and cell polarity. *Development*, *132*, 4421–4436.

Wallingford, J. B., & Harland, R. M. (2002). Neural tube closure requires Dishevelled-dependent convergent extension of the midline. *Development*, *129*, 5815–5825.

Wallingford, J. B., Rowning, B. A., Vogeli, K. M., Rothbacher, U., Fraser, S. E., & Harland, R. M. (2000). Dishevelled controls cell polarity during Xenopus gastrulation. *Nature*, *405*, 81–85.

Wang, J., Etheridge, S. L., & Wynshaw-Boris, A. (2006). The Wnt signaling pathways in mammalian patterning and morphogenesis. In Sergei Sokol (Ed.), *Wnt Signaling in embryonic development. Advances in developmental biology*, Vol. 17, (pp. 111–158) Elsevier.

Wang, Y., Guo, N., & Nathans, J. (2006). The role of Frizzled3 and Frizzled6 in neural tube closure and in the planar polarity of inner-ear sensory hair cells. *The Journal of Neuroscience*, *26*, 2147–2156.

Wang, J., Hamblet, N. S., Mark, S., Dickinson, M., Brinkman, B., Segil, N., et al. (2006). Dishevelled genes mediate a conserved mammalian PCP pathway to regulate convergent extension during neurulation. *Development*, *133*, 1767–1778.

Wang, J., Mark, S., Zhang, X., Qian, D., Yoo, S. J., Radde-Gallwitz, K., et al. (2005). Regulation of polarized extension and planar cell polarity in the cochlea by the vertebrate PCP pathway. *Nature Genetics*, *37*, 980–985.

Wang, Y., & Nathans, J. (2007). Tissue/planar cell polarity in vertebrates: New insights and new questions. *Development*, *134*, 647–658.

Wang, Y., Thekdi, N., Smallwood, P. M., Macke, J. P., & Nathans, J. (2002). Frizzled-3 is required for the development of major fiber tracts in the rostral CNS. *The Journal of Neuroscience*, *22*, 8563–8573.

Wong, G. T., Gavin, B. J., & McMahon, A. P. (1994). Differential transformation of mammary epithelial cells by Wnt genes. *Molecular and Cellular Biology*, *14*, 6278–6286.

Zallen, J. A. (2007). Planar polarity and tissue morphogenesis. *Cell, 129*, 1051–1063.

Zhang, X., Zhu, J., Yang, G. Y., Wang, Q. J., Qian, L., Chen, Y. M., et al. (2007). Dishevelled promotes axon differentiation by regulating atypical protein kinase C. *Nature Cell Biology*, *9*, 743–754.

Zhou, W., Lin, L., Majumdar, A., Li, X., Zhang, X., Liu, W., et al. (2007). Modulation of morphogenesis by non-canonical Wnt signaling requires ATF2-mediated transcriptional activation of TGFb. *Nature Genetics*, *39*, 1225–1234.

CHAPTER TEN

An Expanding Role of Vangl Proteins in Embryonic Development

Elena Torban*, Alexandra Iliescu[†,‡], Philippe Gros[†,‡,1]

*Department of Medicine, McGill University, Montreal, Quebec, Canada
[†]Department of Biochemistry, McGill University, Montreal, Quebec, Canada
[‡]Complex Traits Group, McGill University, Montreal, Quebec, Canada
[1]Corresponding author: e-mail address: philippe.gros@mcgill.ca

Contents

Abstract

The mammalian *Vangl1* and *Vangl2* genes were discovered a decade ago through their association with neural tube defects, in particular the presence of *Vangl2* mutations in independent alleles of the mouse mutant *Loop-tail* (*Lp*), a mouse model of the severe neural tube defect craniorachischisis. Vangl1 and Vangl2 variants have also been detected in familial and sporadic cases of spina bifida. Vangl proteins are highly conserved in evolution with relatives in flies, and distant invertebrates and vertebrates. In these organisms, they play a central role in planar cell polarity (PCP) and convergent extension (CE) movements. Over the past decade, these functional characteristics have also been established for mammalian *Vangl* genes. The careful analysis of mouse *Vangl* genes mutants has showed that these genes and the associated PCP pathway and CE movements are involved in many unexpected developmental processes, from morphogenesis of different tissues, left–right asymmetry, asymmetric cell division, and organization of many epithelial structures, as well as positioning and function of cellular appendages. Genetic studies in double mutants and biochemical studies of interacting proteins have started to elucidate the molecular pathways in which Vangl proteins participate and that regulate these complex events.

Current Topics in Developmental Biology, Volume 101
ISSN 0070-2153
http://dx.doi.org/10.1016/B978-0-12-394592-1.00005-3
© 2012 Elsevier Inc.
All rights reserved.

1. VANGL PROTEINS, NEURULATION, AND NEURAL TUBE DEFECTS

Neurulation, the process by which the brain and spinal cord form during embryonic development, culminates in the formation of the primordial structure of the central nervous system, the neural tube (NT). Aberrations in the formation of the NT result in neural tube defects (NTDs) such as spina bifida and anencephaly, conditions in which the central canal of the malformed brain or spinal cord remains open to the environment. Spina bifida and anencephaly are very frequent in humans (1/1000 live births) and are second only to congenital heart defects as a cause of perinatal mortality. The cellular and molecular mechanisms contributing to neurulation are largely unknown, and multiple factors, both genetic and environmental, have been implicated in the etiology of NTDs underlying their causal heterogeneity (Greene, Stanier, & Copp, 2009; Kibar, Capra, & Gros, 2007).

The mouse is an excellent model to study the formation of NT because of the accessibility of the developing embryo, the abundance of mutants affecting neurulation, and the ease to manipulate environmental and genetic influences on this process (Harris & Juriloff, 2010; Wallingford, 2005). In mice, NT closure is initiated at specific "closure sites" and failure of closure at any of these sites leads to the various NTDs (Fig. 10.1). Failure of closure I results in the most severe NTD, craniorachischisis, when the NT remains completely open from midbrain–hindbrain junction to the tail end. Failure of the caudal spread of fusion from closure site I causes spina bifida, whereas distortion of closure at sites II and III leads to rostral NTD, anencephaly (Copp & Greene, 2010; Fig. 10.1). The identification and characterization of mutations causing NTDs in mice provide an entry point not only for studying the role of these genes, proteins, and cellular mechanisms of neurulation but also for identifying genes involved in the pathogenesis of NTDs in humans.

A large number of mouse mutants harboring different types of NTDs have been described over the years (Harris & Juriloff, 2010). Among them, *Loop-tail* (*Lp*) is a semidominant mutation on chromosome 1 originally described by Strong in 1949 (Strong & Hollander, 1949). The *Lp*/+ heterozygotes display a characteristic "looped" or severely kinked tail (Fig. 10.2) and show wobbly head movements (Strong & Hollander, 1949). The *Lp*/*Lp* homozygotes die *in utero* of a severe NTD, craniorachischisis, that arises from a failure to initiate closure at the closure site I, and NT remains completely

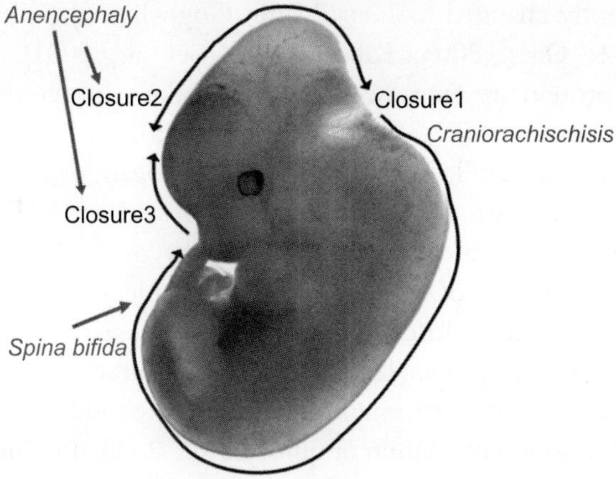

Figure 10.1 Schematic representation of the pattern of anterior neural tube closure. The closure initiation points are shown (closure 1–3), and neural tube defects (anencephaly, spina bifida, craniorachischisis) associated with failure to initiate or propagate closure at these sites are indicated. (For color version of this figure, the reader is referred to the online version of this chapter.)

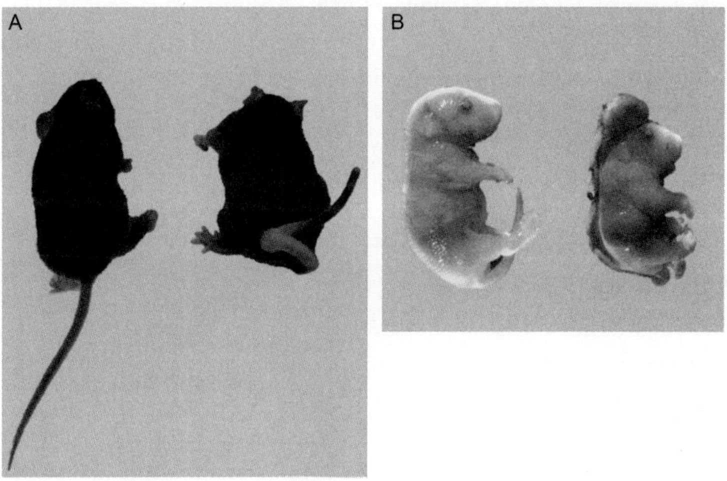

Figure 10.2 Phenotype of Lp/+ and of Lp/Lp mice as compared to WT. (A) Lp/+ heterozygotes are recognizable by their characteristic "looped" tail, while (B) Lp/Lp homozygotes are affected with the NTD craniorachischisis and die *in utero*. (For color version of this figure, the reader is referred to the online version of this chapter.)

open from midbrain–hindbrain juncture to tail end (Greene, Gerrelli, Van Straaten, & Copp, 1998; Fig. 10.2). Using positional cloning and functional complementation in mutant mice, we identified the gene mutated in *Lp* and gave it the provisional appellation *Ltap* (Loop-tail-associated protein), which was subsequently changed to *Vangl2* (Van Gogh-like 2) (Kibar, Gauthier, Lee, Vidal, & Gros, 2003; Kibar, Vogan, et al., 2001). *Ltap/Vangl2* mRNA and protein are expressed throughout the neuroepithelium of the forming NT prior to (E7.5), during (E7.5–8.5), and after (E9.5) NT closure (Kibar, Underhill, et al., 2001; Kibar, Vogan, et al., 2001). Ltap/Vangl2 protein is also developmentally expressed in additional tissues (Torban et al., 2007), and indeed, *Lp/Lp* embryos display defects in the inner ear (organization of hair cells of the cochlea), the heart (outflow tract defects), and the musculoskeletal system (fused ribs) (see below). This suggests that *Ltap/Vangl2* is not only required for successful NT formation but is also instrumental for multiple additional aspects of embryonic development (Montcouquiol et al., 2003; Phillips, Murdoch, Chaudhry, Copp, & Henderson, 2005; Torban et al., 2008).

Ltap/Vangl2 gene encodes a 521-amino acid-residue highly hydrophobic integral membrane protein composed of four transmembrane domains (Kibar, Underhill, et al., 2001; Torban, Wang, Groulx, & Gros, 2004), two solvent accessible extracytoplasmic loops, and intracellular amino and carboxy termini (Iliescu, Gravel, Horth, Apuzzo, & Gros, 2011). The large carboxy-terminal cytoplasmic domain includes a PDZ-binding motif (PBM) at its extremity (Fig. 10.3). We identified independent loss-of-function *Ltap/Vangl2* mutations at the three known *Lp* alleles: the original *Lp* (S464N) and the chemically induced Lp^{m1Jus} (D255E) and Lp^{m2Jus} (R259L) alleles (Guyot et al., 2011; Kibar, Underhill, et al., 2001; Kibar, Vogan, et al., 2001). These mutations are specific for *Lp* mice and affect amino acid residues that are highly conserved throughout evolution. Subcellular localization experiments by immunofluorescence in transfected MDCK cells indicate that while the wild-type protein is expressed at the plasma membrane, *Lp*-associated variants are mistargeted and are retained in the endoplasmic reticulum. *Lp*-associated mutations cause Vangl protein instability, reduced half-life, and increased proteasome-dependent degradation, all possibly secondary to a basic misfolding defect (Gravel, Iliescu, Horth, Apuzzo, & Gros, 2010; Iliescu, Gravel, Horth, Kibar, & Gros, 2011; Torban, Wang, et al., 2004).

Vertebrates have a second *Vangl* gene (*Vangl1*) that encodes a protein sharing ~75% similarity with Vangl2 (Torban et al., 2008). In the developing NT, *Vangl1* mRNA and protein are expressed in the most ventral part of NT (floor plate) and notochord in a pattern complementary to that of Vangl2 (Torban et al., 2008). Embryo homozygotes for a targeted deletion of *Vangl1* ($Vangl1^{gt/gt}$) (Antic et al., 2010) display craniorachischisis at low

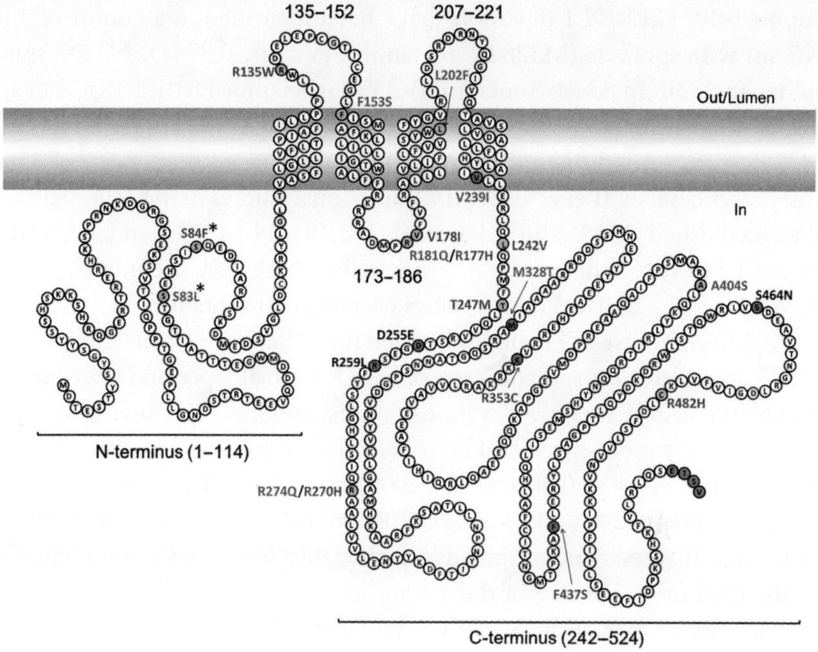

Figure 10.3 Topological model, structural features, and mutant variants of the Vangl protein family. The primary sequence and secondary structure of human VANGL1 protein as established by epitope mapping is presented. The nomenclature adopted in the figure has been standardized to residues in hVANGL1, which is used as molecular backbone. Sequence motifs and individual amino acid residues defining sequence landmarks are depicted in different colors. This includes Vangl protein variants (both human hVANGL1 and hVANGL2 and mouse mVangl2) associated with neural tube defects in mice and humans (orange, pink) and where loss of function has already been demonstrated (pink) or has yet to be characterized (orange). Residues altered in mVangl2 (D255E, R259L, S464N) are depicted in black, hVANGL1 (S83L, F153S, R181Q, L202F, V239I, R274Q, M328T, A404S) are depicted in red, and hVANGL2 (S84F, R135W, R177H, V178I, L242V, T247M, R270H, R353C, F437S, R482H) are depicted in blue. Also depicted is the ETSV sequence corresponding to a PDZ-binding motif (green) and two serine phosphorylation sites for which variants have been detected in human NTD patients (*; S83L in hVANGL1 and S84F in hVANGL2). (See Color Insert.)

frequency. The double heterozygote for *Vangl2* and *Vangl1* mutations (*Vangl2*$^{Lp/+}$:*Vangl1*$^{gt/+}$) (Torban et al., 2008) or homozygous for both *Vangl1* and *Vangl2* (Song et al., 2010) displays craniorachischisis at high frequency. These results point at a genetic interaction between *Vangl1* and *Vangl2* and indicate that both proteins play a critical role in NT formation. Recent studies have also identified a physical interaction between Vangl1 and Vangl2 proteins (Yin, Copley, Goodrich, & Deans, 2012).

Importantly, *VANGL1* mutations have been identified in a cohort of 144 patients with sporadic (M328T) and familial (V239I, R274Q) NTDs (spina bifida), including a *de novo* mutation (V239I) in a familial setting (Kibar et al., 2007). Independently, mutations in human *VANGL2* (S84F, R353C, F437S) were recently identified in stillborn fetuses with anencephaly and holoprosencephaly (Lei et al., 2010). Functional studies in model organisms have established that VANGL1 variants V239I and M328T and VANGL2 variant F347S are functionally inactive (Kibar et al., 2007; Lei et al., 2010; Reynolds et al., 2010). These findings confirm that mutations in *VANGL1* and *VANGL2* cause NTDs in human and mice. Screening of a larger cohort of 673 nonsyndromic sporadic cases of NTDs (cranial, open and closed spinal dysraphisms) in patients of various ethnic origins identified additional independent patient-specific *VANGL1* (S83L, F153S, R181Q, L202F, R274Q, and A404S) (Kibar et al., 2009) and *VANGL2* (S88F, R135W, R177H, V178I, L242V, T247M, R270H, and R482H) (Kibar et al., 2011) mutations. In these studies, putative pathogenic mutations were identified on the basis of (a) absence of the variant in ethnically matched controls, (b) evolutionary conservation of the residue affected, and (c) nonconservative nature of the variant. Of notable interest are Arginine at positions R181 and R274 (VANGL1 numbering) which were found independently mutated to Histidine (H-VANGL2^{R177H}) or Glutamine (Q-VANGL1^{R181Q}) for R181 and mutated to H (VANGL2^{R270H}) or Q (VANGL1^{R274Q}) for R274 in unrelated patients, suggesting a critical role for these two invariant Arginine residues in the function of Vangl proteins. As of today, a total of 22 naturally occurring VANGL protein variants associated with clinical cases of NTDs in humans have been discovered (Fig. 10.3).

2. VANGL PROTEINS, PLANAR CELL POLARITY, AND TISSUE PATTERNING

Vangl proteins are highly conserved in evolution with relatives found in flies (*Vangl/Stbm*), fish (*trilobite/Vangl2*), and frogs (*Xstbm*) (Darken et al., 2002; Jessen et al., 2002; Taylor, Abramova, Charlton, & Adler, 1998; Wolff & Rubin, 1998). The study of these relatives in model organisms has shed considerable light on the function and molecular mechanism of action of mammalian Vangl proteins during development. In flies, mutations in the *Vang* (*Van Gogh*)/*Stbm* (*Strabismus*) gene were initially identified as impairing the planar polarization of several epithelial structures (Gubb & Garcia-Bellido, 1982). In flies, the PCP axis of polarity is visible in the orientation of different structures including the eye unit (ommatidium) and in the uniform polarity of certain appendages such as the hair (trichome) on wing cells. Genetic screens have identified several so-called core PCP genes in which mutations cause defects in polarity of these structures: *frizzled* (*fz*), *disheveled* (*Dsh*), *prickle* (*pk*), *Van Gogh* (*Vang*), *flamingo* (*fmi*), and *diego* (*dgo*) (Gubb & Garcia-Bellido, 1982); extensively reviewed by (McNeill, 2010; Simons & Mlodzik, 2008; Goodrich & Strutt, 2011). Genetic interaction screens *in vivo*, protein:protein interaction experiments *in vitro*, subcellular colocalization studies, and analysis of mosaic clones on fly wings demonstrate that the core PCP proteins assemble into dynamic membrane-bound complexes that rearrange during establishment of PCP in flies. Fz is a 7-transmembrane domains protein that recruits cytoplasmic Dsh to the membrane. Dsh is a modular protein consisting of DIX, PDZ, and DEP domains that can bind Pk, Vang, and Dgo. In flies, Vang can bind Dsh, Pk, and the cytoplasmic ankyrin-repeat protein Dgo. The mechanism by which these proteins establish PCP in flies involves their asymmetric distribution and selective enrichment that culminate in the appearance of two polarized membrane-bound complexes: Dsh/Fz at the distal side and Vang/Pk at the proximal side of wing epithelial cells. This asymmetry in distribution of the core PCP proteins establishes critical positional information. Fmi, a 7-TM protein with a long extracellular domain encoding nine cadherin domains (that enable homotypic interactions), interacts with both Dsh/Fz and Vang/Pk complexes stabilizing and propagating the PCP signal through a "domino" effect (McNeill, 2010; Simons & Mlodzik, 2008; Goodrich & Strutt, 2011). Genetic screens in flies have also identified cadherin proteins Fat (Ft), Dachsous (Ds), and the Golgi kinase Four-jointed (Fj) as additional general PCP genes (Rawls, Guinto, & Wolff, 2002; Yang, Axelrod, & Simon, 2002; Zeidler, Perrimon, & Strutt, 1999) that act to stabilize and propagate the PCP signal. Other evidence suggest that the fat/dachsous pathway may act in parallel to the frizzled/Flamingo pathway (Casal, Lawrence, & Struhl, 2006). Finally, the proteins Fuzzy and Inturned have also been identified as wing-specific PCP "effector" genes in flies (Collier & Gubb, 1997; Gubb & Garcia-Bellido, 1982).

This PCP pathway is conserved in mammals and has been the focus of intense investigation in the past few years. These studies have shown that PCP genes play a critical role in the development and maintenance of broad range of tissues and organs. In mammals, there exist 2 *Vangl* genes, 10 *Frizzled* (*Fz*), 4 *Prickle* (*Pk*), 3 *Dishvelled* (*Dvl*), 3 *Flamingo* (*Celsr*), 1 *Diego*

(*Inversin*) (McNeill, 2010; Simons & Mlodzik, 2008), and 4 *Fat* (Saburi, 2012; Saburi et al., 2008). *Inturned* and *fuzzy* relatives have also been described and studied in vertebrates (Gray et al., 2009; Heydeck, Zeng, & Liu, 2009; Park, Haigo, & Wallingford, 2006). Finally, a number of PCP genes, whose mutations cause defects in planar polarity only in vertebrates, have been identified including *Ptk7* (Lu et al., 2004; Yen et al., 2009), *Scribble* (Murdoch et al., 2003), *Ror2* (Gao et al., 2011), and *Sec24b* (Merte et al., 2010; Wansleeben et al., 2010). The presence of multiple paralogous PCP genes that are expressed at a variable degree of tissue and cell specificity and often display functional redundancy has complicated the functional characterization of individual PCP genes. In addition, embryonic lethality caused by inactivation of some PCP genes (*Vangl2*, *Celsr1*) has until recently precluded a study of their role in the structure, function, and maintenance of adult tissues and possible association with pathologies in the adults.

At the cellular level, Vangl and other mammalian PCP proteins developmentally regulate the structure, positioning, and polarity of cellular appendages in different cell types. They are also involved in the asymmetric morphology, polarization, oriented cell division, migration, and redistribution of many cell populations. Consequently, mutations in PCP genes have been shown to affect biogenesis, structure, and function of a surprisingly large number of tissues and organs (Table 10.1; McNeill, 2010; Simons & Mlodzik, 2008; Tissir & Goffinet, 2010; Wansleeben & Meijlink, 2011). For example, the uniform orientation of stereociliary bundles on outer and inner hair cells of the cochlear duct (organ of Corti) is a classical manifestation of PCP in mammals. Mutations not only in *Vangl1* (Torban et al., 2008) and *Vangl2* (Montcouquiol et al., 2003) but also in other PCP genes such as *Dvl1/Dvl2* (Wang et al., 2005), *Dvl3* (Etheridge et al., 2008), *Fz3/Fz6* (Wang, Guo, & Nathans, 2006), *Fat4* (Saburi et al., 2008), and *Celsr1* (Curtin et al., 2003) disrupt PCP of the stereociliary bundles to a varying degree.

In vertebrates, PCP proteins also participate in morphogenic activity designated convergent extension (CE) movements. CE plays a critical role in several aspects of embryogenesis and this process has been particularly well studied in *Xenopus*. CE is the process in which a tissue or epithelium narrows in one axis and lengthens in a perpendicular axis, allowing tissue remodeling (Keller, 1975, 1976; Wallingford, Fraser, & Harland, 2002). CE plays a critical role during gastrulation, neurulation, and organogenesis. During CE, cells form mediolateral protrusions that generate traction used for

Table 10.1 Phenotype of Vangl2 and Vangl1 mutant mice[a]

Vangl2 (*Loop-tail* mutant) phenotype	
Heterozygote	
– Looped, kinked tail – Imperforate vagina – Enlargement and deformity of lateral brain ventricles – Wobbly head movements – Delayed neural tube closure (remains open at 8–9 somite stage)	Strong and Hollander (1949) Strong and Hollander (1949) Wilson and Center (1977) van Abeelen and Raven (1968) Wang, Hamblet, et al. (2006)
Homozygote[b]	
– Neural tube defects: craniorachischisis (Lp^{D255E} and Lp^{S464N} alleles) or spina bifida (hypomorphic Lp^{R259L} allele)	Strong and Hollander (1949) Guyot et al. (2011) Kibar, Vogan, et al. (2001) Murdoch et al. (2003) Strong and Hollander (1949)
– Abnormal embryo turning	Smith and Stein (1962)
– Open eye leads – PCP defects of hair cell stereociliary bundles in organ of Corti – Reduced size and abnormal morphology of cochlea	Kibar, Vogan, et al. (2001) Montcouquiol et al. (2003) Torban et al. (2008)
– Cardiac defects: abnormal heart looping (E8–E10.5), double outlet right ventricle, ventricular septal defect	Henderson et al. (2001)
– Optic nerve pathfinding defects	Rachel, Murdoch, Beermann, Copp, and Mason (2000)
– Facial branchiomotor neuron migration defects	Glasco et al. (2012)
– Cortical neuronal differentiation defects	Lake and Sokol (2009)

Continued

Table 10.1 Phenotype of Vangl2 and Vangl1 mutant mice[a]—cont'd

– Lung branching morphogenesis defects	Yates, Schnatwinsky, et al. (2010)
– Kidney tubular branching and glomerular maturation defects	Yates, Papakrivopoulos, et al. (2010)
– Musculoskeletal system: abnormal rib morphology (rib fusion), bifurcation, reduced rib number, delayed rib ossification, abnormal vertebrae development/morphology	Stein and Mackensen (1957)
– Reduced fore-stomach size and arch	Matsuyama, Aizawa, and Shimono (2009)
Vangl1 (gene-trap) mutant phenotype	
Heterozygote	
– No phenotype	Torban et al. (2008), Song et al. (2010)
Homozygote	
– No phenotype – Low percent of mice with craniorachischisis	Song et al. (2010) Torban et al. (2008) Antic et al. (2010)
Double Vangl1:Vangl2 mutant phenotype	
Double heterozygote (*Vangl2*$^{Lp/+}$; *Vangl1*$^{gt/+}$)	
– Neural tube defect: craniorachischisis in 60% of mice – PCP defects in the inner ear	Torban et al. (2008)
Double homozygote (*Vangl2*$^{-/-}$; *Vangl2*$^{gt/gt}$)	
– Neural tube defect: craniorachischisis – Laterality defects: embryo turning and heart looping – PCP defects in the inner ear, particularly in the inner hair cells – Randomized position of primary cilia in posterior notochord cells	Song et al. (2010)

[a]This list may not be complete since only the most prominent phenotypes are listed.
[b]Phenotypes in all other organs outside of neural tube are for semidominant Lp^{D255E} and Lp^{S464N} alleles.

mediolateral intercalation. CE movements are responsible for the narrowing and lengthening of the neural plate during NT closure. Homozygosity for mutations in *Vangl1* (Antic et al., 2010), *Vangl2* (Kibar, Underhill, et al., 2001), *Dvl1/Dvl2* (Wang, Guo, et al., 2006), *Dvl2/Dvl3* (Etheridge et al., 2008), *Celsr1* (Curtin et al., 2003), *Fz3/Fz6* (Wang, Guo, et al., 2006), *Scribble* (Murdoch et al., 2003), *Ptk7* (Lu et al., 2004), and *Sec24b* (Merte et al., 2010) causes NTDs of different severity, including the characteristic very severe craniorachischisis. Likewise, severe NTDs (and the PCP defects) are often detected in mutant mice double heterozygous for mutations in *Vangl2* (*Vangl2*$^{Lp/+}$) and other PCP genes from the above-mentioned list, highlighting the key role of *Vangl2* in this process. CE movements are also essential for cardiogenesis; mutations in *Vangl2* and *Fz1/Fz2* cause cardiac defects in the form of aberrant right subclavian artery and double outlet right ventricle (Torban et al., 2008; Yu et al., 2010).

The careful morphological and cytological analyses of anomalies detected in *Lp* mice have shown that, in addition to NT closure and cardiogenesis, Vangl2-dependent PCP and CE activities, are required for patterning of many additional tissues and organs (Table 10.1). *Vangl2* (and *Celsr1*) is expressed in restricted spatial domains of epithelial cells in developing lung. *Lp/Lp* embryos display defects in lung branching morphogenesis manifested as lung hypoplasia, reduction in number and width of airways, thickening of the interstitial mesenchyme, defective saccular formation, and impaired branching in response to FGF10 (Yates, Schnatwinkel, et al., 2010). Vangl2 protein is expressed in developing skin (embryonic epidermis), and homozygous *Lp* embryos have hair follicles that are misangled, with altered anterior–posterior polarization of interfollicular epithelium (Devenport & Fuchs, 2008). Vangl2 is expressed at the lateral edges of the apical pole of uterine epithelial cells, and *Lp* homozygotes display loss of columnar appearance of uterine epithelium. As a result, the *Lp/Lp* and *Lp* heterozygotes both show gross anatomical defects of reproduction system and the *Lp* heterozygotes are often sterile (Vandenberg & Sassoon, 2009). Vangl2 is expressed in the developing gut, including epithelial cells of the developing stomach and duodenum (Torban et al., 2007), and *Lp/Lp* E13.5 embryos display a smaller fore-stomach with a loss of apicobasal polarity and oriented cell division (Matsuyama et al., 2009). Early studies detected *Vangl* mRNA expression in the developing limbs in the E10.5–13.5 embryos (Torban et al., 2008). In addition, Gao et al. (2011) reported asymmetric Vangl2 protein localization in developing limbs chondrocytes. Recent studies identified altered shapes and dimensions of early limb buds in *Lp/Lp* embryos, lack of chondrogenic condensation due to increased cell death in distal limbs, and digits and limb defects including loss of phalanges and nails (Wang, Sinha, Jiao, Serra, & Wang, 2011). *Lp/Lp* embryos also show severe thoracic skeletal defects in the form of solid asymmetrical sternum lacking segmentation; fused ribs and lack of vertebral rib fusion; and abnormal size, shape, and position of vertebrae which are frequently fused (Stein & Mackensen, 1957). In the developing eye at E16.5, Vangl2 protein is expressed in the cornea and in migrating peridermal cells at the time of eyelid closure, immediately prior to keratinization of the fused eyelids (Torban et al., 2007); *Lp/Lp* embryos exhibit a defect in eyelid closure (Greene et al., 1998; Torban et al., 2008).

Vangl2 is also required for asymmetric cell division and specification of different cell types (reviewed in Wansleeben & Meijlink, 2011). For example, examination of *Lp/Lp* brain shows a decrease in the size of the neocortex that is caused by the premature differentiation of the neuronal progenitors at the expense of later born neurons (Lake & Sokol, 2009). These and additional studies *in vitro* demonstrated a role for Vangl2 in suppressing progenitor differentiation and promoting cell fate diversity. Vangl2 appeared to control spindle orientation in neuronal precursors, thereby promoting asymmetric cell division during neuronal fate specification (Lake & Sokol, 2009). Further, Vangl2 is also required for tangential migration of branchiomotor neurons (Bingham, Higashijima, Okamoto, & Chandrasekhar, 2002; Bingham et al., 2010; Jessen et al., 2002), anterior–posterior patterning of the axons of the monoaminergic neurons in brainstem (Fenstermaker et al., 2010), and an establishment of the retinal axon trajectories (Rachel et al., 2000). Elimination of *Vangl2* in *Lp* mice or in other animal models causes alterations in these neuronal populations.

3. ROLE OF VANGL2 IN STRUCTURE AND FUNCTION OF CILIATED CELLS

A role of PCP proteins, in general, and Vangl2 in particular, in the structure and function of motile and sensory cilia has recently emerged (Wallingford & Mitchell, 2011). Motile and sensory cilia project from many cell types. While primary cilia transmit key environmental and mechanical cues to the cells bearing them, motile cilia generate fluid flow of critical physiological importance in the lungs, reproductive tract, and central nervous system. Cilia are essential for normal vertebrate development, as they regulate signal transduction in a number of developmental signaling pathways including Sonic Hedgehog signaling (reviewed in Goetz and Anderson (2010) and Jones and Chen (2008). Defects in the biogenesis or maintenance of cilia cause human pathologies known as "ciliopathies" (Badano, Mitsuma, Beales, & Katsanis, 2006), some of the most frequent manifestations of which include polycystic kidney disease (Goetz & Anderson, 2010; Jones & Chen, 2008) and retinal degeneration (Bramall, Wright, Jacobson, & McInnes, 2010; Ramamurthy & Cayouette, 2009). Cilium is formed by a membrane enclosed, microtubule-based axoneme that protrudes into the extracellular space. It is positioned at the cell apical pole by anchoring to an underlying basal body and transitional fibers (Gerdes, Davis, & Katsanis, 2009). Vangl2 and other PCP proteins have been implicated in several aspects of cilia development, polarized localization, and motile function (Wallingford & Mitchell, 2011). PCP proteins Vangl2 (Guirao et al., 2010; Ross et al., 2005), Dvl (Park, Mitchell, Abitua, Kintner, & Wallingford, 2008; Park et al., 2006), and Fat4 (Saburi et al., 2008) have been detected by double immunofluorescence with cilia-specific markers at the base of cilia in certain ciliated cell types, while in other ciliated cell types, Vangl2 appears to be solely localized to the lateral membrane and not at the base of the cilium (Borovina, Superina, Voskas, & Ciruna, 2010; Song et al., 2010), likely reflecting a different tissue-specific requirements for Vangl2 in the cilia-related processes. Gene silencing experiments and studies with dominant-negative mutant alleles in model organisms such as zebrafish and *Xenopus* have shown that ablation of *trilobite* (*Vangl*) (Antic et al., 2010; May-Simera et al., 2010; Mitchell et al., 2009), *Dvl* (Mitchell et al., 2009; Park et al., 2008), *fuzzy* (Gray et al., 2009; Park et al., 2006), and *inturned* (Park et al., 2006; Zeng, Hoover, & Liu, 2010) causes a general reduction in cilia number and cilia length in different ciliated cell types. Such reduction was linked to the defects in the apical docking of basal bodies and/or defects in the organization of the cortical actin network which supports anchoring of the basal bodies beneath apical surface. Mammalian Vangl2 has been shown to be essential for asymmetric positioning and tilting of primary cilia. Altered positioning of basal body and cilia in posterior notochord cells causes altered nodal flow in embryos devoid of Vangl protein function (homozygous *Vangl1*/*Vangl2* double mutants) (Song et al., 2010), although ciliogenesis *per se* is not affected in these cells. Likewise, ablation of *Vangl2* (Guirao et al., 2010) results in altered ciliary alignment in the maturing ependymal cells lining cerebral ventricles that ultimately causes defective mechanical flow of cerebrospinal fluid. Similar studies of maturing ependymal cells in *Celsr2*$^{-/-}$ mutant mice show reduced number of cilia, abnormal organization of the cilia formed, altered tilting, and additional defects in the apical docking of basal bodies, together leading to little nodal flow and severe hydrocephalus (Tissir & Goffinet, 2010). In *Xenopus* and zebrafish, loss of Vangl2-dependent, cilia-mediated nodal flow results in impaired left/right asymmetry (Antic et al., 2010; Borovina et al., 2010; May-Simera et al., 2010). Together, these studies have identified an important role of Vangl2 in cilial polarity and function in different ciliated cells.

Recent genetic interaction studies and biochemical experiments have suggested that Vangl proteins may be involved in supporting structural integrity and function of adult tissues and organs. However, Vangl2 protein function in adult tissues and adult tissues-related diseases has been difficult to study, as *Vangl2* mouse mutations (*Lp*) are embryonic lethal; the recent description of a conditional mutant allele at *Vangl2* (Song et al., 2010) is likely to provide valuable information on the role of Vangl proteins in adult tissues. Recent studies point toward a possible role for Vangl proteins in postnatal maintenance and functions of renal tissues (Fischer & Pontoglio, 2009). In kidney, cilia are required for normal tubular development and to provide environmental cues (e.g., fluid flow information) to cells bearing them (reviewed by Goetz & Anderson, 2010). Single primary cilia are found at the apex of most renal epithelial cells including the cells of the proximal tubules, the loop of Henle, and the collecting ducts. The axoneme and membrane of the primary cilium are formed via a highly specialized intraflagellar transport (IFT) mechanism (Scholey, 2008). The IFT system relies on kinesin and dynein motors and the associated IFT complexes to organize a bidirectional retrograde and anterograde protein transport. All proteins needed for ciliogenesis and/or cilia maintenance and functions are delivered via a polarized vesicular traffic to and from the basal bodies (Badano et al., 2006; Nachury et al., 2007). Mutations in Bardet–Biedl Syndrome genes encoding proteins involved in vesicular traffic (Jin & Nachury, 2009; Jin et al., 2010; Nachury et al., 2007) or in genes encoding proteins that support integrity of the basal bodies and transitional zone (Gerdes et al., 2009; Williams et al., 2011) cause a group of recessive ciliopathies including Bardet–Biedl syndrome, Nephronophthisis, Joubert syndrome, and Meckel–Gruber syndrome (MKKS). These diseases have a various degree of severity (e.g., MKKS fetuses display lethal cranial NTDs) and affect many tissues; however, all of them are characterized by the presence of cystic kidneys (Hildebrandt, Benzing, & Katsanis, 2011).

Genetic studies have identified a striking reciprocal relationship between PCP genes and genes underlying ciliopathies suggesting that both pathways likely intersect at cilia function: (a) Cystic disease occurs upon loss of cilia function and/or loss of oriented cell division, both of which are regulated by PCP signaling in certain tissues (Fischer et al., 2006); (b) Several PCP proteins including Vangl2 are present at the base of cilia in certain cells (Guirao et al., 2010; Park et al., 2006, 2008; Ross et al., 2005); (c) Mice bearing independent mutations in *Fat4* (Saburi et al., 2008) or *Inversin* (relative of *Diego* that is mutated in the Nephronophthisis type 2 in humans (Simons et al., 2005)) display cystic kidneys. Importantly, the effect of *Fat4* knockdown on renal cystic phenotype is exacerbated by crossing *Fat4* mutant to $Vangl2^{Lp/+}$ (Saburi et al., 2008). Inversin was also shown to form a protein complex with Vangl2 and Dvl2 (Simons et al., 2005); (d) $Bbs1^{-/-}$, $Bbs4^{-/-}$, $Bbs6^{-/-}$, and $Mkks^{-/-}$ mutants display PCP defects in the cochlea (orientation of hair cells), while $Bbs4^{-/-}$ mutants additionally exhibit exencephaly and open eyelids. Notably, all *Bbs* and *Mkks* phenotypes in mice are exacerbated by introducing a single copy of mutated *Vangl2* in $Vangl2^{Lp/+}$ heterozygotes (Ross et al., 2005). In addition, Bbs8 has been shown to physically interact with Vangl2 (May-Simera et al., 2010).

The data obtained in mice with mutations of structural ciliary proteins Ift88 and Kif3a are less clear. Excision of *Kif3a* in renal tubules causes cystic kidney disease, with mutant tubular cells displaying defective orientation of mitotic spindles—a well-established PCP defect (Davenport et al., 2007; Pazour et al., 2000). This suggests that cilia integrity is crucial for normal PCP signaling. However, subcellualr localization of Vangl2 protein is not affected in cochlea of $Ift88^{-/-}$ and $Kif3a^{-/-}$ mouse mutants (Jones et al., 2008), suggesting that the PCP pathway may also act upstream from cilial proteins.

In kidney, Vangl2 is developmentally expressed at E9.5 in mesonephric ducts and tubules. At E16.5, Vangl2 protein is prominently expressed in cells and tubules derived from the ureteric bud (ureteric bud tips, collecting duct), as well as in S-shaped bodies derived from the metanephric mesenchyme, where it is present at cell:cell contacts and at basolateral surface (Babayeva, Zilber, & Torban, 2011; Torban et al., 2007). Vangl2 is also expressed in glomeruli at the basolateral side of podocytes (Babayeva et al., 2011). Studies in *Lp/Lp* mice identify kidney abnormalities: smaller size, abnormal shape, fewer ureteric bud branching tips, hyperplastic medulla with distorted glomeruli, dysmorphic, and dilated tubules with attenuated brush borders (Yates, Papakrivopoulou, et al., 2010), suggesting that Vangl2 may be involved in the physiology of ciliated kidney cells and cystic kidney disease in adult renal tissues.

4. CELLULAR AND MOLECULAR MECHANISM OF ACTION OF VANGL2 IN VERTEBRATES

At the molecular level, the PCP signaling pathway in vertebrates shares structural and mechanistic similarities with its fly counterpart, including the formation of asymmetrically distributed membrane PCP complexes in polarized cells and tissues. Indeed, Vangl1 (Antic et al., 2010; Song et al., 2010), Vangl2 (Devenport & Fuchs, 2008; Guirao et al., 2010; Jones & Chen, 2008; Montcouquiol et al., 2006; Song et al., 2010), Fz3 (Jones & Chen, 2008; Montcouquiol et al., 2006; Wang, Guo, et al., 2006), Fz6 (Devenport & Fuchs, 2008; Wang, Hamblet, et al., 2006), Dvl2 (Darken et al., 2002), Dvl3 (Etheridge et al., 2008), Pk2 (Antic et al., 2010; Borovina et al., 2010), and Celsr1 (Devenport & Fuchs, 2008) have all been described as asymmetrically distributed *in vivo* in polarized and/or ciliated cells of different tissues. These include structures such as the cochlea, epidermis, ependyma, neuroepithelium, and the posterior node in developing embryos. In addition, Vangl2 has been shown to physically interact *in vitro* (pull-down/immunoblotting, immunoprecipitation, yeast two-hybrid system) with other components of the PCP machinery including Dvl1/2/3 (Kibar et al., 2007; Torban et al., 2004), Celsr1 (Devenport & Fuchs, 2008), Scribble (Kallay, McNickle, Brennwald, Hubbard, & Braiterman, 2006; Montcouquiol et al., 2006), and more recently Vangl1 (Yin et al., 2012); the *Lp*-associated inactivating *Vangl2* mutations disrupt these interactions either directly or indirectly due to mistargeting of mutant Vangl2 variants. Further, mutations in mammalian PCP proteins Scrible (Montcouquiol et al., 2006), Celsr1 (Devenport & Fuchs, 2008), and Celsr2 (Tissir & Goffinet, 2010) impair membrane targeting of Vangl2 *in vivo* and/or physical interaction with Vangl2 *in vitro*. These observations have led to a basic model for vertebrate PCP signaling (Tissir & Goffinet, 2010) involving asymmetrically positioned Vangl/Pk versus Fz/Dvl membrane complexes that provide polarity information to cells and cell layers. This polarity information would then be propagated to adjacent cells via homotypic interactions between Celsr proteins expressed by

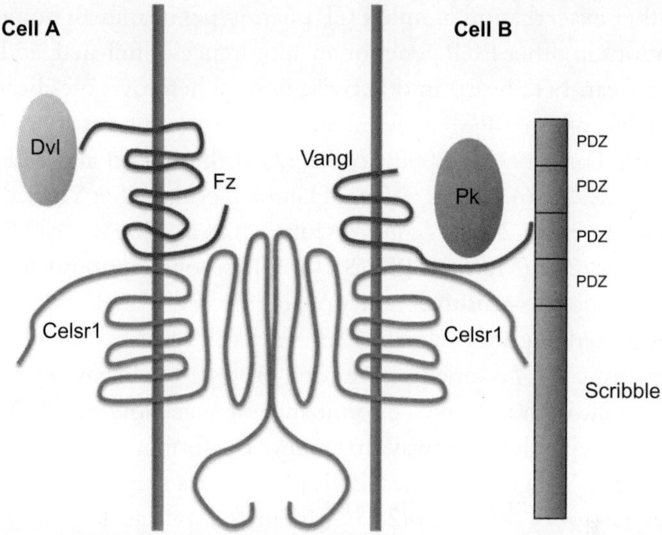

Figure 10.4 Schematic representation of proposed *Vangl2*-dependent PCP complexes in mammalian cells. The intracellular carboxyl terminal domain of Vangl2 has been shown to interact with the cytoplasmic protein Pk and a cytoplasmic segment of membrane-bound Celsr1. The intracellular carboxyl terminal domain of Vangl2 also interacts with the PDZ domains of the cytoplasmic protein Scribble. Cytoplasmic domains of Celsr1 protein molecules from adjacent neighboring cells form homotypic extracellular complexes. (For color version of this figure, the reader is referred to the online version of this chapter.)

neighboring cells (different variations of this model may be active in different cell types) (Fig. 10.4). PCP signaling further requires an interaction between the PDZ-binding domain of Vangl2 and the PDZ-domain of Scribble (Montcouquiol et al., 2006).

A recently discovered Sec24b protein is needed for maturation of Vangl2 positive COPII vesicles and controls membrane targeting of Vangl2 to membrane PCP complexes (Merte et al., 2010; Wansleeben et al., 2010). *Sec24b* mouse mutants display craniorachischisis and inner ear defects, and *Lp*-associated Vangl2 mutant proteins cannot enter COPII vesicles for Sec24b-dependent maturation. In addition, there is an aberrant localization of Vangl2 in Sec24b mutant embryos *in vivo* (Merte et al., 2010; Wansleeben et al., 2010).

Importantly, a unique and critical feature of Vangl2 is that its function represents a key rate-limiting and dosage-sensitive step in the PCP signaling. Heterozygosity for *Vangl2* mutation (*Lp*/+) sensitizes the PCP pathway and causes either exacerbation of mild PCP phenotypes in animals homozygotes for mutations in other PCP genes or an appearance of full-fledged PCP defects (inner ear, NT, heart) in otherwise normal heteroyzgotes bearing one mutated allele of other PCP genes including *Vangl1*(Antic et al., 2010; Song et al., 2010; Torban et al., 2008), *Dvl1/2/3* (Etheridge et al., 2008; Wang, Hamblet, et al., 2006), *Fz1/2* (Wang, Hamblet, et al., 2006; Yu et al., 2010), *Celsr1* (Curtin et al., 2003), *Scribble* (Murdoch et al., 2003), *Ptk7* (Lu et al., 2004), and *Sec24b* (Merte et al., 2010). These observations establish the critical and dosage-sensitive functional role of Vangl2 in molecular processes underlying PCP signaling in many cell types.

Mechanistically, *Lp* variants have been proposed to behave either as partially penetrant with negative codominance or as haploid insufficiency in a gene dosage-dependent pathway. In favor of the former are the observations that (a) several Vangl2-associated PCP phenotypes appear more severe for *Lp* alleles ($Vangl2^{D255E}$, $Vangl2^{S264N}$) than for null allele ($Vangl2^{TM/TM}$) (Yin et al., 2012); (b) Vangl1 and Vangl2 appear to physically interact and, in cotransfection experiments, *Lp* alleles disrupt Vangl1/Vangl2 interactions, as well as trafficking and membrane targeting of Vangl1 and Vangl2 (Yin et al., 2012), and decrease of the posttranslational modification of WT protein (Gao et al., 2011; Song et al., 2010). In support of haploid insufficiency in a gene dosage-dependent pathway are the observations that (a) experimental overexpression or silencing of core PCP genes causes the same phenotype in different animal models tested; (b) all experimentally induced (Gao et al., 2011; Song et al., 2010) or naturally occurring *Vangl2* mutations (Guyot et al., 2011; Kibar, Underhill, et al., 2001; Kibar, Vogan, et al., 2001) described so far show varying degrees of the same phenotype (looped tail, inner ear defects) in heterozygotes and homozygotes *in vivo*; (c) *Lp*-associated Vangl2 protein variants are expressed at lower levels *in vivo* (Montcouquiol et al., 2006; Shafer, Onishi, Lo, Colakoglu, & Zou, 2011; Torban et al., 2007) and display reduced stability and shorter half-life when tested *in vitro* (Gravel et al., 2010; Iliescu, Gravel, Horth, Kibar, et al., 2011); (d) colocalization studies by double immunofluorescence and confocal microscopy in transfected MDCK cells show that expression of $Vangl2^{D255E}$ has no effect on membrane targeting of WT Vangl2 (Gravel et al., 2010). Finally, genetic background has a strong influence on penetrance and expressivity of *Lp*-associated phenotypes (Strong & Hollander, 1949), which complicates analysis of mode of inheritance of *Vangl2* mutations and the associated interpretation of experimental results obtained *in vivo*.

Finally, phosphorylation appears to be an important regulatory mechanism for Vangl protein function. Early work by Kalabis, Rosenberg, and Podolsky (2006) showed that Vangl1 is phosphorylated in response to intestinal trefoil factor (ITF)/TFF3, which reduces Vangl1 plasma membrane localization and may affect cell motility. More recently, Gao et al. (2011) observed that Vangl2 interacts with Ror2 and is phosphorylated both *in vivo* and *in vitro*. In addition, Vangl2 phosphorylation is stimulated by Wnt ligands. A Ser/Thr cluster at Vangl2 amino terminus is targeted for phosphorylation, and the importance of this event is revealed by the presence of hVANGL1 (S83L) and hVANGL2 (S84F) variants at this site in human NTD patients. Finally, coexpression of *Lp*-associated variant Vangl2^{S464N} reduced phosphorylation of the WT protein expressed in the same cell. These results have provided a model in which Vangl2 phosphorylation and associated role in PCP are regulated by Wnt ligands (Gao et al., 2011).

5. CONCLUSIONS

The pivotal role of Vangl2 in NT formation was originally established 10 years ago with the discovery that mutations in *Vangl2* are responsible for the severe NTD craniorachischisis in the well-known mouse mutant *Loop-tail* (Kibar, Vogan, et al., 2001). The careful examination of morphology and dynamic developmental processes altered in *Lp* mice have revealed that Vangl2 plays a similarly important role in patterning and development of a many tissues and organs. The genetic interaction studies in mice heterozygous for *Vangl2*$^{Lp/+}$ and for mutations in other genes have linked Vangl2 and the uniquely Vangl2-sensitive PCP pathway to a number of additional and unsuspected phenomena such as asymmetric cell division, cell migration, and function of key cellular appendages such as cilia. With such a rich spectrum of biological activities essential for normal embryonic development, it is tempting to speculate that Vangl2 also plays an important role in the structural integrity and functional homeostasis of adult organs, tissues, and cells. The recent generation of conditional mutant alleles at *Vangl2* enabling tissue and temporal-specific genetic ablation of *Vangl2* will likely be very informative and may also point at the possible association of *Vangl2* with the diseases of adult tissues.

ACKNOWLEDGMENTS

E. T. and P. G. are supported by grants from the Canadian Institute of Health Research. E. T. is a recipient of the Fonds de Recherche en Santé du Quebec award, and P. G. is a James professor, McGill University.

REFERENCES

Antic, D., Stubbs, J. L., Suyama, K., Kintner, C., Scott, M. P., & Axelrod, J. D. (2010). Planar cell polarity enables posterior localization of nodal cilia and left-right axis determination during mouse and Xenopus embryogenesis. *PloS One*, *5*, e8999.

Babayeva, S., Zilber, Y., & Torban, E. (2011). Planar cell polarity pathway regulates actin rearrangement, cell shape, motility, and nephrin distribution in podocytes. *American Journal of Physiology. Renal Physiology*, *300*, F549–F560.

Badano, J. L., Mitsuma, N., Beales, P. L., & Katsanis, N. (2006). The ciliopathies: An emerging class of human genetic disorders. *Annual Review of Genomics and Human Genetics*, *7*, 125–148.

Bingham, S., Higashijima, S., Okamoto, H., & Chandrasekhar, A. (2002). The Zebrafish trilobite gene is essential for tangential migration of branchiomotor neurons. *Developments in Biologicals*, *242*, 149–160.

Bingham, S. M., Sittaramane, V., Mapp, O., Patil, S., Prince, V. E., & Chandrasekhar, A. (2010). Multiple mechanisms mediate motor neuron migration in the zebrafish hindbrain. *Developmental Neurobiology*, *70*, 87–99.

Borovina, A., Superina, S., Voskas, D., & Ciruna, B. (2010). Vangl2 directs the posterior tilting and asymmetric localization of motile primary cilia. *Nature Cell Biology*, *12*, 407–412.

Bramall, A. N., Wright, A. F., Jacobson, S. G., & McInnes, R. R. (2010). The genomic, biochemical, and cellular responses of the retina in inherited photoreceptor degenerations and prospects for the treatment of these disorders. *Annual Review of Neuroscience*, *33*, 441–472.

Casal, J., Lawrence, P. A., & Struhl, G. (2006). Two separate molecular systems, Dachsous/Fat and Starry night/Frizzled, act independently to confer planar cell polarity. *Development*, *133*, 4561–4572.

Collier, S., & Gubb, D. (1997). Drosophila tissue polarity requires the cell-autonomous activity of the fuzzy gene, which encodes a novel transmembrane protein. *Development*, *124*, 4029–4037.

Copp, A. J., & Greene, N. D. (2010). Genetics and development of neural tube defects. *The Journal of Pathology*, *220*, 217–230.

Curtin, J. A., Quint, E., Tsipouri, V., Arkell, R. M., Cattanach, B., Copp, A. J., et al. (2003). Mutation of Celsr1 disrupts planar polarity of inner ear hair cells and causes severe neural tube defects in the mouse. *Current Biology*, *13*, 1129–1133.

Darken, R. S., Scola, A. M., Rakeman, A. S., Das, G., Mlodzik, M., & Wilson, P. A. (2002). The planar polarity gene strabismus regulates convergent extension movements in Xenopus. *The EMBO Journal*, *21*, 976–985.

Davenport, J. R., Watts, A. J., Roper, V. C., Croyle, M. J., van Groen, T., Wyss, J. M., et al. (2007). Disruption of intraflagellar transport in adult mice leads to obesity and slow-onset cystic kidney disease. *Current Biology*, *17*, 1586–1594.

Devenport, D., & Fuchs, E. (2008). Planar polarization in embryonic epidermis orchestrates global asymmetric morphogenesis of hair follicles. *Nature Cell Biology*, *10*, 1257–1268.

Etheridge, S. L., Ray, S., Li, S., Hamblet, N. S., Lijam, N., Tsang, M., et al. (2008). Murine dishevelled 3 functions in redundant pathways with dishevelled 1 and 2 in normal cardiac outflow tract, cochlea, and neural tube development. *PLoS Genetics*, *4*, e1000259.

Fenstermaker, A. G., Prasad, A. A., Bechara, A., Adolfs, Y., Tissir, F., Goffinet, A., et al. (2010). Wnt/planar cell polarity signaling controls the anterior-posterior organization of monoaminergic axons in the brainstem. *The Journal of Neuroscience*, *30*, 16053–16064.

Fischer, E., Legue, E., Doyen, A., Nato, F., Nicolas, J. F., Torres, V., et al. (2006). Defective planar cell polarity in polycystic kidney disease. *Nature Genetics*, *38*, 21–23.

Fischer, E., & Pontoglio, M. (2009). Planar cell polarity and cilia. *Seminars in Cell & Developmental Biology*, *20*, 998–1005.

Gao, B., Song, H., Bishop, K., Elliot, G., Garrett, L., English, M. A., et al. (2011). Wnt signaling gradients establish planar cell polarity by inducing Vangl2 phosphorylation through Ror2. *Developmental Cell*, *20*, 163–176.

Gerdes, J. M., Davis, E. E., & Katsanis, N. (2009). The vertebrate primary cilium in development, homeostasis, and disease. *Cell*, *137*, 32–45.

Glasco, D. M., Sittaramane, V., Bryant, W., Fritzsch, B., Sawant, A., Paudyal, A., et al. (2012). The mouse Wnt/PCP protein Vangl2 is necessary for migration of facial branchiomotor neurons, and functions independently of Dishevelled. *Developments in Biologicals*, *369*(2), 211–222.

Goetz, S. C., & Anderson, K. V. (2010). The primary cilium: A signalling centre during vertebrate development. *Nature Reviews Genetics*, *11*, 331–344.

Goodrich, L. V., & Strutt, D. (2011). Principles of planar polarity in animal development. *Development*, *138*, 1877–1892.

Gravel, M., Iliescu, A., Horth, C., Apuzzo, S., & Gros, P. (2010). Molecular and cellular mechanisms underlying neural tube defects in the loop-tail mutant mouse. *Biochemistry*, *49*, 3445–3455.

Gray, R. S., Abitua, P. B., Wlodarczyk, B. J., Szabo-Rogers, H. L., Blanchard, O., Lee, I., et al. (2009). The planar cell polarity effector Fuz is essential for targeted membrane trafficking, ciliogenesis and mouse embryonic development. *Nature Cell Biology*, *11*, 1225–1232.

Greene, N. D., Gerrelli, D., Van Straaten, H. W., & Copp, A. J. (1998). Abnormalities of floor plate, notochord and somite differentiation in the loop-tail (Lp) mouse: A model of severe neural tube defects. *Mechanisms of Development*, *73*, 59–72.

Greene, N. D., Stanier, P., & Copp, A. J. (2009). Genetics of human neural tube defects. *Human Molecular Genetics*, *18*, R113–R129.

Gubb, D., & Garcia-Bellido, A. (1982). A genetic analysis of the determination of cuticular polarity during development in Drosophila melanogaster. *Journal of Embryology and Experimental Morphology*, *68*, 37–57.

Guirao, B., Meunier, A., Mortaud, S., Aguilar, A., Corsi, J. M., Strehl, L., et al. (2010). Coupling between hydrodynamic forces and planar cell polarity orients mammalian motile cilia. *Nature Cell Biology*, *12*, 341–350.

Guyot, M. C., Bosoi, C. M., Kharfallah, F., Reynolds, A., Drapeau, P., Justice, M., et al. (2011). A novel hypomorphic Looptail allele at the planar cell polarity Vangl2 gene. *Developmental Dynamics*, *240*, 839–849.

Harris, M. J., & Juriloff, D. M. (2010). An update to the list of mouse mutants with neural tube closure defects and advances toward a complete genetic perspective of neural tube closure. *Birth Defects Research. Part A, Clinical and Molecular Teratology*, *88*, 653–669.

Henderson, D. J., Conway, S. J., Greene, N. D., Gerrelli, D., Murdoch, J. N., Anderson, R. H., et al. (2001). Cardiovascular defects associated with abnormalities in midline development in the Loop-tail mouse mutant. *Circulation Research*, *89*, 6–12.

Heydeck, W., Zeng, H., & Liu, A. (2009). Planar cell polarity effector gene Fuzzy regulates cilia formation and Hedgehog signal transduction in mouse. *Developmental Dynamics*, *238*, 3035–3042.

Hildebrandt, F., Benzing, T., & Katsanis, N. (2011). Ciliopathies. *The New England Journal of Medicine*, *364*, 1533–1543.

Iliescu, A., Gravel, M., Horth, C., Apuzzo, S., & Gros, P. (2011). Transmembrane topology of mammalian planar cell polarity protein Vangl1. *Biochemistry*, *50*, 2274–2282.

Iliescu, A., Gravel, M., Horth, C., Kibar, Z., & Gros, P. (2011). Loss of membrane targeting of Vangl proteins causes neural tube defects. *Biochemistry*, *50*, 795–804.

Jessen, J. R., Topczewski, J., Bingham, S., Sepich, D. S., Marlow, F., Chandrasekhar, A., et al. (2002). Zebrafish trilobite identifies new roles for Strabismus in gastrulation and neuronal movements. *Nature Cell Biology*, *4*, 610–615.

Jin, H., & Nachury, M. V. (2009). The BBSome. *Current Biology*, *19*, R472–R473.

Jin, H., White, S. R., Shida, T., Schulz, S., Aguiar, M., Gygi, S. P., et al. (2010). The conserved Bardet-Biedl syndrome proteins assemble a coat that traffics membrane proteins to cilia. *Cell*, *141*, 1208–1219.

Jones, C., & Chen, P. (2008). Primary cilia in planar cell polarity regulation of the inner ear. *Current Topics in Developmental Biology*, *85*, 197–224.

Jones, C., Roper, V. C., Foucher, I., Qian, D., Banizs, B., Petit, C., et al. (2008). Ciliary proteins link basal body polarization to planar cell polarity regulation. *Nature Genetics*, *40*, 69–77.

Kalabis, J., Rosenberg, I., & Podolsky, D. K. (2006). Vangl1 protein acts as a downstream effector of intestinal trefoil factor (ITF)/TFF3 signaling and regulates wound healing of intestinal epithelium. *The Journal of Biological Chemistry*, *281*, 6434–6441.

Kallay, L. M., McNickle, A., Brennwald, P. J., Hubbard, A. L., & Braiterman, L. T. (2006). Scribble associates with two polarity proteins, Lgl2 and Vangl2, via distinct molecular domains. *Journal of Cellular Biochemistry*, *99*, 647–664.

Keller, R. E. (1975). Vital dye mapping of the gastrula and neurula of Xenopus laevis. I. Prospective areas and morphogenetic movements of the superficial layer. *Developments in Biologicals*, *42*, 222–241.

Keller, R. E. (1976). Vital dye mapping of the gastrula and neurula of Xenopus laevis. II. Prospective areas and morphogenetic movements of the deep layer. *Developments in Biologicals*, *51*, 118–137.

Kibar, Z., Bosoi, C. M., Kooistra, M., Salem, S., Finnell, R. H., De Marco, P., et al. (2009). Novel mutations in VANGL1 in neural tube defects. *Human Mutation*, *30*, E706–E715.

Kibar, Z., Capra, V., & Gros, P. (2007). Toward understanding the genetic basis of neural tube defects. *Clinical Genetics*, *71*, 295–310.

Kibar, Z., Gauthier, S., Lee, S. H., Vidal, S., & Gros, P. (2003). Rescue of the neural tube defect of loop-tail mice by a BAC clone containing the Ltap gene. *Genomics*, *82*, 397–400.

Kibar, Z., Salem, S., Bosoi, C. M., Pauwels, E., De Marco, P., Merello, E., et al. (2011). Contribution of VANGL2 mutations to isolated neural tube defects. *Clinical Genetics*, *80*, 76–82.

Kibar, Z., Underhill, D. A., Canonne-Hergaux, F., Gauthier, S., Justice, M. J., & Gros, P. (2001). Identification of a new chemically induced allele (Lp(m1Jus)) at the loop-tail locus: Morphology, histology, and genetic mapping. *Genomics*, *72*, 331–337.

Kibar, Z., Vogan, K. J., Groulx, N., Justice, M. J., Underhill, D. A., & Gros, P. (2001). Ltap, a mammalian homolog of Drosophila Strabismus/Van Gogh, is altered in the mouse neural tube mutant Loop-tail. *Nature Genetics*, *28*, 251–255.

Lake, B. B., & Sokol, S. Y. (2009). Strabismus regulates asymmetric cell divisions and cell fate determination in the mouse brain. *The Journal of Cell Biology*, *185*, 59–66.

Lei, Y. P., Zhang, T., Li, H., Wu, B. L., Jin, L., & Wang, H. Y. (2010). VANGL2 mutations in human cranial neural-tube defects. *The New England Journal of Medicine*, *362*, 2232–2235.

Lu, X., Borchers, A. G., Jolicoeur, C., Rayburn, H., Baker, J. C., & Tessier-Lavigne, M. (2004). PTK7/CCK-4 is a novel regulator of planar cell polarity in vertebrates. *Nature*, *430*, 93–98.

Matsuyama, M., Aizawa, S., & Shimono, A. (2009). Sfrp controls apicobasal polarity and oriented cell division in developing gut epithelium. *PLoS Genetics*, *5*, e1000427.

May-Simera, H. L., Kai, M., Hernandez, V., Osborn, D. P., Tada, M., & Beales, P. L. (2010). Bbs8, together with the planar cell polarity protein Vangl2, is required to establish left-right asymmetry in zebrafish. *Developments in Biologicals*, *345*, 215–225.

McNeill, H. (2010). Planar cell polarity: Keeping hairs straight is not so simple. *Cold Spring Harbor Perspectives in Biology*, *2*, a003376.

Merte, J., Jensen, D., Wright, K., Sarsfield, S., Wang, Y., Schekman, R., et al. (2010). Sec24b selectively sorts Vangl2 to regulate planar cell polarity during neural tube closure. *Nature Cell Biology*, *12*, 41–46 Suppl. pp. 1–8.

Mitchell, B., Stubbs, J. L., Huisman, F., Taborek, P., Yu, C., & Kintner, C. (2009). The PCP pathway instructs the planar orientation of ciliated cells in the Xenopus larval skin. *Current Biology*, *19*, 924–929.

Montcouquiol, M., Rachel, R. A., Lanford, P. J., Copeland, N. G., Jenkins, N. A., & Kelley, M. W. (2003). Identification of Vangl2 and Scrb1 as planar polarity genes in mammals. *Nature*, *423*, 173–177.

Montcouquiol, M., Sans, N., Huss, D., Kach, J., Dickman, J. D., Forge, A., et al. (2006). Asymmetric localization of Vangl2 and Fz3 indicate novel mechanisms for planar cell polarity in mammals. *The Journal of Neuroscience*, *26*, 5265–5275.

Murdoch, J. N., Henderson, D. J., Doudney, K., Gaston-Massuet, C., Phillips, H. M., Paternotte, C., et al. (2003). Disruption of scribble (Scrb1) causes severe neural tube defects in the circletail mouse. *Human Molecular Genetics*, *12*, 87–98.

Nachury, M. V., Loktev, A. V., Zhang, Q., Westlake, C. J., Peranen, J., Merdes, A., et al. (2007). A core complex of BBS proteins cooperates with the GTPase Rab8 to promote ciliary membrane biogenesis. *Cell*, *129*, 1201–1213.

Park, T. J., Haigo, S. L., & Wallingford, J. B. (2006). Ciliogenesis defects in embryos lacking inturned or fuzzy function are associated with failure of planar cell polarity and Hedgehog signaling. *Nature Genetics*, *38*, 303–311.

Park, T. J., Mitchell, B. J., Abitua, P. B., Kintner, C., & Wallingford, J. B. (2008). Dishevelled controls apical docking and planar polarization of basal bodies in ciliated epithelial cells. *Nature Genetics*, *40*, 871–879.

Pazour, G. J., Dickert, B. L., Vucica, Y., Seeley, E. S., Rosenbaum, J. L., Witman, G. B., et al. (2000). Chlamydomonas IFT88 and its mouse homologue, polycystic kidney disease gene tg737, are required for assembly of cilia and flagella. *The Journal of Cell Biology*, *151*, 709–718.

Phillips, H. M., Murdoch, J. N., Chaudhry, B., Copp, A. J., & Henderson, D. J. (2005). Vangl2 acts via RhoA signaling to regulate polarized cell movements during development of the proximal outflow tract. *Circulation Research*, *96*, 292–299.

Rachel, R. A., Murdoch, J. N., Beermann, F., Copp, A. J., & Mason, C. A. (2000). Retinal axon misrouting at the optic chiasm in mice with neural tube closure defects. *Genesis*, *27*, 32–47.

Ramamurthy, V., & Cayouette, M. (2009). Development and disease of the photoreceptor cilium. *Clinical Genetics*, *76*, 137–145.

Rawls, A. S., Guinto, J. B., & Wolff, T. (2002). The cadherins fat and dachsous regulate dorsal/ventral signaling in the Drosophila eye. *Current Biology*, *12*, 1021–1026.

Reynolds, A., McDearmid, J. R., Lachance, S., De Marco, P., Merello, E., Capra, V., et al. (2010). VANGL1 rare variants associated with neural tube defects affect convergent extension in zebrafish. *Mechanisms of Development*, *127*, 385–392.

Ross, A. J., May-Simera, H., Eichers, E. R., Kai, M., Hill, J., Jagger, D. J., et al. (2005). Disruption of Bardet-Biedl syndrome ciliary proteins perturbs planar cell polarity in vertebrates. *Nature Genetics*, *37*, 1135–1140.

Saburi, S., Hester, I., Fischer, E., Pontoglio, M., Eremina, V., Gessler, M., et al. (2008). Loss of Fat4 disrupts PCP signaling and oriented cell division and leads to cystic kidney disease. *Nature Genetics*, *40*, 1010–1015.

Scholey, J. M. (2008). Intraflagellar transport motors in cilia: Moving along the cell's antenna. *The Journal of Cell Biology*, *180*, 23–29.

Shafer, B., Onishi, K., Lo, C., Colakoglu, G., & Zou, Y. (2011). Vangl2 promotes Wnt/planar cell polarity-like signaling by antagonizing Dvl1-mediated feedback inhibition in growth cone guidance. *Developmental Cell*, *20*, 177–191.

Simons, M., Gloy, J., Ganner, A., Bullerkotte, A., Bashkurov, M., Kronig, C., et al. (2005). Inversin, the gene product mutated in nephronophthisis type II, functions as a molecular switch between Wnt signaling pathways. *Nature Genetics*, *37*, 537–543.

Simons, M., & Mlodzik, M. (2008). Planar cell polarity signaling: From fly development to human disease. *Annual Review of Genetics*, *42*, 517–540.

Smith, L. J., & Stein, K. F. (1962). Planar cell polarity signaling: From fly development to human disease. *Journal of Embryology and Experimental Morphology*, *10*, 73–87.

Song, H., Hu, J., Chen, W., Elliott, G., Andre, P., Gao, B., et al. (2010). Planar cell polarity breaks bilateral symmetry by controlling ciliary positioning. *Nature*, *466*, 378–382.

Stein, K. F., & Mackensen, J. A. (1957). Abnormal development of the thoracic skeleton in mice homozygous for the gene for looped-tail. *The American Journal of Anatomy*, *100*, 205–223.

Strong, L. C., & Hollander, F. (1949). Hereditary Loop-tail in the house mouse: Accompanied by imperforate vagina and with lethal craniorachischisis when homozygous. *The Journal of Heredity*, *40*, 329–334.

Taylor, J., Abramova, N., Charlton, J., & Adler, P. N. (1998). Van Gogh: A new Drosophila tissue polarity gene. *Genetics*, *150*, 199–210.

Tissir, F., & Goffinet, A. M. (2010). Planar cell polarity signaling in neural development. *Current Opinion in Neurobiology*, *20*, 572–577.

Torban, E., Kor, C., & Gros, P. (2004). Van Gogh-like2 (Strabismus) and its role in planar cell polarity and convergent extension in vertebrates. *Trends in Genetics*, *20*, 570–577.

Torban, E., Patenaude, A. M., Leclerc, S., Rakowiecki, S., Gauthier, S., Andelfinger, G., et al. (2008). Genetic interaction between members of the Vangl family causes neural tube defects in mice. *Proceedings of the National Academy of Sciences of the United States of America*, *105*, 3449–3454.

Torban, E., Wang, H. J., Groulx, N., & Gros, P. (2004). Independent mutations in mouse Vangl2 that cause neural tube defects in looptail mice impair interaction with members of the Dishevelled family. *The Journal of Biological Chemistry*, *279*, 52703–52713.

Torban, E., Wang, H. J., Patenaude, A. M., Riccomagno, M., Daniels, E., Epstein, D., et al. (2007). Tissue, cellular and sub-cellular localization of the Vangl2 protein during embryonic development: Effect of the Lp mutation. *Gene Expression Patterns*, *7*, 346–354.

van Abeelen, J. H., & Raven, S. M. (1968). Enlarged ventricles in the cerebrum of loop-tail mice. *Experientia*, *24*, 191–192.

Vandenberg, A. L., & Sassoon, D. A. (2009). Non-canonical Wnt signaling regulates cell polarity in female reproductive tract development via van gogh-like 2. *Development*, *136*, 1559–1570.

Wallingford, J. B. (2005). Neural tube closure and neural tube defects: Studies in animal models reveal known knowns and known unknowns. *American Journal of Medical Genetics. Part C, Seminars in Medical Genetics*, *135C*, 59–68.

Wallingford, J. B., Fraser, S. E., & Harland, R. M. (2002). Convergent extension: The molecular control of polarized cell movement during embryonic development. *Developmental Cell*, *2*, 695–706.

Wallingford, J. B., & Mitchell, B. (2011). Strange as it may seem: The many links between Wnt signaling, planar cell polarity, and cilia. *Genes & Development*, *25*, 201–213.

Wang, Y., Guo, N., & Nathans, J. (2006). The role of Frizzled3 and Frizzled6 in neural tube closure and in the planar polarity of inner-ear sensory hair cells. *The Journal of Neuroscience*, *26*, 2147–2156.

Wang, J., Hamblet, N. S., Mark, S., Dickinson, M. E., Brinkman, B. C., Segil, N., et al. (2006). Dishevelled genes mediate a conserved mammalian PCP pathway to regulate convergent extension during neurulation. *Development*, *133*, 1767–1778.

Wang, J., Mark, S., Zhang, X., Qian, D., Yoo, S. J., Radde-Gallwitz, K., et al. (2005). Regulation of polarized extension and planar cell polarity in the cochlea by the vertebrate PCP pathway. *Nature Genetics*, *37*, 980–985.

Wang, B., Sinha, T., Jiao, K., Serra, R., & Wang, J. (2011). Disruption of PCP signaling causes limb morphogenesis and skeletal defects and may underlie Robinow syndrome and brachydactyly type B. *Human Molecular Genetics*, *20*, 271–285.

Wansleeben, C., Feitsma, H., Montcouquiol, M., Kroon, C., Cuppen, E., & Meijlink, F. (2010). Planar cell polarity defects and defective Vangl2 trafficking in mutants for the COPII gene Sec24b. *Development*, *137*, 1067–1073.

Wansleeben, C., & Meijlink, F. (2011). The planar cell polarity pathway in vertebrate development. *Developmental Dynamics*, *240*, 616–626.

Williams, C. L., Li, C., Kida, K., Inglis, P. N., Mohan, S., Semenec, L., et al. (2011). MKS and NPHP modules cooperate to establish basal body/transition zone membrane associations and ciliary gate function during ciliogenesis. *The Journal of Cell Biology*, *192*, 1023–1041.

Wilson, D. B., & Center, E. M. (1977). Differences in cerebral morphology in 2 stocks of mutant mice heterozygous for the loop-tail (Lp)-gene. *Experientia*, *33*, 1502–1503.

Wolff, T., & Rubin, G. M. (1998). Strabismus, a novel gene that regulates tissue polarity and cell fate decisions in Drosophila. *Development*, *125*, 1149–1159.

Yang, C. H., Axelrod, J. D., & Simon, M. A. (2002). Regulation of Frizzled by fat-like cadherins during planar polarity signaling in the Drosophila compound eye. *Cell*, *108*, 675–688.

Yates, L. L., Papakrivopoulou, J., Long, D. A., Goggolidou, P., Connolly, J. O., Woolf, A. S., et al. (2010). The planar cell polarity gene Vangl2 is required for mammalian kidney-branching morphogenesis and glomerular maturation. *Human Molecular Genetics*, *19*, 4663–4676.

Yates, L. L., Schnatwinkel, C., Murdoch, J. N., Bogani, D., Formstone, C. J., Townsend, S., et al. (2010). The PCP genes Celsr1 and Vangl2 are required for normal lung branching morphogenesis. *Human Molecular Genetics*, *19*, 2251–2267.

Yen, W. W., Williams, M., Periasamy, A., Conaway, M., Burdsal, C., Keller, R., et al. (2009). PTK7 is essential for polarized cell motility and convergent extension during mouse gastrulation. *Development*, *136*, 2039–2048.

Yin, H., Copley, C. O., Goodrich, L. V., & Deans, M. R. (2012). Comparison of phenotypes between different vangl2 mutants demonstrates dominant effects of the Looptail mutation during hair cell development. *PLoS One*, *7*, e31988.

Yu, H., Smallwood, P. M., Wang, Y., Vidaltamayo, R., Reed, R., & Nathans, J. (2010). Frizzled 1 and frizzled 2 genes function in palate, ventricular septum and neural tube closure: General implications for tissue fusion processes. *Development*, *137*, 3707–3717.

Zeidler, M. P., Perrimon, N., & Strutt, D. I. (1999). The four-jointed gene is required in the Drosophila eye for ommatidial polarity specification. *Current Biology*, *9*, 1363–1372.

Zeng, H., Hoover, A. N., & Liu, A. (2010). PCP effector gene Inturned is an important regulator of cilia formation and embryonic development in mammals. *Developments in Biologicals*, *339*, 418–428.

CHAPTER ELEVEN

Wnt Regulation of Planar Cell Polarity (PCP)

Bo Gao[1]
National Human Genome Research Institute, Bethesda, Maryland, USA
[1]Corresponding author: e-mail address: gaob@mail.nih.gov

Contents

Abstract

Planar cell polarity (PCP), a process controlling coordinated, uniformly polarized cellular behaviors in a field of cells, has been identified to be critically required for many fundamental developmental processes. However, a global directional cue that establishes PCP in a three-dimensional tissue or organ with respect to the body axes remains elusive. In vertebrate, while Wnt-secreted signaling molecules have been implicated in regulating PCP in a β-catenin-independent manner, whether they function permissively or act as a global cue to convey directional information is not clearly defined. In addition, the underlying molecular mechanism by which Wnt signal is transduced to core PCP proteins is largely unknown. In this chapter, I review the roles of Wnt signaling in regulating PCP during vertebrate development and update our knowledge of its regulatory mechanism.

Current Topics in Developmental Biology, Volume 101 2012, Published by Elsevier Inc.
ISSN 0070-2153
http://dx.doi.org/10.1016/B978-0-12-394592-1.00008-9

1. INTRODUCTION

A fundamental process in the development of a complex multicellular organism is spatial and temporal control of cell proliferation, differentiation, survival, and polarity. In the past, our understanding about regulatory mechanisms of cell proliferation, apoptosis, and cell-fate determination has been greatly advanced, while relatively little attention was given to cell polarity. Most cell types exhibit some kind of polarity, for instance, epithelial cells show apical–basal polarities while neurons are polarized into domains specialized for either receiving (dendrites) or transmitting (axons) cellular signals (Craig & Banker, 1994; Rodriguez-Boulan & Powell, 1992; Simons & Fuller, 1985). Epithelial cells lining internal and external surface of animal bodies are characterized by apical–basal polarities with apical membrane facing outside of the body or internal lumen and basal membrane attaching to the underlying cells (Simons & Fuller, 1985). Mesenchymal cells, such as fibroblasts, also have polarities. In order to move in certain directions, these cells must form leading and trailing edges. Such front–rear polarity is required for directional migration (Ridley et al., 2003). These types of cell polarities have been extensively studied and continue to be active research fields of cell biology. However, all these polarities are studied mostly at single-cell level. The mechanisms that establish and coordinate cell polarities with other cellular processes in a complex multicellular organism when tissues and organs are shaped still remain largely unknown. An emerging mechanism whereby directional morphogenesis is regulated is planar cell polarity (PCP). PCP, which was originally identified in *Drosophila* as the polarity of epithelial cells within the plane orthogonal to their apical–basal axis, controls coordinated, uniformly polarized cellular behaviors in a field of cells (Gubb & Garcia-Bellido, 1982). More importantly, regulation by PCP has been also extended to mesenchymal cells (Simons & Mlodzik, 2008). With new findings made in recent years, PCP has emerged to be critical for many basic developmental and physiological processes as it plays important roles in morphogenesis of most developing organs and tissues. However, fundamental questions still remained to be answered. This review aims to summarize recent advances in our understanding of how PCP is established, highlighting the role of global cue in initiating and establishing PCP and potential future research directions.

2. Wnt AND PLANAR CELL POLARITY

PCP was first discovered in the insects *Oncopeltus fasciatus* (Lawrence & Shelton, 1975) and then extensively studied in *Drosophila melanogaster*, in which a group of core PCP genes were identified (McNeill, 2010; Seifert & Mlodzik, 2007; Tree, Ma, & Axelrod, 2002; Vladar, Antic, & Axelrod, 2009). Disruption of these core PCP proteins results in misorientation of the bristles in the *Drosophila* wing and disarrangement of the ommatidia in the *Drosophila* compound eye. In vertebrate embryos, such as frog (*Xenopus laevis*) and zebrafish (*Danio rerio*), the same group of PCP proteins regulates convergent extension (CE), a process coordinating extension of the anterior–posterior (A–P) body axis while the mediolateral (ML) axis was narrowed. The zebrafish PCP mutants failed to extend their A–P axis properly leading to a shortened and broadened body axis (Gray, Roszko, & Solnica-Krezel, 2011; Wang & Nathans, 2007). Using CE as a functional PCP readout in vertebrate, frog and zebrafish have become popular model organisms to test genes regulating PCP. In mammals, PCP regulates a diverse array of essential developmental processes, ranging from neural tube closure to determining left–right (L–R) asymmetry, indicating its fundamental roles in the development of vertebrates (Gray et al., 2011). Not surprisingly, more and more PCP mutations have been identified in various human diseases in recent years (Bosoi et al., 2011; De Marco et al., 2012; Frojmark et al., 2011; Kibar et al., 2009; Kibar, Capra, & Gros, 2007; Kibar et al., 2011; Kibar, Torban, et al., 2007; Lei et al., 2010; Robinson et al., 2012; Tao et al., 2011). In spite of much progress made in different organisms, the question why hundreds of cells can adopt the uniform planar polarity still remains to be answered. Although a long-range global cue that initiates and establishes PCP was predicted, its nature and regulatory mechanism is not yet clear. Therefore, the molecular nature and functional mechanisms of the global cue are still one of the most important issues in the PCP field.

2.1. The PCP pathway

The framework of PCP was largely built from studies in *Drosophila*, in which distally oriented bristles in the wing or ommatidia organized in a crystalline-like array in the compound eye are most frequently studied as models of planar polarity. Polarity in other places, such as bristles on the fly abdomen and thorax, has been also explored (Lawrence, Casal, & Struhl, 2002, 2004;

Olguin, Glavic, & Mlodzik, 2011). Genetic screening in *Drosophila* identified a set of core PCP proteins, including seven-pass transmembrane proteins Frizzled (Fz) (Adler, Krasnow, & Liu, 1997; Vinson & Adler, 1987) and Flamingo (Fmi) (Chae et al., 1999; Usui et al., 1999), a four-pass transmembrane protein Van Gogh (Vang) (Taylor, Abramova, Charlton, & Adler, 1998; Wolff & Rubin, 1998), and cytoplasmic proteins Disheveled (Dsh) (Theisen et al., 1994), Prickle (Pk) (Gubb et al., 1999), and Diego (Dgo) (Feiguin, Hannus, Mlodzik, & Eaton, 2001). The key characteristic of these proteins is that they exhibit asymmetrical subcellular localization once PCP is established. In the *Drosophila* wing cells, all these proteins initially distribute evenly on the entire cell membrane, but during establishment of PCP, Fz, Dsh, and Dgo preferentially accumulate to the distal and Vang and Pk to the proximal side of cell membrane. The atypical cadherin protein Fmi localizes both distally and proximally (Axelrod, 2001; Chen et al., 2008; Lawrence et al., 2004; Strutt, 2001; Tree et al., 2002; Usui et al., 1999). The fly harboring mutations in any one of these core proteins displayed typical PCP phenotype in the wing, which is randomized hair orientation. Since there are many excellent reviews written on PCP in *Drosophila* (Adler, 2002; Lawrence, Struhl, & Casal, 2007; McNeill, 2010; Seifert & Mlodzik, 2007; Strutt, 2009; Tree et al., 2002; Vladar et al., 2009), details on this subject will not be further discussed.

In lower vertebrate model organisms, such as *Xenopus* and zebrafish, same set of PCP proteins regulates a key process of gastrulation, CE (Ciruna, Jenny, Lee, Mlodzik, & Schier, 2006; Wallingford & Harland, 2001). In this process, ML-elongated mesenchymal cells intercalate each other and move toward the midline forcing the embryo to narrow (converge) along the ML axis and elongate (extend) along the A–P axis concurrently (Keller, 2002; Keller et al., 2000; Wallingford, Fraser, & Harland, 2002). For example, the zebrafish *trilobite* mutant caused by Vangl2 (vertebrate homologue of *Drosophila* Vang) mutations exhibited convergence and extension defects with shorter and broader body axis (Jessen et al., 2002). Although the cellular signals required for CE are complicated and not yet fully understood, impaired CE is now often viewed as an evidence of vertebrate PCP defect.

In mammals, disrupted CE movements during mammalian neurulation lead to failure of neural tube closure (Tissir & Goffinet, 2010; Ueno & Greene, 2003). Neurulation is a process in which the neural plate bends up and later fuses to form the hollow tube that will eventually differentiate into the brain and the spinal cord of the central nervous system. Incomplete neural tube closure causes the condition of spina bifida, the most common disabling birth defect in the United States (Northrup & Volcik, 2000). Indeed, mutations in core PCP proteins (Vangl1, Vangl2, Prickle1, Frizzled6, and Celsr1) have been recently identified in patients with various neural tube defects (Bosoi et al., 2011; De Marco et al., 2012; Kibar et al., 2009; Kibar et al., 2007; Kibar et al., 2011; Kibar, Torban, et al., 2007; Lei et al., 2010; Robinson et al., 2012). Neural tube closure is initiated at several places along the A–P axis forming two open ends of the neural tube, cranial (head) end, and caudal (tail) end, also called the anterior neuropore and the posterior neuropore. The neuropores will ultimately close so that the neural tube becomes an irregular tube sealed at both ends. Failure to close the anterior end or posterior end results in anencephaly or spina bifida, respectively. If the entire neural tube fails to close, it is referred to as craniorachischisis, a condition that can be caused by core PCP protein mutations in mice (Gilbert, 2000). While neural tube defects have been widely accepted as one of major PCP phenotypes in mammals, caution must be taken. Other signaling pathways, for instance, Sonic hedgehog signaling pathway is also essential for mammalian neural tube closure (Copp & Greene, 2010; Copp, Greene, & Murdoch, 2003). Perturbation of this pathway leads to exencephaly (early stage of anencephaly) or holoprosencephaly. Another mammalian developmental process requiring PCP is the orientation of sensory hair cells in the cochlea of the inner ear (Rida & Chen, 2009; Wang & Nathans, 2007). The sensory epithelium has four rows of hair cells, three outer rows, and one inner row. In wild-type mouse embryo, all hair cells orient unidirectionally. But in PCP mutants, the orientation of hair cells becomes randomized. Both hair cell orientation of the inner ear and neural tube closure are referred as robust and sensitive readouts for PCP in mammalian development.

The tissue morphogenetic processes found to require PCP have expanded in recent years (Goodrich & Strutt, 2011; Gray et al., 2011; Zallen, 2007). With the generation of null mutants of both Vangl1 and Vangl2 (Song et al., 2010), it is now possible to completely inactive the Vangl-mediated PCP signaling and test the roles of PCP in these mutants. The first directional morphogenetic process that requires PCP is the establishment of L–R asymmetry during gastrulation (Song et al., 2010). PCP is required to establish a unidirectional leftward nodal flow when breaking the bilateral symmetry through controlling cilium positioning in the node cells (Song et al., 2010). The leftward nodal flow across posterior notochord (PNC) is the earliest event in the *de novo* formation of L–R asymmetry establishing L–R asymmetry, and this unidirectional nodal flow is generated by the posteriorly localized motile cilia in the PNC (Hamada, Meno, Watanabe, & Saijoh, 2002; Hirokawa, Tanaka, Okada, & Takeda, 2006; Nonaka, Shiratori, Saijoh, & Hamada, 2002; Nonaka et al., 2005; Shiratori & Hamada, 2006). When the core PCP proteins Vangl1 and Vangl2 were genetically removed, cilia are randomly positioned in the PNC cells leading to turbulent nodal flow and randomized L–R asymmetry (Song et al., 2010). Similar to that observed in *Drosophila* wing, the mouse hairs also orient predominantly in the anterior to posterior direction (proximal to distal direction in the limb). The mouse hair is derived from precisely orientated hair follicles determined by planar-polarized basal epidermal cells. Disruption of core PCP proteins, for instance, Vangl2 or Frizzled6, leads to global hair orientation defects, with waves, whorls, and tufts over the body surface (Devenport & Fuchs, 2008; Guo, Hawkins, & Nathans, 2004; Wang, Chang, & Nathans, 2010; Wang, Guo, & Nathans, 2006).

During CE, PCP regulates cell polarity in mesenchymal cells. Recently, it is found that in the developing limb, chondrocytes are also polarized along the proximal–distal (P–D) axis of the limb and PCP is required for limb elongation along the P–D axis (Gao et al., 2011; Wang, Sinha, Jiao, Serra, & Wang, 2011), possibly through a process similar to CE (Gao et al., 2011). Notably, PCP has been found to play important roles in various developmental and physiological processes in recently years, including elongation of craniofacial processes, axon guidance, and organogenesis of heart, lung, kidney, and eye. PCP may also play a role in tumor formation/metastasis (Carroll & Das, 2011; Henderson & Chaudhry, 2011; Sugiyama, Lovicu, & McAvoy, 2011; Tissir & Goffinet, 2010; Topczewski, Dale, & Sisson, 2011; Wang, 2009; Yates & Dean, 2011).

2.2. Global cues and PCP

Over the past 10 years, despite a growing list of diverse roles of PCP genes in development and disease has been shown, little is known about the mechanism of PCP initiation and establishment. Two classes of models, a global module and a local module, mostly based on studies in *Drosophila*, have been proposed (Axelrod, 2009; Axelrod & Tomlin, 2011; Blair, 2012; Strutt, 2009; Wu & Mlodzik, 2009). The local module describes the interaction between core PCP proteins across cell membrane and how such cell–cell interaction forms a positive-feedback loop to reinforce and amplify the polarization of these core PCP proteins within individual cells (Axelrod, 2009; Chen et al., 2008; Wu & Mlodzik, 2008). However, prior to the amplification of polarization, the symmetrical localization of core PCP proteins must be altered. More importantly, PCP must be established in a field of cells in a three-dimensional tissue or organ. A distinct character of PCP is uniformity in a large filed of cells, which requires not only local coordination between cells but also a global directional cue. For instance, it was unknown why mesenchymal cells in the limb establish PCP along P–D, not A–P axis; why skin hairs orient away from the head and why cilia are localized posteriorly in the node. The global module introduces mechanisms to link the global direction of PCP to the tissue axes. Among these mechanisms, Ft (Fat)/Ds (Daschous)/Fj (Four-jointed) system has been proposed to act upstream of PCP to provide a directional cue and these work is mostly done in the context of the *Drosophila* wing hair polarity (Adler, Charlton, & Liu, 1998; Casal, Struhl, & Lawrence, 2002; Ma, Yang, McNeill, Simon, & Axelrod, 2003; Yang, Axelrod, & Simon, 2002; Zeidler, Perrimon, & Strutt, 2000). This system consists of proximally expressed protocadherin Ft, protocadherin Ds, and distally expressed Golgi-resident kinase Fj. Ft and Ds form heterodimers across cell membrane, and Fj phosphorylates them to modulate their binding. The graded expression of Ds and Fj results in the biased Ft/Ds heterodimer formation along P–D axis in the fly wing, providing an instructive cue to convey the directional information from Ds/Fj transcriptional gradient to downstream PCP components. However, how Ds/Fj signal is transmitted to core PCP proteins is completely unknown and other evidence supported a model in which Ft/Ds/Fj system and core PCP proteins function in parallel (Casal, Lawrence, & Struhl, 2006; Lawrence et al., 2007). In addition, it is unknown what sets up the graded expression of Ds/Fj. In mammals, mutants of Ft/Ds/Fj components display either no or mild PCP phenotypes (Ciani, Patel, Allen, & ffrench-Constant, 2003; Mao et al., 2011; Probst, Rock, Gessler, Vortkamp, & Puschel, 2007; Saburi et al., 2008; Saburi, Hester, Goodrich, & McNeill, 2012). Notably, this system may also regulate another fundamentally important signaling pathway, the Hippo pathway, which controls organ size through regulating cell proliferation and apoptosis. Therefore, it becomes challenging to clearly separate their roles in PCP and Hippo (Matakatsu & Blair, 2012).

Another proposed mechanism is Frizzled gradient model, in which a gradient of unknown signal sets up a Frizzled activity gradient across a field of cells. Then, through the cross-cell Flamingo homodimers, each cell compares the level of Frizzled activity with its adjacent cells and orients itself toward its neighbor with lower Frizzled activity (Lawrence et al., 2004). However, the nature of proposed upstream signal establishing Frizzled activity gradient is not defined, either. In both models, it is unknown what provides upstream global cues to break the symmetry of PCP protein localization and initiate PCP. Since secreted molecules generated from a localized source can form morphogen gradients across a field of cells, morphogens that regulate formation of body axes have been proposed to also act as global cues for PCP (Lawrence et al., 2007). While major morphogens, such as Hedgehog (Hh) and Decapentaplegic (Dpp), do not directly regulate PCP in *Drosophila* (Strutt, 2009), Wingless (Wg) has been hypothesized to regulate PCP as its receptor, Frizzled (Fz), is one of the core PCP proteins. However, extensive studies in fly so far failed to prove a role of Wg in PCP (Wu & Mlodzik, 2009). Nevertheless, it remains a formal possibility that Wg may exert its PCP regulatory function within a small time window in which Wg's role in PCP and Wg/Armadillo signaling can be cleanly separated.

In vertebrates, however, Wnt5b and Wnt11 are clearly required for CE during gastrulation, a process regulated by PCP (Heisenberg et al., 2000; Rauch et al., 1997). Zebrafish mutants *pipetail* (*ppt*, *wnt5b*) and *silberblick* (*slb*, *wnt11*) exhibit compromised CE movements with broadened and shortened body axis, similar to the *trilobite* (*tri*) mutant caused by mutations in the core PCP protein Vangl2 (Heisenberg et al., 2000; Jessen et al., 2002; Rauch et al., 1997). In mammals, some Wnts, like Wnt5a and Wnt11, transduce their signals predominantly through noncanonical β-catenin-independent pathways (Schneider & Mercola, 2001; Topol et al., 2003; Westfall et al., 2003). Importantly, Wnt5a genetically interacts with core PCP protein, Vangl2, in sensory hair cell orientation in inner ear and neural tube closure, strongly supporting its role in regulating PCP (Qian et al., 2007). The role of Wnt5a required for establishing PCP is demonstrated in the developing limb, in which Wnt5a regulates PCP in the forming chondrocytes along the P–D axis (Fig. 11.1A) and likely in other places such as the frontonasal processes and CE along A–P axis (Fig. 11.1B) (Gao et al., 2011). In chick embryos, skeletal muscle fibers are well organized along the body A–P axis. Elongated myocytes are aligned

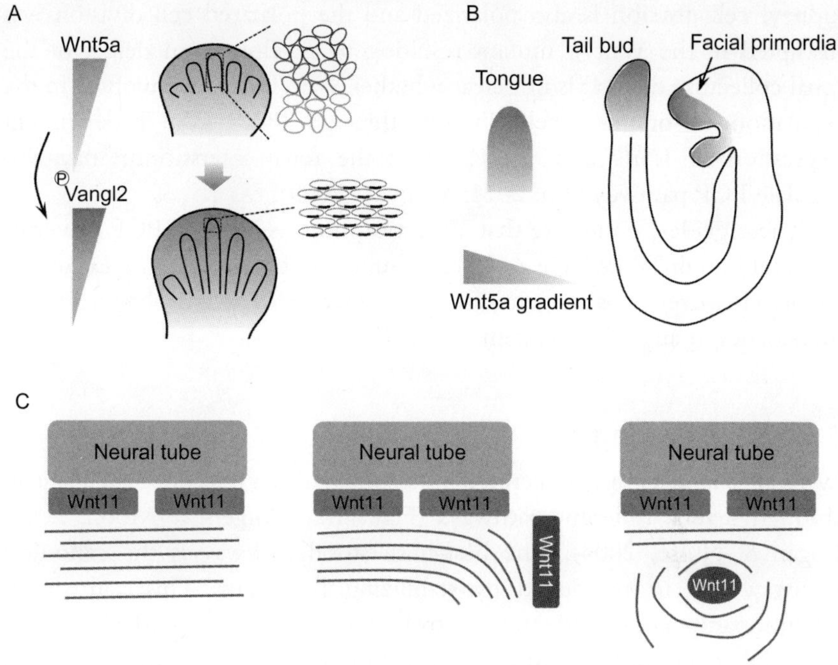

Figure 11.1 *Instructive role of Wnt morphogen gradient in establishing tissue polarity.* (A) In the developing limb bud, by inducing different levels of Vangl2 phosphorylation (blue), Wnt5a gradient (orange) is translated into an activity gradient of Vangl2. Such activity gradient then induces its protein asymmetrical localization (blue) leading to the downstream polarized cell behaviors. (B) Wnt5a is expressed in a gradient in the branchial arches, tail bud and tongue tip. Wnt5a may play similar roles in elongation of these tissues as *Wnt5a*$^{-/-}$ embryos exhibited PCP defects in these places. (C) Muscle fibers parallel and align with neural tube and the Wnt11 expression area. If Wnt11-expressing cells were placed between two somites or within the myotome (purple), muscle fibers will be reoriented to the ectopic Wnt11. (For interpretation of the references to color in this figure legend, the reader is referred to the online version of this chapter.)

parallel to the neural tube. It is shown that Wnt11 expressed in the chick neural tube is necessary and sufficient to orient muscle fibers through core PCP proteins (Fig. 11.1C) (Gros, Serralbo, & Marcelle, 2009). It has been demonstrated that PCP controls oriented cell division in tissue elongation (Gong, Mo, & Fraser, 2004). In mouse kidney morphogenesis, Wnt9b is found to be necessary for CE processes and polarized divisions of kidney epithelial cells that determine the tubule diameter. Attenuation of Wnt9b signaling affects the PCP leading to significantly increased diameter of kidney tubules (Karner et al., 2009). In the collecting duct epithelium of kidney, cell division is also polarized and the polarized cell division was disrupted in the Wnt7b mutant resulting in an elongation defect of the renal collecting duct. It is not clear whether Wnt7b itself is involved in the regulation of oriented cell division through PCP. Wnt7b-dependent expression of *Wnt4* and/or *Wnt11* in the renal interstitium may also regulate PCP pathway (Yu, 2011; Yu et al., 2009).

These evidence indicate that Wnt morphogens regulate PCP in vertebrate, although Wnts have not been found to control PCP in *Drosophila* so far. However, it is not entirely clear whether Wnt morphogens act as global cues in any of the system.

2.3. Wnt signaling

Wnts are a large family of secreted molecules that can transduce their signals through several different pathways (Fig. 11.2) (Angers & Moon, 2009; Logan & Nusse, 2004). The best-understood pathway is the canonical pathway that is transduced by stabilizing β-catenin. This pathway is evolutionarily conserved and controls many processes of development and adult tissue homeostasis (Clevers, 2006; MacDonald, Tamai, & He, 2009). In this pathway, Wnt proteins bind to their coreceptors Frizzleds and Lrp5/6 on the cell surface leading to phosphorylation of Lrp5/6, which then inactivates the β-catenin destruction complex containing Axin/GSK3/APC by recruiting Axin to the cell membrane (MacDonald et al., 2009). The stabilized β-catenin accumulates in the cytoplasma and then enters the nucleus to form a complex with TCF/Lef and activate target genes expression. In the absence of Wnt ligands, β-catenin is phosphorylated by GSK3 in the destruction complex and degraded (Macdonald, Semenov, & He, 2007; MacDonald et al., 2009). According to the ability of Wnts to induce β-catenin accumulation in the nucleus, Wnt family members were divided into two distinct classes. For example, Wnt1, Wnt3a, and Wnt8 (canonical Wnts) strongly stabilize β-catenin but other Wnts like Wnt5a and Wnt11 (noncanonical Wnts) could not. Instead, Wnt5a represses canonical Wnt signaling (Mikels & Nusse, 2006; Topol et al., 2003; Torres et al., 1996; Westfall et al., 2003). However, other evidence showed that context-dependent combination of Wnt receptors, rather than Wnt proteins themselves, may play more important roles in determining which branch, "canonical" or "noncanonical," Wnt ligands signal through (van Amerongen, Mikels, & Nusse, 2008). It is likely that all Wnts can signal through both the "canonical" and the

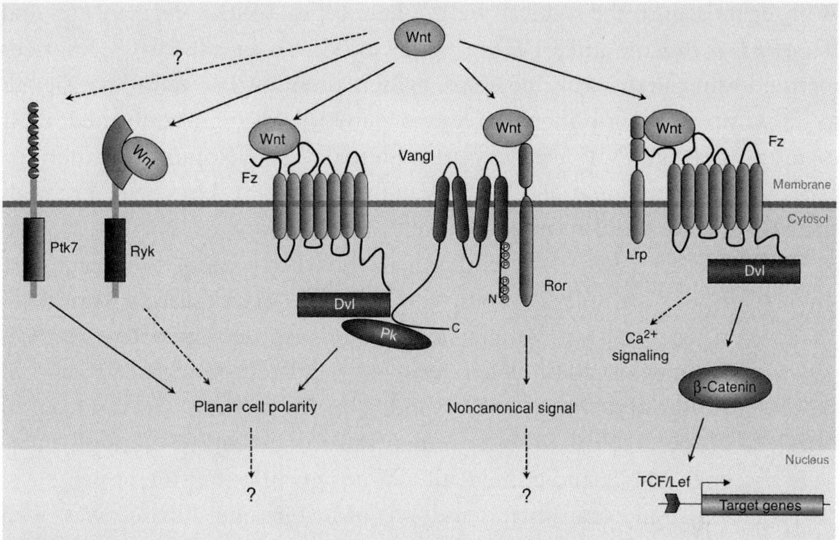

Figure 11.2 *Wnt signaling pathways.* In the canonical Wnt/β-catenin pathway (right panel), secreted Wnt proteins bind the Frizzled (Fz) and low-density lipoprotein receptor-related proteins 5 and 6 (LRP) coreceptors to activate Disheveled (Dvl) leading to the stabilization of β-catenin, which then enters into the nucleus, where it binds T cell factor/the lymphoid enhancer factor (TCF/Lef) to transactivate target genes expression. In the Wnt/PCP pathway (middle panel), Wnt proteins also bind Fz, which recruits Dvl. On the other hand, Wnt proteins bind receptor tyrosine kinase orphan receptor (Ror) to form a complex with Vangl, which recruits Prickle (PK). As a result, Vangl is phosphorylated in a Wnt dose-dependent manner. It is proposed that PK and Dvl bind and antagonize each other to generate asymmetrical protein localization, a critical regulatory event of PCP. Other receptor tyrosine kinases including PTK7 and Ryk are also involved in PCP (left panel). (See Color Insert.)

"noncanonical" pathways and that, in a particular cell, the specific combination of receptors determines that one pathway dominates over the other. While the term "noncanonical Wnt pathway" is being widely used in the literature, its definition is quite loose (Semenov, Habas, Macdonald, & He, 2007). Among of them, Wnt/calcium and Wnt/PCP pathways have been studied more extensively (Kohn & Moon, 2005; Veeman, Axelrod, & Moon, 2003). However, the *in vivo* function of Wnt/Calcium pathway still awaits rigorous genetic tests in mammals.

While the *in vivo* evidence in mouse embryos has shown that Wnt signaling is required in PCP pathway, the underlying mechanism by which Wnt signaling regulates PCP remains largely unknown. The Frizzled family proteins are assumed to mediate Wnt signaling due to their abilities to bind

Wnt ligands and the typical PCP phenotypes of the *Frizzled1/2* and *Frizzled3/6* mutant mice (Wang et al., 2006; Yu et al., 2010). Frizzleds recruit Disheveled family proteins, which then further transduce signals to downstream components. Elegant genetics analysis confirmed their essential roles in PCP pathway in mammalian development (Etheridge et al., 2008; Hamblet et al., 2002; Wang et al., 2006). However, Frizzleds and Disheveleds are not only dedicated to PCP pathway, but they are also indispensable for Wnt/β-catenin signaling, which hampers mechanistic studies of the Wnt/PCP pathway through manipulating Frizzled/Disheveled. In contrast, Vangl1 and Vangl2 are dedicated to the PCP pathway (Song et al., 2010). Therefore, the regulatory mechanisms of PCP can be readily studied in *Vangl1* and *Vangl2* mutants. In addition to Frizzleds, some atypical Wnt receptors, including Ror2, Ryk, and Ptk7, have also been implicated in mediating Wnt signaling (Fig. 11.2).

Ror2, a single transmembrane receptor tyrosine kinase, was well established to mediate Wnt5a signal both *in vitro* and *in vivo* (Green, Kuntz, & Sternberg, 2008; Minami, Oishi, Endo, & Nishita, 2010). Wnt5a inhibits Wnt/β-catenin signals through Ror2 (Mikels & Nusse, 2006), and *Ror2*$^{-/-}$ mice is similar to, although less severe than, the *Wnt5a*$^{-/-}$ mutant. Both mutant mice display similar phenotypes of dwarfism, flat face, short limbs and tails, and some other defects (DeChiara et al., 2000; Gao et al., 2011; Oishi et al., 2003; Takeuchi et al., 2000; Yamaguchi, Bradley, McMahon, & Jones, 1999; Yang, Topol, Lee, & Wu, 2003). In addition, mutations in both *ROR2* and *WNT5A* are identified to cause Robinow syndrome, which is characterized by short-limbed dwarfism, abnormalities in the head, face, and external genitalia, as well as vertebral segmentation (Afzal & Jeffery, 2003; Afzal et al., 2000; Person et al., 2010; van Bokhoven et al., 2000). Not surprisingly, Ror2, like its ligand Wnt5a, also genetically interacts with Vangl2 (Gao et al., 2011; Qian et al., 2007). When Vangl2 was further removed in the Ror2-null background, double knockout mice exhibited nearly identical phenotypes compared to the *Wnt5a*$^{-/-}$ mutant in many places, including shortened limbs, tails, and craniofacial processes (Gao et al., 2011). These defects together with biochemical studies which have identified Wnt5a-induced complex formation between Ror2 and Vangl2 indicated that Wnt5a transduces its signal through Ror2 and Vangl2 to regulate PCP (Gao et al., 2011). A very recent report showed that *Ror1* and *Ror2* double knockout mice displayed exencephaly (indicating a PCP phenotype) at a low penetrance and died at E15.5. The phenotypes are more severe than those of the *Wnt5a*$^{-/-}$ mutant, suggesting that Ror1/2 may also transduce signals of other Wnt proteins. However, in the limb, distal digits of forelimbs of the *Ror1*$^{-/-}$ and *Ror2*$^{-/-}$ double mutant embryo still form. This phenotype is much less severe than that of the *Wnt5a*$^{-/-}$ mutant, suggesting that there are some other receptors mediating the Wnt5a signal (Ho et al., 2012).

Ryk, another single transmembrane receptor tyrosine kinase with unfunctional kinase domain (Halford & Stacker, 2001), binds to Disheveled. It potentiates Wnt1 and Wnt3a-dependent gene expression and is required for axon guidance and Wnt3a-induced neurite outgrowth (Lu, Yamamoto, Ortega, & Baltimore, 2004). Several groups showed that Ryk is also required for Wnt5a-mediated axon guidance in both *Drosophila* and mammals (Li, Hutchins, & Kalil, 2009; Yoshikawa, McKinnon, Kokel, & Thomas, 2003). In addition, Wnt5b has been identified to provide an instructive cue to regulate zebrafish gastrulation movements through Ryk (Lin, Baye, Westfall, & Slusarski, 2010). Therefore, Ryk appears to mediate both β-catenin-dependent and β-catenin-independent pathways. Very recently, Stacker's group demonstrated that Ryk is involved in PCP pathway in vertebrate. They found Ryk genetically interacts with Wnt11 during CE movement in zebrafish. In mouse, *Ryk*$^{-/-}$ mutant exhibited sensory hair cell orientation defects in inner ear cochlea and Ryk genetically interacts with Vangl2 (Macheda et al., 2012).

The tyrosine kinase receptor Ptk7 with a defective kinase domain was originally identified as a PCP protein because its mutation disrupts neural tube closure and inner ear hair cell polarity in mouse, and it genetically interacts with Vangl2 (Lu et al., 2004). In *Xenopus*, Ptk7 was shown to be required for Frizzled7-mediated Disheveled localization and regulate neural crest migration (Shnitsar & Borchers, 2008). While the Ptk7 was strongly implicated in Wnt/PCP pathway, its role in canonical Wnt signaling seems to be controversial. Puppo et al. (2011) reported Ptk7 positively regulates Wnt/β-catenin signaling upstream from GSK3. However, Peradziryi et al. showed Ptk7/Otk inhibits canonical Wnt signaling. They also showed Ptk7 coimmunoprecipitates with Wnt3a and Wnt8, but not Wnt5a and Wnt11 (Peradziryi et al., 2011). Nevertheless, whether Ptk7, which does not contain any known Wnt proteins binding domain, is a true Wnt receptor remains to be determined.

Despite innumerable successful gene discoveries in the Wnt/β-catenin signaling pathway, which is mainly based on the convenient TCF/LEF luciferase assay (Topflash), advances in finding regulatory components of

Wnt/PCP pathway still largely rely on genetic screening in *Drosophila* or findings from forward or reverse mouse genetics, as the most reliable functional readout of PCP is still based on *in vivo* phenotypic analysis. The Rho family GTPases (Rho, Rac, Cdc42) and JNK pathway have been demonstrated to be the downstream transducers of core PCP proteins to regulate cell morphology (Kikuchi, Yamamoto, & Kishida, 2007; Schlessinger, Hall, & Tolwinski, 2009; Tada & Kai, 2009). Because of these, people have attempted to use AP1- or ATF2-luciferase reporter (downstream transcription factors of JNK pathway) or c-Jun/JNK phosphorylation as readouts for Wnt/PCP pathway (Ohkawara & Niehrs, 2011; Park & Moon, 2002; Shafer, Onishi, Lo, Colakoglu, & Zou, 2011). However, it is still largely unclear whether JNK is a major mediator of PCP in vertebrates *in vivo* as JNK activity can be regulated by many other signaling pathways (Weston & Davis, 2002, 2007). Furthermore, it is still questionable whether these cell morphology regulators should be placed into the core PCP pathway. Importantly, those proteins which disruption changed cell morphology should not be simply interpreted as PCP proteins/effectors. They could be two distinct pathways both regulating cell morphology (Lapebie, Borchiellini, & Houliston, 2011). Lack of a simple and reliable molecular readout of Wnt/PCP pathway has severely hindered our understanding of signal transduction in the Wnt/PCP pathway.

Establishment of PCP is likely to be regulated at the protein level by asymmetrical protein localization, not at the transcriptional level as none of the core PCP proteins is a transcription factor. Recently, a novel signaling mechanism underlying Wnt5a-dependent PCP establishment in mouse embryonic development has been identified (Gao et al., 2011). Vangl2, a core PCP protein, is phosphorylated at two highly conserved clusters of serines/threonines in response to Wnt5a in a Wnt5a-dose-dependent manner. As Vangl2 phosphorylation is required for Wnt/PCP signal transduction, it is likely that a Wnt5a signaling gradient can theoretically orient the cells it signals to by regulating the Vangl2 protein activity in each cell. These findings have opened new doors to further understanding Wnt/PCP signal transduction.

2.4. Wnt signaling gradients establish PCP

Although Wnt proteins, especially Wnt5 and Wnt11, have been strongly implicated in regulating PCP in vertebrate, a big question remained: do they indeed provide directional cues to initiate PCP or are they only required for PCP establishment as permissive signals?

In an *in vitro* culture system, Witze et al. found that Wnt5a plays a permissive role in promoting directional migration of melanoma cells toward a gradient of the chemokine CXCL12 by inducing an asymmetrically distributed Frizzled-containing receptor–actin–myosin polarity (W-RAMP) structure (Witze et al., 2008). It should be pointed out that the study and observation are made at single-cell level *in vitro* and it is not clear whether the phenomenon is truly related to PCP. In addition, regulation of PCP requires close cell–cell contact in a field of cells. In the zebrafish model, injecting *wnt11* mRNA into one-cell stage embryos can rescue *slb* mutant phenotype, suggesting its role in regulating CE is permissive (Heisenberg et al., 2000). However, notably, the mutant embryos were not fully rescued (57% *wnt11*-injected *slb* exhibited other abnormal phenotype) (Heisenberg et al., 2000). Additionally, *ppt* mutant phenotype could not be rescued by misexpressing *Wnt5b*. Instead, wnt5b-injected *ppt* embryos exhibited variable malformations (Kilian et al., 2003). All these suggest that the instructive role of Wnt5b or Wnt11 cannot be ruled out and maybe a better assay system and/or readout is needed in addressing this issue. Indeed, more and more evidence in different organisms indicates that Wnt gradients do provide directional information to regulate polarized cell behaviors in a group of cells. A study in *Caenorhabditis elegans* found that the ground polarity is established by instructive Wnt/EGL-20 activity via CAM-1 (a Ror receptor tyrosine kinase) and Van Gogh/VANG-1 (Green, Inoue, & Sternberg, 2008). In zebrafish, it has been shown that Wnt5b regulates gastrulation movements by acting as an instructive cue through Ryk (Lin et al., 2010). In the mouse developing palate, an essential role of Wnt5a in regulating directional cell migration through Ror2 was demonstrated (He et al., 2008). Recently, it is shown that Wnt5a controls directional outgrowth and A–P guidance of commissural axons through core PCP proteins (Shafer et al., 2011).

Wnt11 has also been shown to act as an instructive cue. In the chick embryo, Wnt11 signaling in early muscle fibers regulates their directional elongation and orientation through the PCP pathway (Fig. 11.1C) (Gros et al., 2009). The neural tube is necessary and sufficient to polarize muscle fibers, and the myocytes are aligned in parallel to each other and along with the A–P axis of the embryo. *Wnt11* expression at the medial border of somites is induced by secreted Wnt1 and/or Wnt3a from the dorsal neural tube and is required to control the polarized elongation of myocytes through the PCP components. When the elongating myocytes are exposed to a localized exogenous source of Wnt11, myocyte elongation was redirected to be around the localized exogenous Wnt11. When *Wnt11* is expressed in myocytes uniformly, myofibers become disorganized. These data argue strongly for an instructive role of Wnt ligands in the PCP pathway. However, it is still questionable what is the relationship between muscle fiber orientation and PCP, as polarized localization of core PCP protein, the hallmark of PCP, in myocytes has not been identified. Furthermore, it is still unclear whether Wnt11 truly regulates PCP in mammals. Although *Wnt11*$^{-/-}$ mutant mice displayed multiple defects in kidney development and cardiogenesis, no obvious PCP phenotypes were reported (Cohen, Miller, Wang, Moon, & Morrisey, 2012; Flaherty & Dawn, 2008; Majumdar, Vainio, Kispert, McMahon, & McMahon, 2003; Uysal-Onganer & Kypta, 2012; Zhou et al., 2007). In contrast to Wnt5a, Wnt11 does not genetically interact with core PCP protein, Vangl2 (H. Song & Y. Yang, unpublished data). Direct evidence that Wnt11 regulates PCP is still lacking.

During development, it is an intriguing question why some of the organs and tissues preferentially extend in one direction? For instance, the limbs preferentially elongate along the P–D axis. It was proposed that Fgf (fibroblast growth factors) family members secreted from the apical ectodermal ridge play a central role in limb P–D elongation through setting up a gradient of proliferation rate (higher at the distal, but lower at the proximal part of limb buds), which then drives oriented limb elongation along P–D axis (Ede & Law, 1969; Fernandez-Teran, Hinchliffe, & Ros, 2006; Hopyan, Sharpe, & Yang, 2011; Niswander, Tickle, Vogel, Booth, & Martin, 1993; Reiter & Solursh, 1982). However, elegant experimental and computational studies based on three-dimensional analysis of developing limb buds did not support this "proliferation gradient model," and suggested directional cell activities, rather than graded proliferation rate, are the major driving force behind the early limb bud outgrowth (Boehm et al., 2010; Gros et al., 2010). Studies in our lab also showed that when chondrocytes are forming, they are indeed polarized along the P–D axis, indicated by asymmetrically localized Vangl2 protein, a hall marker of PCP. Interestingly, those limb mesenchymal cells that are not differentiating into chondrocytes in the interdigital region did not show Vangl2 asymmetrical localization. As Vangl2 asymmetrical localization, as well as directional outgrowth and extension of cartilage in wild-type embryo, is abolished in the *Wnt5a*$^{-/-}$ mutant, Wnt5a, which is expressed in the distal limb bud and forms a gradient from distal to proximal limb, controls elongation of the cartilage along the P–D axis in limb development through a process that requires PCP (Gao et al., 2011). It is possible that limb mesenchymal cells gain directional information from the Wnt5a morphogen gradient to extended cartilage along the P–D axis. Besides the developing limb, it is very likely that in other developmental processes such as the frontonasal elongation, A–P body axis extension, and secondary palate fusion, Wnt5a gradient may also define a unidirectional polarity (Fig. 11.1B). In all these places, *Wnt5a* is expressed in a graded fashion (branchial arches, tail bud, and palatal mesenchyme) (He et al., 2008; Liu et al., 2012; Yamaguchi et al., 1999). *Wnt5a* and *Vangl2* mutant mice exhibited similar defects in these places (Gao et al., 2011; Liu et al., 2012, 2009; Yamaguchi et al., 1999). Biochemically, our finding of a Wnt5a gradient leading to a Vangl2 phosphorylation gradient provides a mechanism whereby Wnt5a signaling gradient elicits a graded response (Gao et al., 2011), supporting the instructive role of Wnts (Fig. 11.1A). Although the graded *Wnt5a* expression may be functionally important in regulating PCP, more rigorous genetic assays are needed to directly test whether Wnt5a acts instructively or permissively in controlling PCP.

Secreted frizzled-related protein (Sfrp) family members are homologous to the extracellular domains of Frizzleds. They bind to and modulate activity of Wnt proteins including Wnt5a. Disruption of certain Sfrp proteins causes PCP-like phenotype in multiple tissues and they genetically interact with Vangl2 (Matsuyama, Aizawa, & Shimono, 2009; Satoh, Matsuyama, Takemura, Aizawa, & Shimono, 2008; Sugiyama et al., 2010). In addition, when Sfrp2 that potentially disturbs all Wnt signals is misexpressed, it disrupted global alignment and orientation of lens fibers in the eye, suggesting that Wnt ligands act as an extracellular directional signal to polarize fibers in the eye lens (Sugiyama et al., 2010). However, while the Vangl2 is asymmetrically localized to the anterior pole side of lens cells, the Wnt signal gradient is not well defined in the mammalian eye development.

In the developing cochlea of the inner ear, four rows of sensory hair cells reorient themselves from E17 to P10 to achieve a final uniform polarity. *Wnt7a* is found to be highly expressed in the developing cochlea, and by E16.5, it becomes restricted to pillar cells, which are supporting cells between three outer rows and one inner row of hair cells. When *in vitro*-cultured cochlea explants are exposed to Wnt7a conditional medium, which should theoretically disturb the potential morphogen gradient generated by asymmetrically expressed endogenous *Wnt7a* in pillar cells, the reorientation process of outer hair cells is inhibited. However, *Wnt7a*$^{-/-}$ mutant itself has no obvious orientation defects, suggesting possible functional redundancy from other *Wnt*s expressed in the cochlea (*Wnt5a*, *Wnt2*, *Wnt10b*, or more)

(Dabdoub et al., 2003). Consistently, treatment with Sfrp1 or Wif1 (Wnt inhibitory factor 1), which neutralizes all endogenous Wnt proteins, also disrupts the outer hair cell reorientation (Dabdoub et al., 2003). These evidence suggest a role of Wnt7a as an instructive cue for reorientation of sensory hair cells by regulating PCP during inner ear cochlea development. However, to distinguish the instructive or permissive role of Wnt morphogen gradient in regulating sensory hair cell orientation in the inner ear, artificial Wnt gradient should be generated to test whether it is sufficient to redirect the polarity of sensory hair cells (Dabdoub & Kelley, 2005). In addition, it is puzzling why treatment with Wn7a conditional medium, Sfrp1, or Wif1 has no effect on inner hair cell orientation, which indicates a possible Wnt-independent mechanism (Dabdoub et al., 2003). Since sensory hair cell orientation is a complicated model system to study mammalian PCP, genetic manipulation in other systems and *in vitro* culture assay will help us to further dissect the role of Wnt morphogen gradient in PCP, whether the role is instructive or permissive.

Taken together, it appears that more evidence supports an instructive, rather than permissive role of Wnt morphogen in establishing PCP in vertebrates. However, there is so far no direct genetic evidence to demonstrate the instructive model or rule out the permissive model. The mechanism whereby Wnt gradient sets up the PCP or asymmetrically localized PCP proteins needs to be further investigated. The findings of Wnt-regulated Vangl2 phosphorylation may provide more insight to this issue.

3. MISSING LINK BETWEEN Wnt AND PCP

From the discovery of *Wnt1* in mouse mammary gland tumors (Nusse & Varmus, 1982) and the findings of major downstream components, there is a great advance in our understandings of canonical Wnt signal transductions in both development and disease (Klaus & Birchmeier, 2008). In contrast, while genetic studies of PCP or tissue polarity also emerged around 30 years ago (Gubb & Garcia-Bellido, 1982) and a number of PCP core proteins and effectors have been identified afterward, the mechanisms whereby PCP is regulated still remain to be identified. One of the challenges is to understand how Wnt ligands regulate PCP in vertebrates.

Secreted signaling molecules such as Fgf, Tgf-β, Hedgehog, or Wnt family members can govern tissue patterning by forming extracellular protein gradients. It was well understood that such gradients provide quantitative information to generate different cell types and control cell proliferation

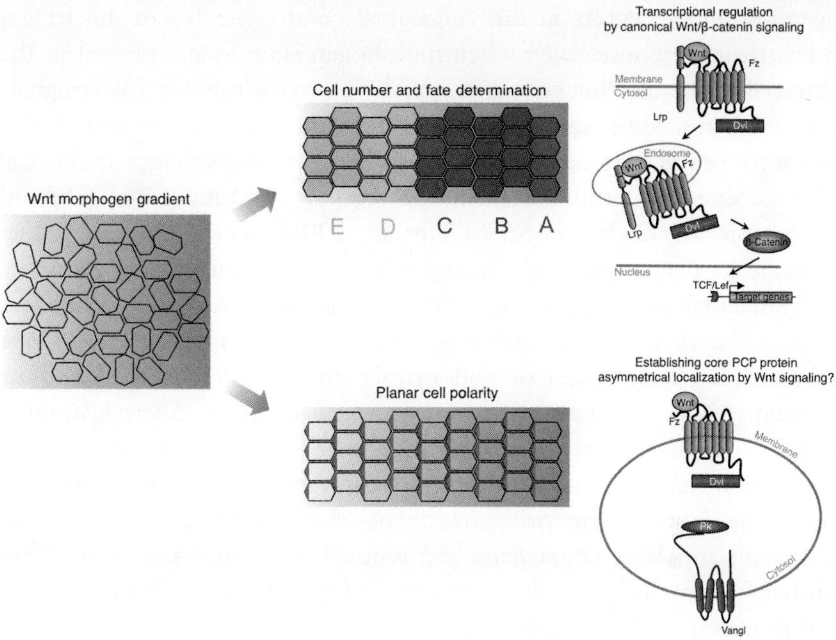

Figure 11.3 *Dual function of Wnt morphogen gradient*. First, Wnt signaling gradient provides quantitative information to control cell proliferation and cell differentiation through canonical Wnt/β-catenin pathway, which is achieved via endocytosis of key regulatory components, β-catenin stabilization and nuclear translocation, and, finally, transactivation of target genes. Second, Wnt morphogen acts as a global cue to establish PCP to convey directional information, which is evidenced by asymmetrical localization of core PCP proteins through an unknown mechanism. (See Color Insert.)

in a specific spatial order at distinct threshold concentration (Fig. 11.3). These are regulated by a series of events including receptor binding, downstream signal transduction in the cytoplasm, and finally transcriptional responses. Recent studies have extended the roles of Wnt morphogen to regulate polarity by providing directional information. However, it is still largely unknown how responding cells interpret Wnt morphogen gradient and then use the information to instruct polarized cellular behaviors (Fig. 11.3).

Following ligand binding, internalization of receptors has been found to be critical for signal transduction in many pathways (Seto, Bellen, & Lloyd, 2002). For instance, Fgf8 morphogen gradient is interpreted in receiving cells by endocytic trafficking (Nowak, Machate, Yu, Gupta, & Brand, 2011). Furthermore, the duration and extent of signaling exposure to morphogen gradients can be memorized by receiving cells, and ligand-bound receptors in the endosomal compartments can still trigger downstream responses even when morphogen is no longer present in the extracellular matrix (Jullien & Gurdon, 2005). In the canonical Wnt signaling pathway, while early genetic analysis in *Drosophila* indicated that endocytic trafficking degrades Wingless protein, supporting a traditional view of signaling termination through receptor downregulation (Dubois, Lecourtois, Alexandre, Hirst, & Vincent, 2001), later studies show that endocytosis plays a positive role and suggest that maximal signal activities are generated from endosomes (Blitzer & Nusse, 2006; Seto & Bellen, 2006). Either knocking down Dynamin or Rab5 (a small GTPase required for early stages of endocytosis) in *Drosophila* S2R+ cells or overexpressing dominant-negative forms of them in *Drosophila* wing imaginal discs reduced Wingless signaling (Seto & Bellen, 2006). In a mammalian cell culture system, blocking endocytosis by interfering Dynamin function or drug treatment also decreased the levels of β-catenin and Wnt target gene expression when more exogenous Wnt proteins accumulated on the membrane (Blitzer & Nusse, 2006). Studies on β-arrestin, which links receptor proteins to clathrin and clathrin adaptor AP-2 to promote endocytosis, supported the positive role of clathrin-mediated endocytosis in Wnt/β-catenin signaling (Bryja, Gradl, Schambony, Arenas, & Schulte, 2007). In addition, another major endocytic route, caveolin-mediated endocytosis, was found to be necessary for canonical Wnt signal transduction as well (Yamamoto, Komekado, & Kikuchi, 2006). In the Wnt/PCP pathway, important role of intracellular trafficking has also been implicated. Sec24b, a member of COPII coat protein complex that transports proteins from endoplasmic reticulum to Golgi, was identified to selectively transport core PCP protein Vangl2. *Sec24b* mutant mice exhibited craniorachischisis and decreased Vangl2 membrane localization (Merte et al., 2010; Wansleeben et al., 2010). Involvement of endocytosis in PCP was first suggested by Eaton's group (Classen, Anderson, Marois, & Eaton, 2005). During development, *Drosophila* wing epithelial cells gradually become hexagonally packed shortly before hair formation in a PCP-dependent manner. They proposed a model in which PCP proteins polarize cadherin endocytosis and recycling leading to the cellular packing geometry in the fly wing (Classen et al., 2005). Later on, it is found that interactions between Disheveled and AP-2 appear to be required for PCP activity through engaging Wnt5a-dependent internalization of Frizzled4 (Yu et al., 2007). Along this line, it has been shown before that

Disheveled 2 can recruit β-arrestin 2 to trigger Frizzled4 endocytosis upon Wnt5a stimulation (Chen et al., 2003). In addition, gain- and loss-of-function studies of β-arrestin 2 in *Xenopus* show that its endocytic activity is essential for proper CE movements without affecting cell-fate determination, suggesting the essential role of endocytosis in regulating PCP (Kim & Han, 2007). By *in vivo* time-lapse imaging on *Drosophila* wing, Uemura's group observed that Frizzled- and Flamingo-containing intracellular particles preferentially move toward distal cell boundaries along P–D-oriented microtubules prior to prehair formation (Shimada, Yonemura, Ohkura, Strutt, & Uemura, 2006). Therefore, polarized vesicular trafficking of Frizzled proteins was proposed to be one of the possible mechanisms underlying PCP establishment (Gagliardi, Piddini, & Vincent, 2008). Recently, there is more evidence to indicate that endocytosis is essential for PCP establishment. Rab4- and Rab11-mediated Frizzled recycling favors PCP over canonical Wnt signaling (Purvanov, Koval, & Katanaev, 2010). In addition, a Rab5 effector, Rabenosyn-5 (Rbsn-5), is required for PCP in the fly wing epithelia cells through regulating endocytic trafficking of Flamingo (Mottola, Classen, Gonzalez-Gaitan, Eaton, & Zerial, 2010). In the *Drosophila* eye, internalization of Flamingo requires AP-2 bound Echinoid (Ed), an adherens junction-associated cell adhesion molecule. *Ed* mutant exhibited more Flamingo proteins on the membrane of nonrotating interommatidial cells resulting in misorientation of neighboring photoreceptors in ommatidia (Ho et al., 2010). Apart from core PCP proteins, Syndecan 4-mediated R-spondin 3 signal can induce Wnt/PCP through clathrin-mediated endocytosis (Ohkawara, Glinka, & Niehrs, 2011). Furthermore, Wnt11 controls mesendodermal cell cohesion and migration in zebrafish gastrulation through regulating Rab5-mediated endocytosis of E-cadherin (Ulrich et al., 2005). In another report, Ryk is found to act as a coreceptor of Frizzled 7 to promote Wnt11-stimulated endocytosis of Disheveled in *Xenopus* CE movements (Kim, Her, & Han, 2008). Together, endocytosis of core PCP proteins, especially Frizzleds, plays a key role in establishing PCP and Wnt proteins may control PCP-like processes through regulating endocytosis. However, how Wnts regulate endocytosis and how endocytosis leads to asymmetrically localized core PCP protein still remain to be elucidated. Interestingly, a recent paper suggested a correlation between Frizzled 3 phosphorylation and its endocytosis, although whether Wnt proteins regulate phosphorylation of Frizzled 3 is unknown (Shafer et al., 2011). It will be interesting to investigate whether Wnt-regulated Vangl2 phosphorylation also regulates endocytosis and contributes to PCP establishment (Gao et al., 2011).

An interesting pattern of asymmetrical localization of PCP components is that Vangl and Prickle are always localized opposite to the direction of increasing Wnt dosages. But Disheveled, probably Frizzled, is often localized to the plasma membrane along the direction of increasing Wnt dosages. For instance, in the early developing limb, there is a distal to proximal Wnt5a gradient (Yamaguchi et al., 1999) and Vangl2 is localized to the proximal side of the plasma membrane (Gao et al., 2011). In the mouse node, Vangl1, Vangl2, and Prickle2 are preferentially localized on the anterior side of cell membrane (Antic et al., 2010; Song et al., 2010) opposite to both Wnt3a and Wnt5a gradient from posterior tail bud. But Disheveled proteins are localized to the posterior edge of node cells (Hashimoto et al., 2010). Similarly, in the zebrafish gastrulae, PK and Dvl are enriched near anterior and posterior cell membranes, respectively (Yin, Kiskowski, Pouille, Farge, & Solnica-Krezel, 2008). In the basal layer of mouse back skins, Vangl2 is also asymmetrically accumulated on the anterior side of cell membrane, although the identity of Wnt gradient is unknown in this context (Devenport & Fuchs, 2008; Devenport, Oristian, Heller, & Fuchs, 2011). In the mouse eye, Vangl2 is predominantly localized to the anterior pole side of lens cells, but positions of other PCP proteins (Frizzled6, Disheveled2/3, and Prickle1) appear quite complicated (Sugiyama et al., 2010). Interestingly, although Wnt protein expression is dynamic in the developing lens, their expression is predominantly restricted to the cells at the lens equator at later embryonic states (Ang, Stump, Lovicu, & McAvoy, 2004), suggesting there might have been a Wnt gradient from equator to anterior pole. Therefore, it will be interesting to identify the mechanism whereby the direction of Wnt gradient determines the pattern of PCP protein localization, a critical aspect of Wnt regulation of PCP as global cue.

4. CONCLUSION

In the past, the efforts focused on identifying new PCP components and regulatory mechanisms of PCP have been most productive in *Drosophila*. The functions and the underlying mechanisms of PCP in different organs and tissues of vertebrates are also being revealed. However, a critical question that remains unanswered is how a large group of cells orient in the same direction with respect to the body axes, mainly because of difficulties in identifying a global cue in *Drosophila*. Instead of a single global cue, recent reports indicated several local cues as well as mechanical forces during growth and morphogenesis can polarize core PCP proteins and regulate planar polarity in *Drosophila* wing epithelium (Aigouy et al., 2010; Sagner et al., 2012). In this review, I have highlighted the role of Wnt morphogen in establishing PCP in vertebrate. Recent progress on this issue has taken us a step closer to understanding Wnt-initiated PCP establishment. However, more efforts should be aimed at exploiting the precise regulatory mechanism of Wnt signal inputs on PCP as we have achieved in canonical Wnt/β-catenin signaling pathway. Since it is likely that such regulation does not occur at transcriptional level in Wnt/PCP pathway, it has been challenging to devise a relevant and productive assay system to study PCP *in vitro*. Another challenge is understanding how the morphogen gradient instructs asymmetrical distribution of core PCP proteins. Advanced live imaging techniques will be essential on top of an *in vitro* PCP assay system in mammalian cells. Another interesting question is how PCP is maintained during regeneration. Studies on proliferating skins gave some clues. Eevenport and her colleagues showed that in mouse basal epithelia cells, mitotic endocytosis of core PCP proteins (Flamingo, Frizzled, and Vangl) and selectively recycling back to the cell surface (Frizzled to posterior side and Vangl to anterior side) will reestablish PCP (Devenport et al., 2011). Whether secreted Wnt proteins have impact on this process is unknown. Probably not, since evidence has been shown that such process is reliant on neighboring polarized cells (Devenport et al., 2011). Nevertheless, how endocytic vesicles containing Frizzled or Vangl interpret directional cues from neighboring cells and recycle selectively remains unknown. Similarly, during initial PCP establishment in embryonic development, how Wnt morphogen regulates polarized intracellular trafficking of core PCP proteins is also unidentified. Answering the above questions will provide much more mechanistic insight to our understanding of PCP establishment and maintenance.

ACKNOWLEDGMENT

The author is supported by the Intramural Research Program of the National Human Genome Institute of the National Institutes of Health. I thank Dr. Yingzi Yang for critical discussions in writing this chapter.

REFERENCES

Adler, P. N. (2002). Planar signaling and morphogenesis in Drosophila. *Developmental Cell, 2*, 525–535.

Adler, P. N., Charlton, J., & Liu, J. (1998). Mutations in the cadherin superfamily member gene dachsous cause a tissue polarity phenotype by altering frizzled signaling. *Development, 125*, 959–968.

Adler, P. N., Krasnow, R. E., & Liu, J. (1997). Tissue polarity points from cells that have higher Frizzled levels towards cells that have lower Frizzled levels. *Current Biology, 7*, 940–949.

Afzal, A. R., & Jeffery, S. (2003). One gene, two phenotypes: ROR2 mutations in autosomal recessive Robinow syndrome and autosomal dominant brachydactyly type B. *Human Mutation, 22*, 1–11.

Afzal, A. R., Rajab, A., Fenske, C. D., Oldridge, M., Elanko, N., Ternes-Pereira, E., et al. (2000). Recessive Robinow syndrome, allelic to dominant brachydactyly type B, is caused by mutation of ROR2. *Nature Genetics, 25*, 419–422.

Aigouy, B., Farhadifar, R., Staple, D. B., Sagner, A., Roper, J. C., Julicher, F., et al. (2010). Cell flow reorients the axis of planar polarity in the wing epithelium of Drosophila. *Cell, 142*, 773–786.

Ang, S. J., Stump, R. J., Lovicu, F. J., & McAvoy, J. W. (2004). Spatial and temporal expression of Wnt and Dickkopf genes during murine lens development. *Gene Expression Patterns, 4*, 289–295.

Angers, S., & Moon, R. T. (2009). Proximal events in Wnt signal transduction. *Nature Reviews. Molecular Cell Biology, 10*, 468–477.

Antic, D., Stubbs, J. L., Suyama, K., Kintner, C., Scott, M. P., & Axelrod, J. D. (2010). Planar cell polarity enables posterior localization of nodal cilia and left-right axis determination during mouse and Xenopus embryogenesis. *PLoS One, 5*, e8999.

Axelrod, J. D. (2001). Unipolar membrane association of Dishevelled mediates Frizzled planar cell polarity signaling. *Genes & Development, 15*, 1182–1187.

Axelrod, J. D. (2009). Progress and challenges in understanding planar cell polarity signaling. *Seminars in Cell & Developmental Biology, 20*, 964–971.

Axelrod, J. D., & Tomlin, C. J. (2011). Modeling the control of planar cell polarity. *Wiley Interdisciplinary Reviews. Systems Biology and Medicine, 3*, 588–605.

Blair, S. S. (2012). Cell polarity: Overdosing on PCPs. *Current Biology, 22*, R567–R569.

Blitzer, J. T., & Nusse, R. (2006). A critical role for endocytosis in Wnt signaling. *BMC Cell Biology, 7*, 28.

Boehm, B., Westerberg, H., Lesnicar-Pucko, G., Raja, S., Rautschka, M., Cotterell, J., et al. (2010). The role of spatially controlled cell proliferation in limb bud morphogenesis. *PLoS Biology, 8*, e1000420.

Bosoi, C. M., Capra, V., Allache, R., Trinh, V. Q., De Marco, P., Merello, E., et al. (2011). Identification and characterization of novel rare mutations in the planar cell polarity gene PRICKLE1 in human neural tube defects. *Human Mutation, 32*, 1371–1375.

Bryja, V., Gradl, D., Schambony, A., Arenas, E., & Schulte, G. (2007). Beta-arrestin is a necessary component of Wnt/beta-catenin signaling in vitro and in vivo. *Proceedings of the National Academy of Sciences of the United States of America, 104*, 6690–6695.

Carroll, T. J., & Das, A. (2011). Planar cell polarity in kidney development and disease. *Organogenesis, 7*, 180–190.

Casal, J., Lawrence, P. A., & Struhl, G. (2006). Two separate molecular systems, Dachsous/Fat and Starry night/Frizzled, act independently to confer planar cell polarity. *Development, 133*, 4561–4572.

Casal, J., Struhl, G., & Lawrence, P. A. (2002). Developmental compartments and planar polarity in Drosophila. *Current Biology, 12*, 1189–1198.

Chae, J., Kim, M. J., Goo, J. H., Collier, S., Gubb, D., Charlton, J., et al. (1999). The Drosophila tissue polarity gene starry night encodes a member of the protocadherin family. *Development, 126*, 5421–5429.

Chen, W. S., Antic, D., Matis, M., Logan, C. Y., Povelones, M., Anderson, G. A., et al. (2008). Asymmetric homotypic interactions of the atypical cadherin flamingo mediate intercellular polarity signaling. *Cell, 133*, 1093–1105.

Chen, W., ten Berge, D., Brown, J., Ahn, S., Hu, L. A., Miller, W. E., et al. (2003). Dishevelled 2 recruits beta-arrestin 2 to mediate Wnt5A-stimulated endocytosis of Frizzled 4. *Science, 301*, 1391–1394.

Ciani, L., Patel, A., Allen, N. D., & ffrench-Constant, C. (2003). Mice lacking the giant protocadherin mFAT1 exhibit renal slit junction abnormalities and a partially penetrant cyclopia and anophthalmia phenotype. *Molecular and Cellular Biology, 23*, 3575–3582.

Ciruna, B., Jenny, A., Lee, D., Mlodzik, M., & Schier, A. F. (2006). Planar cell polarity signalling couples cell division and morphogenesis during neurulation. *Nature, 439*, 220–224.

Classen, A. K., Anderson, K. I., Marois, E., & Eaton, S. (2005). Hexagonal packing of Drosophila wing epithelial cells by the planar cell polarity pathway. *Developmental Cell, 9*, 805–817.

Clevers, H. (2006). Wnt/beta-catenin signaling in development and disease. *Cell, 127*, 469–480.

Cohen, E. D., Miller, M. F., Wang, Z., Moon, R. T., & Morrisey, E. E. (2012). Wnt5a and Wnt11 are essential for second heart field progenitor development. *Development, 139*, 1931–1940.

Copp, A. J., & Greene, N. D. (2010). Genetics and development of neural tube defects. *The Journal of Pathology, 220*, 217–230.

Copp, A. J., Greene, N. D., & Murdoch, J. N. (2003). The genetic basis of mammalian neurulation. *Nature Reviews. Genetics, 4*, 784–793.

Craig, A. M., & Banker, G. (1994). Neuronal polarity. *Annual Review of Neuroscience, 17*, 267–310.

Dabdoub, A., Donohue, M. J., Brennan, A., Wolf, V., Montcouquiol, M., Sassoon, D. A., et al. (2003). Wnt signaling mediates reorientation of outer hair cell stereociliary bundles in the mammalian cochlea. *Development, 130*, 2375–2384.

Dabdoub, A., & Kelley, M. W. (2005). Planar cell polarity and a potential role for a Wnt morphogen gradient in stereociliary bundle orientation in the mammalian inner ear. *Journal of Neurobiology, 64*, 446–457.

DeChiara, T. M., Kimble, R. B., Poueymirou, W. T., Rojas, J., Masiakowski, P., Valenzuela, D. M., et al. (2000). Ror2, encoding a receptor-like tyrosine kinase, is required for cartilage and growth plate development. *Nature Genetics, 24*, 271–274.

De Marco, P., Merello, E., Rossi, A., Piatelli, G., Cama, A., Kibar, Z., et al. (2012). FZD6 is a novel gene for human neural tube defects. *Human Mutation, 33*, 384–390.

Devenport, D., & Fuchs, E. (2008). Planar polarization in embryonic epidermis orchestrates global asymmetric morphogenesis of hair follicles. *Nature Cell Biology, 10*, 1257–1268.

Devenport, D., Oristian, D., Heller, E., & Fuchs, E. (2011). Mitotic internalization of planar cell polarity proteins preserves tissue polarity. *Nature Cell Biology, 13*, 893–902.

Dubois, L., Lecourtois, M., Alexandre, C., Hirst, E., & Vincent, J. P. (2001). Regulated endocytic routing modulates wingless signaling in Drosophila embryos. *Cell, 105*, 613–624.

Ede, D. A., & Law, J. T. (1969). Computer simulation of vertebrate limb morphogenesis. *Nature, 221*, 244–248.

Etheridge, S. L., Ray, S., Li, S., Hamblet, N. S., Lijam, N., Tsang, M., et al. (2008). Murine dishevelled 3 functions in redundant pathways with dishevelled 1 and 2 in normal cardiac outflow tract, cochlea, and neural tube development. *PLoS Genetics, 4*, e1000259.

Feiguin, F., Hannus, M., Mlodzik, M., & Eaton, S. (2001). The ankyrin repeat protein Diego mediates Frizzled-dependent planar polarization. *Developmental Cell*, *1*, 93–101.

Fernandez-Teran, M. A., Hinchliffe, J. R., & Ros, M. A. (2006). Birth and death of cells in limb development: A mapping study. *Developmental Dynamics*, *235*, 2521–2537.

Flaherty, M. P., & Dawn, B. (2008). Noncanonical Wnt11 signaling and cardiomyogenic differentiation. *Trends in Cardiovascular Medicine*, *18*, 260–268.

Frojmark, A. S., Schuster, J., Sobol, M., Entesarian, M., Kilander, M. B., Gabrikova, D., et al. (2011). Mutations in Frizzled 6 cause isolated autosomal-recessive nail dysplasia. *American Journal of Human Genetics*, *88*, 852–860.

Gagliardi, M., Piddini, E., & Vincent, J. P. (2008). Endocytosis: A positive or a negative influence on Wnt signalling? *Traffic*, *9*, 1–9.

Gao, B., Song, H., Bishop, K., Elliot, G., Garrett, L., English, M. A., et al. (2011). Wnt signaling gradients establish planar cell polarity by inducing Vangl2 phosphorylation through Ror2. *Developmental Cell*, *20*, 163–176.

Gilbert, S. F. (2000). *Developmental biology* (6th ed.). Sunderland: Sinauer Associates.

Gong, Y., Mo, C., & Fraser, S. E. (2004). Planar cell polarity signalling controls cell division orientation during zebrafish gastrulation. *Nature*, *430*, 689–693.

Goodrich, L. V., & Strutt, D. (2011). Principles of planar polarity in animal development. *Development*, *138*, 1877–1892.

Gray, R. S., Roszko, I., & Solnica-Krezel, L. (2011). Planar cell polarity: Coordinating morphogenetic cell behaviors with embryonic polarity. *Developmental Cell*, *21*, 120–133.

Green, J. L., Inoue, T., & Sternberg, P. W. (2008). Opposing Wnt pathways orient cell polarity during organogenesis. *Cell*, *134*, 646–656.

Green, J. L., Kuntz, S. G., & Sternberg, P. W. (2008). Ror receptor tyrosine kinases: Orphans no more. *Trends in Cell Biology*, *18*, 536–544.

Gros, J., Hu, J. K., Vinegoni, C., Feruglio, P. F., Weissleder, R., & Tabin, C. J. (2010). WNT5A/JNK and FGF/MAPK pathways regulate the cellular events shaping the vertebrate limb bud. *Current Biology*, *20*, 1993–2002.

Gros, J., Serralbo, O., & Marcelle, C. (2009). WNT11 acts as a directional cue to organize the elongation of early muscle fibres. *Nature*, *457*, 589–593.

Gubb, D., & Garcia-Bellido, A. (1982). A genetic analysis of the determination of cuticular polarity during development in Drosophila melanogaster. *Journal of Embryology and Experimental Morphology*, *68*, 37–57.

Gubb, D., Green, C., Huen, D., Coulson, D., Johnson, G., Tree, D., et al. (1999). The balance between isoforms of the prickle LIM domain protein is critical for planar polarity in Drosophila imaginal discs. *Genes & Development*, *13*, 2315–2327.

Guo, N., Hawkins, C., & Nathans, J. (2004). Frizzled6 controls hair patterning in mice. *Proceedings of the National Academy of Sciences of the United States of America*, *101*, 9277–9281.

Halford, M. M., & Stacker, S. A. (2001). Revelations of the RYK receptor. *Bioessays*, *23*, 34–45.

Hamada, H., Meno, C., Watanabe, D., & Saijoh, Y. (2002). Establishment of vertebrate left-right asymmetry. *Nature Reviews. Genetics*, *3*, 103–113.

Hamblet, N. S., Lijam, N., Ruiz-Lozano, P., Wang, J., Yang, Y., Luo, Z., et al. (2002). Dishevelled 2 is essential for cardiac outflow tract development, somite segmentation and neural tube closure. *Development*, *129*, 5827–5838.

Hashimoto, M., Shinohara, K., Wang, J., Ikeuchi, S., Yoshiba, S., Meno, C., et al. (2010). Planar polarization of node cells determines the rotational axis of node cilia. *Nature Cell Biology*, *12*, 170–176.

He, F., Xiong, W., Yu, X., Espinoza-Lewis, R., Liu, C., Gu, S., et al. (2008). Wnt5a regulates directional cell migration and cell proliferation via Ror2-mediated noncanonical pathway in mammalian palate development. *Development*, *135*, 3871–3879.

Heisenberg, C. P., Tada, M., Rauch, G. J., Saude, L., Concha, M. L., Geisler, R., et al. (2000). Silberblick/Wnt11 mediates convergent extension movements during zebrafish gastrulation. *Nature, 405*, 76–81.

Henderson, D. J., & Chaudhry, B. (2011). Getting to the heart of planar cell polarity signaling. *Birth Defects Research. Part A, Clinical and Molecular Teratology, 91*, 460–467.

Hirokawa, N., Tanaka, Y., Okada, Y., & Takeda, S. (2006). Nodal flow and the generation of left-right asymmetry. *Cell, 125*, 33–45.

Ho, Y. H., Lien, M. T., Lin, C. M., Wei, S. Y., Chang, L. H., & Hsu, J. C. (2010). Echinoid regulates Flamingo endocytosis to control ommatidial rotation in the Drosophila eye. *Development, 137*, 745–754.

Ho, H. Y., Susman, M. W., Bikoff, J. B., Ryu, Y. K., Jonas, A. M., Hu, L., et al. (2012). Wnt5a-Ror-Dishevelled signaling constitutes a core developmental pathway that controls tissue morphogenesis. *Proceedings of the National Academy of Sciences of the United States of America, 109*, 4044–4051.

Hopyan, S., Sharpe, J., & Yang, Y. (2011). Budding behaviors: Growth of the limb as a model of morphogenesis. *Developmental Dynamics, 240*, 1054–1062.

Jessen, J. R., Topczewski, J., Bingham, S., Sepich, D. S., Marlow, F., Chandrasekhar, A., et al. (2002). Zebrafish trilobite identifies new roles for Strabismus in gastrulation and neuronal movements. *Nature Cell Biology, 4*, 610–615.

Jullien, J., & Gurdon, J. (2005). Morphogen gradient interpretation by a regulated trafficking step during ligand-receptor transduction. *Genes & Development, 19*, 2682–2694.

Karner, C. M., Chirumamilla, R., Aoki, S., Igarashi, P., Wallingford, J. B., & Carroll, T. J. (2009). Wnt9b signaling regulates planar cell polarity and kidney tubule morphogenesis. *Nature Genetics, 41*, 793–799.

Keller, R. (2002). Shaping the vertebrate body plan by polarized embryonic cell movements. *Science, 298*, 1950–1954.

Keller, R., Davidson, L., Edlund, A., Elul, T., Ezin, M., Shook, D., et al. (2000). Mechanisms of convergence and extension by cell intercalation. *Philosophical Transactions of the Royal Society of London. Series B, Biological Sciences, 355*, 897–922.

Kibar, Z., Bosoi, C. M., Kooistra, M., Salem, S., Finnell, R. H., De Marco, P., et al. (2009). Novel mutations in VANGL1 in neural tube defects. *Human Mutation, 30*, E706–E715.

Kibar, Z., Capra, V., & Gros, P. (2007). Toward understanding the genetic basis of neural tube defects. *Clinical Genetics, 71*, 295–310.

Kibar, Z., Salem, S., Bosoi, C. M., Pauwels, E., De Marco, P., Merello, E., et al. (2011). Contribution of VANGL2 mutations to isolated neural tube defects. *Clinical Genetics, 80*, 76–82.

Kibar, Z., Torban, E., McDearmid, J. R., Reynolds, A., Berghout, J., Mathieu, M., et al. (2007). Mutations in VANGL1 associated with neural-tube defects. *The New England Journal of Medicine, 356*, 1432–1437.

Kikuchi, A., Yamamoto, H., & Kishida, S. (2007). Multiplicity of the interactions of Wnt proteins and their receptors. *Cellular Signalling, 19*, 659–671.

Kilian, B., Mansukoski, H., Barbosa, F. C., Ulrich, F., Tada, M., & Heisenberg, C. P. (2003). The role of Ppt/Wnt5 in regulating cell shape and movement during zebrafish gastrulation. *Mechanisms of Development, 120*, 467–476.

Kim, G. H., & Han, J. K. (2007). Essential role for beta-arrestin 2 in the regulation of Xenopus convergent extension movements. *The EMBO Journal, 26*, 2513–2526.

Kim, G. H., Her, J. H., & Han, J. K. (2008). Ryk cooperates with Frizzled 7 to promote Wnt11-mediated endocytosis and is essential for Xenopus laevis convergent extension movements. *The Journal of Cell Biology, 182*, 1073–1082.

Klaus, A., & Birchmeier, W. (2008). Wnt signalling and its impact on development and cancer. *Nature Reviews. Cancer, 8*, 387–398.

Kohn, A. D., & Moon, R. T. (2005). Wnt and calcium signaling: Beta-catenin-independent pathways. *Cell Calcium*, *38*, 439–446.

Lapebie, P., Borchiellini, C., & Houliston, E. (2011). Dissecting the PCP pathway: One or more pathways? Does a separate Wnt-Fz-Rho pathway drive morphogenesis? *Bioessays*, *33*, 759–768.

Lawrence, P. A., Casal, J., & Struhl, G. (2002). Towards a model of the organisation of planar polarity and pattern in the Drosophila abdomen. *Development*, *129*, 2749–2760.

Lawrence, P. A., Casal, J., & Struhl, G. (2004). Cell interactions and planar polarity in the abdominal epidermis of Drosophila. *Development*, *131*, 4651–4664.

Lawrence, P. A., & Shelton, P. M. (1975). The determination of polarity in the developing insect retina. *Journal of Embryology and Experimental Morphology*, *33*, 471–486.

Lawrence, P. A., Struhl, G., & Casal, J. (2007). Planar cell polarity: One or two pathways? *Nature Reviews. Genetics*, *8*, 555–563.

Lei, Y. P., Zhang, T., Li, H., Wu, B. L., Jin, L., & Wang, H. Y. (2010). VANGL2 mutations in human cranial neural-tube defects. *The New England Journal of Medicine*, *362*, 2232–2235.

Li, L., Hutchins, B. I., & Kalil, K. (2009). Wnt5a induces simultaneous cortical axon outgrowth and repulsive axon guidance through distinct signaling mechanisms. *The Journal of Neuroscience*, *29*, 5873–5883.

Lin, S., Baye, L. M., Westfall, T. A., & Slusarski, D. C. (2010). Wnt5b-Ryk pathway provides directional signals to regulate gastrulation movement. *The Journal of Cell Biology*, *190*, 263–278.

Liu, H. X., Grosse, A. S., Iwatsuki, K., Mishina, Y., Gumucio, D. L., & Mistretta, C. M. (2012). Separate and distinctive roles for Wnt5a in tongue, lingual tissue and taste papilla development. *Developmental Biology*, *361*, 39–56.

Liu, H. X., Grosse, A. M., Walton, K. D., Saims, D. A., Gumucio, D. L., & Mistretta, C. M. (2009). WNT5a in tongue and fungiform Papilla development. *Annals of the New York Academy of Sciences*, *1170*, 11–17.

Logan, C. Y., & Nusse, R. (2004). The Wnt signaling pathway in development and disease. *Annual Review of Cell and Developmental Biology*, *20*, 781–810.

Lu, X., Borchers, A. G., Jolicoeur, C., Rayburn, H., Baker, J. C., & Tessier-Lavigne, M. (2004). PTK7/CCK-4 is a novel regulator of planar cell polarity in vertebrates. *Nature*, *430*, 93–98.

Lu, W., Yamamoto, V., Ortega, B., & Baltimore, D. (2004). Mammalian Ryk is a Wnt coreceptor required for stimulation of neurite outgrowth. *Cell*, *119*, 97–108.

Ma, D., Yang, C. H., McNeill, H., Simon, M. A., & Axelrod, J. D. (2003). Fidelity in planar cell polarity signalling. *Nature*, *421*, 543–547.

Macdonald, B. T., Semenov, M. V., & He, X. (2007). SnapShot: Wnt/beta-catenin signaling. *Cell*, *131*, 1204.

MacDonald, B. T., Tamai, K., & He, X. (2009). Wnt/beta-catenin signaling: Components, mechanisms, and diseases. *Developmental Cell*, *17*, 9–26.

Macheda, M. L., Sun, W. W., Kugathasan, K., Hogan, B. M., Bower, N. I., Halford, M. M., et al. (2012). The Wnt receptor Ryk plays a role in mammalian planar cell polarity signaling. *The Journal of Biological Chemistry*, *287*, 29312–29323.

Majumdar, A., Vainio, S., Kispert, A., McMahon, J., & McMahon, A. P. (2003). Wnt11 and Ret/Gdnf pathways cooperate in regulating ureteric branching during metanephric kidney development. *Development*, *130*, 3175–3185.

Mao, Y., Mulvaney, J., Zakaria, S., Yu, T., Morgan, K. M., Allen, S., et al. (2011). Characterization of a Dchs1 mutant mouse reveals requirements for Dchs1-Fat4 signaling during mammalian development. *Development*, *138*, 947–957.

Matakatsu, H., & Blair, S. S. (2012). Separating planar cell polarity and Hippo pathway activities of the protocadherins Fat and Dachsous. *Development*, *139*, 1498–1508.

Matsuyama, M., Aizawa, S., & Shimono, A. (2009). Sfrp controls apicobasal polarity and oriented cell division in developing gut epithelium. *PLoS Genetics*, *5*, e1000427.

McNeill, H. (2010). Planar cell polarity: Keeping hairs straight is not so simple. *Cold Spring Harbor Perspectives in Biology*, *2*, a003376.

Merte, J., Jensen, D., Wright, K., Sarsfield, S., Wang, Y., Schekman, R., et al. (2010). Sec24b selectively sorts Vangl2 to regulate planar cell polarity during neural tube closure. *Nature Cell Biology*, *12*, 41–46, sup pp. 41–48.

Mikels, A. J., & Nusse, R. (2006). Purified Wnt5a protein activates or inhibits beta-catenin-TCF signaling depending on receptor context. *PLoS Biology*, *4*, e115.

Minami, Y., Oishi, I., Endo, M., & Nishita, M. (2010). Ror-family receptor tyrosine kinases in noncanonical Wnt signaling: Their implications in developmental morphogenesis and human diseases. *Developmental Dynamics*, *239*, 1–15.

Mottola, G., Classen, A. K., Gonzalez-Gaitan, M., Eaton, S., & Zerial, M. (2010). A novel function for the Rab5 effector Rabenosyn-5 in planar cell polarity. *Development*, *137*, 2353–2364.

Niswander, L., Tickle, C., Vogel, A., Booth, I., & Martin, G. R. (1993). FGF-4 replaces the apical ectodermal ridge and directs outgrowth and patterning of the limb. *Cell*, *75*, 579–587.

Nonaka, S., Shiratori, H., Saijoh, Y., & Hamada, H. (2002). Determination of left-right patterning of the mouse embryo by artificial nodal flow. *Nature*, *418*, 96–99.

Nonaka, S., Yoshiba, S., Watanabe, D., Ikeuchi, S., Goto, T., Marshall, W. F., et al. (2005). De novo formation of left-right asymmetry by posterior tilt of nodal cilia. *PLoS Biology*, *3*, e268.

Northrup, H., & Volcik, K. A. (2000). Spina bifida and other neural tube defects. *Current Problems in Pediatrics*, *30*, 313–332.

Nowak, M., Machate, A., Yu, S. R., Gupta, M., & Brand, M. (2011). Interpretation of the FGF8 morphogen gradient is regulated by endocytic trafficking. *Nature Cell Biology*, *13*, 153–158.

Nusse, R., & Varmus, H. E. (1982). Many tumors induced by the mouse mammary tumor virus contain a provirus integrated in the same region of the host genome. *Cell*, *31*, 99–109.

Ohkawara, B., Glinka, A., & Niehrs, C. (2011). Rspo3 binds syndecan 4 and induces Wnt/PCP signaling via clathrin-mediated endocytosis to promote morphogenesis. *Developmental Cell*, *20*, 303–314.

Ohkawara, B., & Niehrs, C. (2011). An ATF2-based luciferase reporter to monitor non-canonical Wnt signaling in Xenopus embryos. *Developmental Dynamics*, *240*, 188–194.

Oishi, I., Suzuki, H., Onishi, N., Takada, R., Kani, S., Ohkawara, B., et al. (2003). The receptor tyrosine kinase Ror2 is involved in non-canonical Wnt5a/JNK signalling pathway. *Genes to Cells*, *8*, 645–654.

Olguin, P., Glavic, A., & Mlodzik, M. (2011). Intertissue mechanical stress affects Frizzled-mediated planar cell polarity in the Drosophila notum epidermis. *Current Biology*, *21*, 236–242.

Park, M., & Moon, R. T. (2002). The planar cell-polarity gene stbm regulates cell behaviour and cell fate in vertebrate embryos. *Nature Cell Biology*, *4*, 20–25.

Peradziryi, H., Kaplan, N. A., Podleschny, M., Liu, X., Wehner, P., Borchers, A., et al. (2011). PTK7/Otk interacts with Wnts and inhibits canonical Wnt signalling. *The EMBO Journal*, *30*, 3729–3740.

Person, A. D., Beiraghi, S., Sieben, C. M., Hermanson, S., Neumann, A. N., Robu, M. E., et al. (2010). WNT5A mutations in patients with autosomal dominant Robinow syndrome. *Developmental Dynamics*, *239*, 327–337.

Probst, B., Rock, R., Gessler, M., Vortkamp, A., & Puschel, A. W. (2007). The rodent Four-jointed ortholog Fjx1 regulates dendrite extension. *Developmental Biology*, *312*, 461–470.

Puppo, F., Thome, V., Lhoumeau, A. C., Cibois, M., Gangar, A., Lembo, F., et al. (2011). Protein tyrosine kinase 7 has a conserved role in Wnt/beta-catenin canonical signalling. *EMBO Reports*, *12*, 43–49.

Purvanov, V., Koval, A., & Katanaev, V. L. (2010). A direct and functional interaction between Go and Rab5 during G protein-coupled receptor signaling. *Science Signaling*, *3*, ra65.

Qian, D., Jones, C., Rzadzinska, A., Mark, S., Zhang, X., Steel, K. P., et al. (2007). Wnt5a functions in planar cell polarity regulation in mice. *Developmental Biology*, *306*, 121–133.

Rauch, G. J., Hammerschmidt, M., Blader, P., Schauerte, H. E., Strahle, U., Ingham, P. W., et al. (1997). Wnt5 is required for tail formation in the zebrafish embryo. *Cold Spring Harbor Symposia on Quantitative Biology*, *62*, 227–234.

Reiter, R. S., & Solursh, M. (1982). Mitogenic property of the apical ectodermal ridge. *Developmental Biology*, *93*, 28–35.

Rida, P. C., & Chen, P. (2009). Line up and listen: Planar cell polarity regulation in the mammalian inner ear. *Seminars in Cell & Developmental Biology*, *20*, 978–985.

Ridley, A. J., Schwartz, M. A., Burridge, K., Firtel, R. A., Ginsberg, M. H., Borisy, G., et al. (2003). Cell migration: Integrating signals from front to back. *Science*, *302*, 1704–1709.

Robinson, A., Escuin, S., Doudney, K., Vekemans, M., Stevenson, R. E., Greene, N. D., et al. (2012). Mutations in the planar cell polarity genes CELSR1 and SCRIB are associated with the severe neural tube defect craniorachischisis. *Human Mutation*, *33*, 440–447.

Rodriguez-Boulan, E., & Powell, S. K. (1992). Polarity of epithelial and neuronal cells. *Annual Review of Cell Biology*, *8*, 395–427.

Saburi, S., Hester, I., Fischer, E., Pontoglio, M., Eremina, V., Gessler, M., et al. (2008). Loss of Fat4 disrupts PCP signaling and oriented cell division and leads to cystic kidney disease. *Nature Genetics*, *40*, 1010–1015.

Saburi, S., Hester, I., Goodrich, L., & McNeill, H. (2012). Functional interactions between Fat family cadherins in tissue morphogenesis and planar polarity. *Development*, *139*, 1806–1820.

Sagner, A., Merkel, M., Aigouy, B., Gaebel, J., Brankatschk, M., Julicher, F., et al. (2012). Establishment of global patterns of planar polarity during growth of the Drosophila Wing epithelium. *Current Biology*, *22*, 1296–1301.

Satoh, W., Matsuyama, M., Takemura, H., Aizawa, S., & Shimono, A. (2008). Sfrp1, Sfrp2, and Sfrp5 regulate the Wnt/beta-catenin and the planar cell polarity pathways during early trunk formation in mouse. *Genesis*, *46*, 92–103.

Schlessinger, K., Hall, A., & Tolwinski, N. (2009). Wnt signaling pathways meet Rho GTPases. *Genes & Development*, *23*, 265–277.

Schneider, V. A., & Mercola, M. (2001). Wnt antagonism initiates cardiogenesis in Xenopus laevis. *Genes & Development*, *15*, 304–315.

Seifert, J. R., & Mlodzik, M. (2007). Frizzled/PCP signalling: A conserved mechanism regulating cell polarity and directed motility. *Nature Reviews. Genetics*, *8*, 126–138.

Semenov, M. V., Habas, R., Macdonald, B. T., & He, X. (2007). SnapShot: Noncanonical Wnt signaling pathways. *Cell*, *131*, 1378.

Seto, E. S., & Bellen, H. J. (2006). Internalization is required for proper Wingless signaling in Drosophila melanogaster. *The Journal of Cell Biology*, *173*, 95–106.

Seto, E. S., Bellen, H. J., & Lloyd, T. E. (2002). When cell biology meets development: Endocytic regulation of signaling pathways. *Genes & Development*, *16*, 1314–1336.

Shafer, B., Onishi, K., Lo, C., Colakoglu, G., & Zou, Y. (2011). Vangl2 promotes Wnt/planar cell polarity-like signaling by antagonizing Dvl1-mediated feedback inhibition in growth cone guidance. *Developmental Cell*, *20*, 177–191.

Shimada, Y., Yonemura, S., Ohkura, H., Strutt, D., & Uemura, T. (2006). Polarized transport of Frizzled along the planar microtubule arrays in Drosophila wing epithelium. *Developmental Cell*, *10*, 209–222.

Shiratori, H., & Hamada, H. (2006). The left-right axis in the mouse: From origin to morphology. *Development, 133*, 2095–2104.

Shnitsar, I., & Borchers, A. (2008). PTK7 recruits dsh to regulate neural crest migration. *Development, 135*, 4015–4024.

Simons, K., & Fuller, S. D. (1985). Cell surface polarity in epithelia. *Annual Review of Cell Biology, 1*, 243–288.

Simons, M., & Mlodzik, M. (2008). Planar cell polarity signaling: From fly development to human disease. *Annual Review of Genetics, 42*, 517–540.

Song, H., Hu, J., Chen, W., Elliott, G., Andre, P., Gao, B., et al. (2010). Planar cell polarity breaks bilateral symmetry by controlling ciliary positioning. *Nature, 466*, 378–382.

Strutt, D. I. (2001). Asymmetric localization of frizzled and the establishment of cell polarity in the Drosophila wing. *Molecular Cell, 7*, 367–375.

Strutt, D. (2009). Gradients and the specification of planar polarity in the insect cuticle. *Cold Spring Harbor Perspectives in Biology, 1*, a000489.

Sugiyama, Y., Lovicu, F. J., & McAvoy, J. W. (2011). Planar cell polarity in the mammalian eye lens. *Organogenesis, 7*, 191–201.

Sugiyama, Y., Stump, R. J., Nguyen, A., Wen, L., Chen, Y., Wang, Y., et al. (2010). Secreted frizzled-related protein disrupts PCP in eye lens fiber cells that have polarised primary cilia. *Developmental Biology, 338*, 193–201.

Tada, M., & Kai, M. (2009). Noncanonical Wnt/PCP signaling during vertebrate gastrulation. *Zebrafish, 6*, 29–40.

Takeuchi, S., Takeda, K., Oishi, I., Nomi, M., Ikeya, M., Itoh, K., et al. (2000). Mouse Ror2 receptor tyrosine kinase is required for the heart development and limb formation. *Genes to Cells, 5*, 71–78.

Tao, H., Manak, J. R., Sowers, L., Mei, X., Kiyonari, H., Abe, T., et al. (2011). Mutations in prickle orthologs cause seizures in flies, mice, and humans. *American Journal of Human Genetics, 88*, 138–149.

Taylor, J., Abramova, N., Charlton, J., & Adler, P. N. (1998). Van Gogh: A new Drosophila tissue polarity gene. *Genetics, 150*, 199–210.

Theisen, H., Purcell, J., Bennett, M., Kansagara, D., Syed, A., & Marsh, J. L. (1994). dishevelled is required during wingless signaling to establish both cell polarity and cell identity. *Development, 120*, 347–360.

Tissir, F., & Goffinet, A. M. (2010). Planar cell polarity signaling in neural development. *Current Opinion in Neurobiology, 20*, 572–577.

Topczewski, J., Dale, R. M., & Sisson, B. E. (2011). Planar cell polarity signaling in craniofacial development. *Organogenesis, 7*, 255–259.

Topol, L., Jiang, X., Choi, H., Garrett-Beal, L., Carolan, P. J., & Yang, Y. (2003). Wnt-5a inhibits the canonical Wnt pathway by promoting GSK-3-independent beta-catenin degradation. *The Journal of Cell Biology, 162*, 899–908.

Torres, M. A., Yang-Snyder, J. A., Purcell, S. M., DeMarais, A. A., McGrew, L. L., & Moon, R. T. (1996). Activities of the Wnt-1 class of secreted signaling factors are antagonized by the Wnt-5A class and by a dominant negative cadherin in early Xenopus development. *The Journal of Cell Biology, 133*, 1123–1137.

Tree, D. R., Ma, D., & Axelrod, J. D. (2002). A three-tiered mechanism for regulation of planar cell polarity. *Seminars in Cell & Developmental Biology, 13*, 217–224.

Tree, D. R., Shulman, J. M., Rousset, R., Scott, M. P., Gubb, D., & Axelrod, J. D. (2002). Prickle mediates feedback amplification to generate asymmetric planar cell polarity signaling. *Cell, 109*, 371–381.

Ueno, N., & Greene, N. D. (2003). Planar cell polarity genes and neural tube closure. *Birth Defects Research. Part C, Embryo Today, 69*, 318–324.

Ulrich, F., Krieg, M., Schotz, E. M., Link, V., Castanon, I., Schnabel, V., et al. (2005). Wnt11 functions in gastrulation by controlling cell cohesion through Rab5c and E-cadherin. *Developmental Cell, 9*, 555–564.

Usui, T., Shima, Y., Shimada, Y., Hirano, S., Burgess, R. W., Schwarz, T. L., et al. (1999). Flamingo, a seven-pass transmembrane cadherin, regulates planar cell polarity under the control of Frizzled. *Cell, 98*, 585–595.

Uysal-Onganer, P., & Kypta, R. M. (2012). Wnt11 in 2011—The regulation and function of a non-canonical Wnt. *Acta Physiologica (Oxford, England), 204*, 52–64.

van Amerongen, R., Mikels, A., & Nusse, R. (2008). Alternative wnt signaling is initiated by distinct receptors. *Science Signaling, 1*, re9.

van Bokhoven, H., Celli, J., Kayserili, H., van Beusekom, E., Balci, S., Brussel, W., et al. (2000). Mutation of the gene encoding the ROR2 tyrosine kinase causes autosomal recessive Robinow syndrome. *Nature Genetics, 25*, 423–426.

Veeman, M. T., Axelrod, J. D., & Moon, R. T. (2003). A second canon. Functions and mechanisms of beta-catenin-independent Wnt signaling. *Developmental Cell, 5*, 367–377.

Vinson, C. R., & Adler, P. N. (1987). Directional non-cell autonomy and the transmission of polarity information by the frizzled gene of Drosophila. *Nature, 329*, 549–551.

Vladar, E. K., Antic, D., & Axelrod, J. D. (2009). Planar cell polarity signaling: The developing cell's compass. *Cold Spring Harbor Perspectives in Biology, 1*, a002964.

Wallingford, J. B., Fraser, S. E., & Harland, R. M. (2002). Convergent extension: The molecular control of polarized cell movement during embryonic development. *Developmental Cell, 2*, 695–706.

Wallingford, J. B., & Harland, R. M. (2001). Xenopus Dishevelled signaling regulates both neural and mesodermal convergent extension: Parallel forces elongating the body axis. *Development, 128*, 2581–2592.

Wang, Y. (2009). Wnt/Planar cell polarity signaling: A new paradigm for cancer therapy. *Molecular Cancer Therapeutics, 8*, 2103–2109.

Wang, Y., Chang, H., & Nathans, J. (2010). When whorls collide: The development of hair patterns in frizzled 6 mutant mice. *Development, 137*, 4091–4099.

Wang, Y., Guo, N., & Nathans, J. (2006). The role of Frizzled3 and Frizzled6 in neural tube closure and in the planar polarity of inner-ear sensory hair cells. *The Journal of Neuroscience, 26*, 2147–2156.

Wang, J., Hamblet, N. S., Mark, S., Dickinson, M. E., Brinkman, B. C., Segil, N., et al. (2006). Dishevelled genes mediate a conserved mammalian PCP pathway to regulate convergent extension during neurulation. *Development, 133*, 1767–1778.

Wang, Y., & Nathans, J. (2007). Tissue/planar cell polarity in vertebrates: New insights and new questions. *Development, 134*, 647–658.

Wang, B., Sinha, T., Jiao, K., Serra, R., & Wang, J. (2011). Disruption of PCP signaling causes limb morphogenesis and skeletal defects and may underlie Robinow syndrome and brachydactyly type B. *Human Molecular Genetics, 20*, 271–285.

Wansleeben, C., Feitsma, H., Montcouquiol, M., Kroon, C., Cuppen, E., & Meijlink, F. (2010). Planar cell polarity defects and defective Vangl2 trafficking in mutants for the COPII gene Sec24b. *Development, 137*, 1067–1073.

Westfall, T. A., Brimeyer, R., Twedt, J., Gladon, J., Olberding, A., Furutani-Seiki, M., et al. (2003). Wnt-5/pipetail functions in vertebrate axis formation as a negative regulator of Wnt/beta-catenin activity. *The Journal of Cell Biology, 162*, 889–898.

Weston, C. R., & Davis, R. J. (2002). The JNK signal transduction pathway. *Current Opinion in Genetics and Development, 12*, 14–21.

Weston, C. R., & Davis, R. J. (2007). The JNK signal transduction pathway. *Current Opinion in Cell Biology, 19*, 142–149.

Witze, E. S., Litman, E. S., Argast, G. M., Moon, R. T., & Ahn, N. G. (2008). Wnt5a control of cell polarity and directional movement by polarized redistribution of adhesion receptors. *Science, 320*, 365–369.

Wolff, T., & Rubin, G. M. (1998). Strabismus, a novel gene that regulates tissue polarity and cell fate decisions in Drosophila. *Development, 125*, 1149–1159.

Wu, J., & Mlodzik, M. (2008). The frizzled extracellular domain is a ligand for Van Gogh/Stbm during nonautonomous planar cell polarity signaling. *Developmental Cell, 15*, 462–469.

Wu, J., & Mlodzik, M. (2009). A quest for the mechanism regulating global planar cell polarity of tissues. *Trends in Cell Biology, 19*, 295–305.

Yamaguchi, T. P., Bradley, A., McMahon, A. P., & Jones, S. (1999). A Wnt5a pathway underlies outgrowth of multiple structures in the vertebrate embryo. *Development, 126*, 1211–1223.

Yamamoto, H., Komekado, H., & Kikuchi, A. (2006). Caveolin is necessary for Wnt-3a-dependent internalization of LRP6 and accumulation of beta-catenin. *Developmental Cell, 11*, 213–223.

Yang, C. H., Axelrod, J. D., & Simon, M. A. (2002). Regulation of Frizzled by fat-like cadherins during planar polarity signaling in the Drosophila compound eye. *Cell, 108*, 675–688.

Yang, Y., Topol, L., Lee, H., & Wu, J. (2003). Wnt5a and Wnt5b exhibit distinct activities in coordinating chondrocyte proliferation and differentiation. *Development, 130*, 1003–1015.

Yates, L. L., & Dean, C. H. (2011). Planar polarity: A new player in both lung development and disease. *Organogenesis*, 7, 209–216.

Yin, C., Kiskowski, M., Pouille, P. A., Farge, E., & Solnica-Krezel, L. (2008). Cooperation of polarized cell intercalations drives convergence and extension of presomitic mesoderm during zebrafish gastrulation. *The Journal of Cell Biology, 180*, 221–232.

Yoshikawa, S., McKinnon, R. D., Kokel, M., & Thomas, J. B. (2003). Wnt-mediated axon guidance via the Drosophila Derailed receptor. *Nature, 422*, 583–588.

Yu, J. (2011). Wnt signaling and renal medulla formation. *Pediatric Nephrology, 26*, 1553–1557.

Yu, J., Carroll, T. J., Rajagopal, J., Kobayashi, A., Ren, Q., & McMahon, A. P. (2009). A Wnt7b-dependent pathway regulates the orientation of epithelial cell division and establishes the cortico-medullary axis of the mammalian kidney. *Development, 136*, 161–171.

Yu, A., Rual, J. F., Tamai, K., Harada, Y., Vidal, M., He, X., et al. (2007). Association of Dishevelled with the clathrin AP-2 adaptor is required for Frizzled endocytosis and planar cell polarity signaling. *Developmental Cell, 12*, 129–141.

Yu, H., Smallwood, P. M., Wang, Y., Vidaltamayo, R., Reed, R., & Nathans, J. (2010). Frizzled 1 and frizzled 2 genes function in palate, ventricular septum and neural tube closure: General implications for tissue fusion processes. *Development, 137*, 3707–3717.

Zallen, J. A. (2007). Planar polarity and tissue morphogenesis. *Cell, 129*, 1051–1063.

Zeidler, M. P., Perrimon, N., & Strutt, D. I. (2000). Multiple roles for four-jointed in planar polarity and limb patterning. *Developmental Biology, 228*, 181–196.

Zhou, W., Lin, L., Majumdar, A., Li, X., Zhang, X., Liu, W., et al. (2007). Modulation of morphogenesis by noncanonical Wnt signaling requires ATF/CREB family-mediated transcriptional activation of TGFbeta2. *Nature Genetics, 39*, 1225–1234.

INDEX

Note: Page numbers followed by "*f*" indicate figures, and "*t*" indicate tables.

D

E

F

CONTENTS OF PREVIOUS VOLUMES

Volume 47

Volume 48

Volume 49

Volume 50

Volume 51

Volume 52

Volume 53

Volume 54

Volume 55

Volume 56

Volume 57

Volume 58

Volume 59

Volume 60

Volume 61

Volume 62

Volume 63

Volume 64

Volume 65

Volume 66

Volume 67

Volume 68

Volume 69

Volume 70

Volume 71

Volume 72

Volume 73

Volume 74

Volume 75

Volume 76

Volume 77

Volume 78

Volume 79

Volume 80

Volume 81

Volume 82

Volume 83

Volume 84

Volume 85

Volume 86

Volume 87

Volume 88

Volume 91

Volume 92

Volume 93

Volume 94

Volume 95

Volume 96

Volume 97

Volume 98

Volume 99

Volume 100

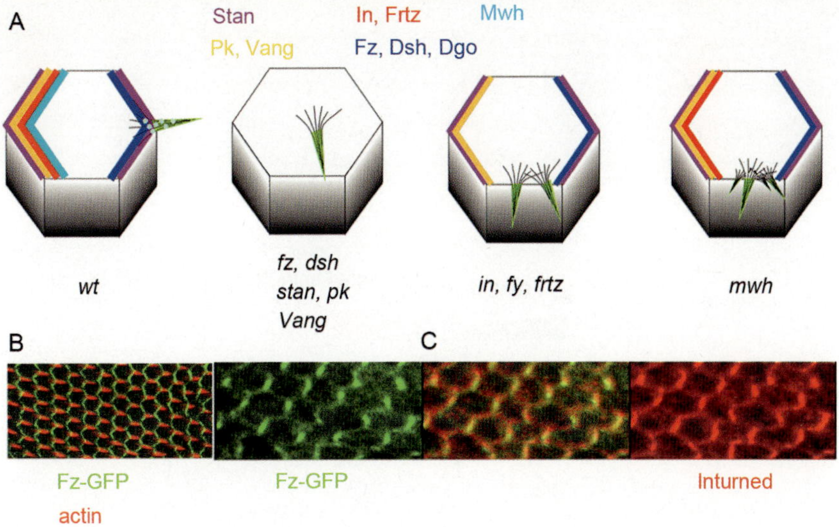

Paul N. Adler, Figure 1.2 Polarized protein accumulation and mutant phenotypes. A shows cartoons of wild-type and mutant cells. In the wild-type cell, all of the *fz/stan* pathway proteins accumulate asymmetrically and hairs are formed at the distal vertex. In cells, mutant for *fz* or other core genes hairs form at a relatively central location and no proteins accumulate asymmetrically. In cells, mutant for *in* or any of the other PPE genes multiple hairs form at an abnormal location on the cell periphery. The core proteins accumulate asymmetrically but the PPE proteins and Mwh do not. In cells, mutant for *mwh* genes multiple hairs form at an abnormal location on the cell periphery. The core and PPE proteins accumulate asymmetrically but Mwh does not. B shows the accumulation of Fz-GFP by direct imaging of GFP (green) (no immunostaining) and actin in red (hairs). C shows the coordinate asymmetric accumulation of Fz (green) and In (red).

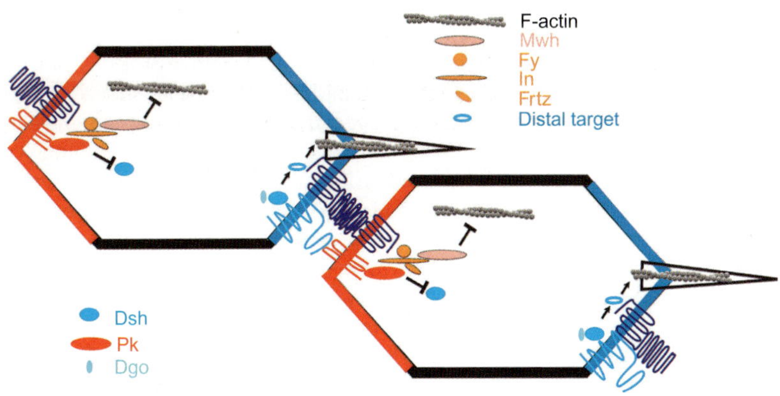

Paul N. Adler, Figure 1.6 A possible model for PCP protein action in wing cells. Two neighboring cells are diagramed. Proteins on the distal membrane of one cell (Fz and Stan) interact with proteins on the proximal membrane of the neighboring cell (Vang and Stan). The core transmembrane proteins recruit the cytoplasmic core proteins (Dsh, Dgo, and Pk). The PPE proteins are recruited to the proximal side where In recruits Mwh. Mwh inhibits the actin cytoskeleton in the distal part of the cell. Dsh activates an activator of the actin cytoskeleton distally. For simplicity, only one of the feedback interactions between cytoplasmic proteins is drawn (Pk inhibiting Dsh on the distal side).

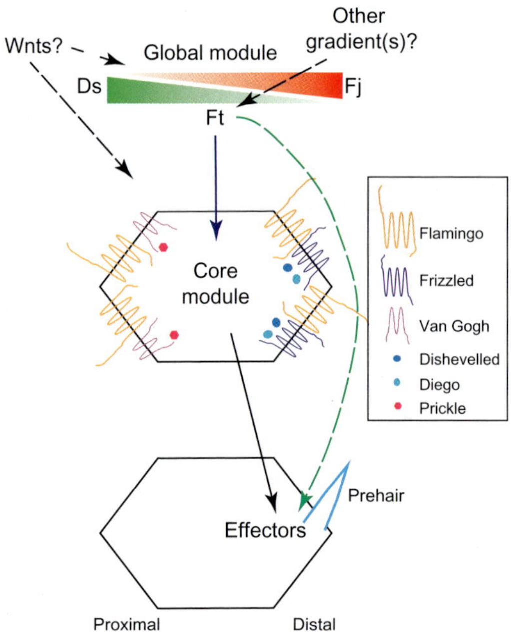

Ying Peng and Jeffrey D. Axelrod, Figure 2.1 Hierarchical model of the PCP signaling pathway. The pathway consists of three modules, the global-, core-, and tissue-specific effector modules. According to the series model, the global module provides directional input to the core module (blue arrow) that establishes and amplifies subcellular asymmetry. This subcellular asymmetry is used to direct tissue-specific effector module function within the cell. According to the parallel model, the global module communicates directly with the tissue-specific effector module (green dashed arrow), without signaling to the core module. Directional information for the global module comes from tissue-level expression gradients of Ds and Fj, but it is likely that other gradients are also important, at least in the wing (black dashed arrow). While Wnt proteins seem not to play a direct role in PCP signaling in *Drosophila*, they appear to do so in vertebrates (gray dashed arrows). Precisely, how Wnts affect vertebrate PCP is unclear. Asymmetrically segregated core PCP proteins are shown. Various effector modules produce different tissue-specific responses. Here, effectors establish the distal location for growth of a prehair, as on the wing and abdomen.

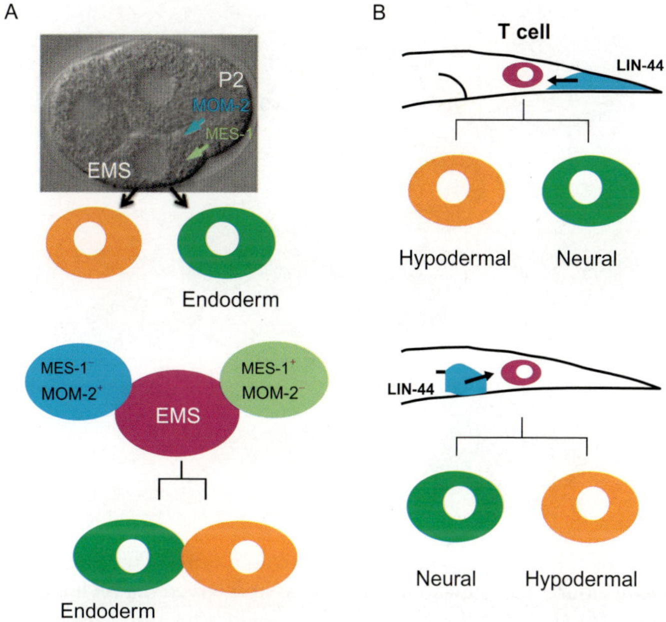

Hitoshi Sawa, Figure 3.3 Wnts function as polarity cues. (A) The EMS polarity is controlled by the MES-1 and MOM-2/Wnt signals from the P2 cell. When isolated wild-type EMS is combined with two P2 cell mutants of *mes-1* or *mom-2*, endoderm produced from an EMS daughter produced on the side with *mes-1* P2 was attached. (B) The T-cell polarity is controlled by LIN-44 from its posterior neighbor cell. When LIN-44 is expressed on the anterior side of the T cell in *lin-44* mutants, the T-cell polarity is nearly completely reversed.

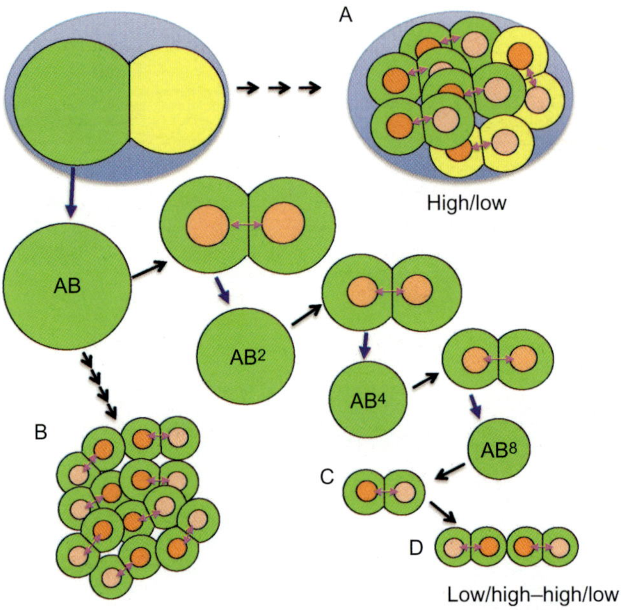

Hitoshi Sawa, Figure 3.5 Regulation of POP-1 asymmetry in the AB lineage. In normal development, (A) POP-1 asymmetry is first observed after the division of AB granddaughters (AB^4). (B) When P1 is isolated and cultured alone, POP-1 asymmetry is observed after the division of AB^4, but the daughter proximal to the cell cluster tends to have higher POP-1 signal than the distal one that does not contact with the cluster. (C) The sequentially isolated AB^8 cell shows POP-1 asymmetry. (D) In the next division, AB^8 daughters show low/high–high/low POP-1 asymmetry. Green cells, pink double arrows, blue arrows, and black arrows represent the AB lineage, daughter pairs, cell isolation, and cell divisions, respectively.

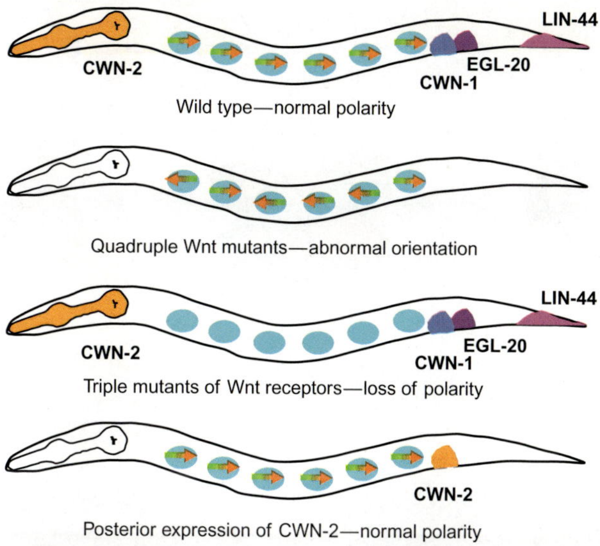

Hitoshi Sawa, Figure 3.6 Regulation of polarity orientation in seam cells by multiple Wnts. In wild type, Wnts are expressed anteriorly (CWN-2) or posteriorly (CWN-1, EGL-20, LIN-44) to seam cells. In quadruple Wnt mutants as in quintuple ones, polarity orientations were randomized. In contrast, after knock down of three Wnt receptors, cells were not polarized. Posterior expression of CWN-2 rescued polarity defects in a compound Wnt mutant.

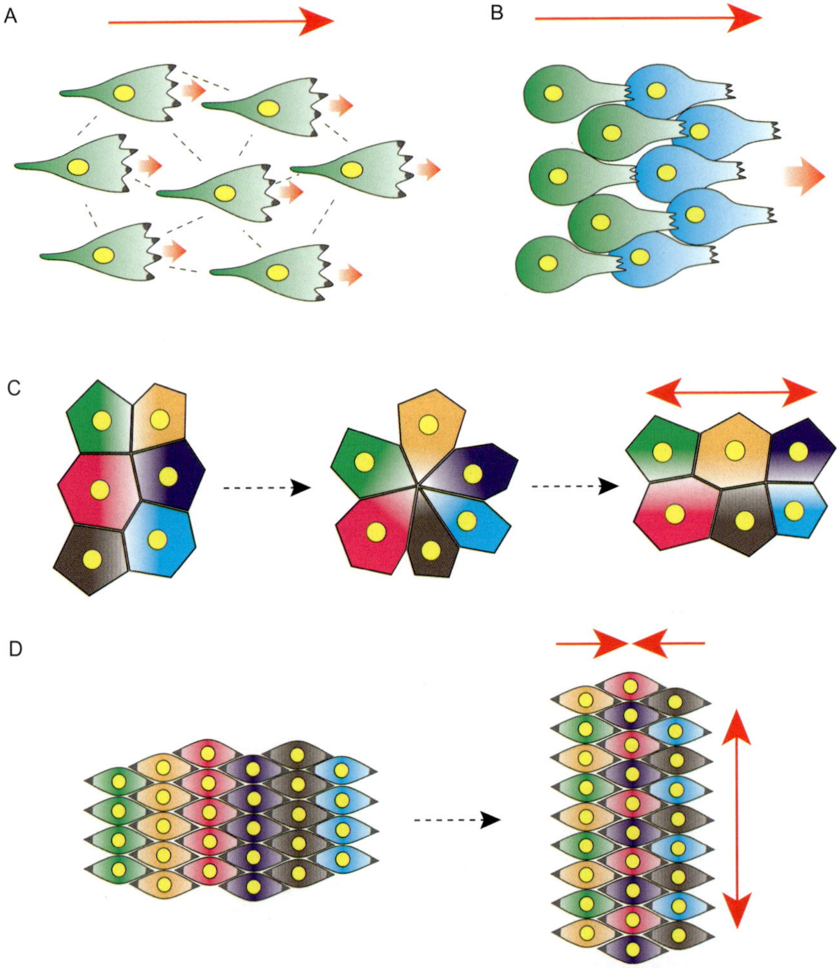

Masazumi Tada and Masatake Kai, Figure 4.1 Different types of cell movements. (A) Collective cell migration in which cells move individually (e.g., neural crest cells). Cells are loosely associated with each other and migrate in a uniform direction. (B) Collective cell migration as a cohesive clump (e.g., prechordal plate progenitors). Cells are tightly packed and migrate in one direction without changing neighbors. Cells at the leading edge (blue) form actin-based protrusions (shown in gray). (C) Cell intercalation mediated by formation and resolving of rosette (e.g., *Drosophila* germband). Five or more cells meet at one point ("rosette" formation) and resolve to make a new cell–cell boundary in relation to the direction of their movement. (D) Cell intercalation in convergent extension (e.g., notochord progenitors). Cells intercalate mediolaterally to simultaneously narrow and elongate the tissue along the anteroposterior axis.

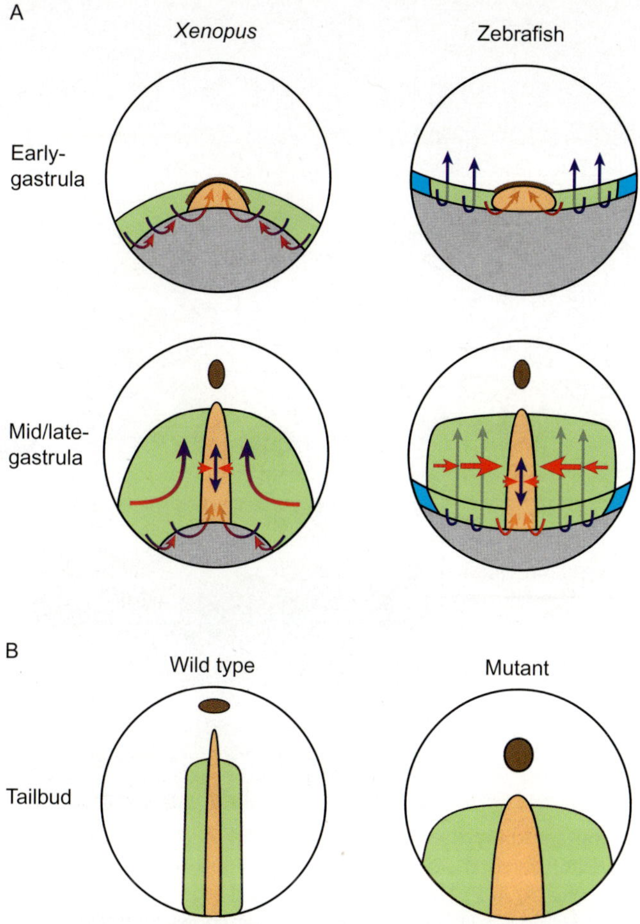

Masazumi Tada and Masatake Kai, Figure 4.2 Schematic illustrations of cell movements during gastrulation in *Xenopus* and zebrafish. (A) At the early-gastrula stage in *Xenopus*, notochord (shown in orange) and somite (green) progenitors involute at the marginal zone and simultaneously converge (blue to red arrows) toward the dorsal midline by cell intercalation. At the mid/late-gastrula stage, extensive mediolateral intercalation drives convergent extension (red and blue arrows) of the axial and lateral mesoderm (orange and green). In zebrafish, somite progenitors (green) internalize at the germ ring (green and light blue belt) and migrate anteriorly (blue arrows) at the early-gastrula stage. Cells at the dorsal-most marginal zone internalize and converge (red arrows) to create the physical shield (orange). After 75% epiboly (mid-gasatrula stage), somite progenitors (green) initiate to converge toward the midline, with increasing their speed and directionality (red arrows). The presumptive notochord (orange) elongates and narrows by mediolateral cell intercalation (red and blue arrows). Prechordal plate progenitors (brown) migrate anteriorly as a cohesive cluster in both *Xenopus* and zebrafish. (B) Wild-type and typical mutant phenotypes at the tailbud stage in zebrafish. In PCP mutants, the presumptive notochord (orange) and somites (green) are widened laterally and are shorter along the anterior–posterior axis than in wild-type embryos. This is accompanied with compromised anterior migration of the prospective prechordal plate (brown).

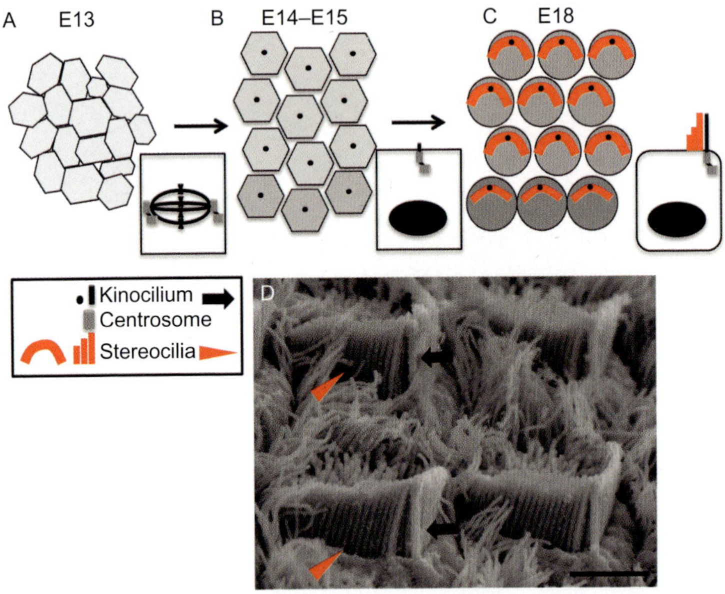

Helen May-Simera and Matthew W. Kelley, Figure 5.2 The kinocilium during hair cell development. Extension of the cochlea duct occurs concurrently with the polarization of individual hair cells. (A) At E13, the precursors for the organ of Corti are still undergoing mitosis and the cells are still actively dividing. Inset shows lateral view. (B) As the hair cells emerge from terminal mitosis, the basal body/centrosome migrates to the apex of the cell and the kinocilium begins to protrude. Between E15 and E16, the kinocilium begins to migrate toward the lateral edge of the luminal hair cell surface and adjacent microvilli elongate to form stereocilia. Inset shows location of basal body and extending kinocilium in the lateral view. (C) By E18, the initial migration of the kinocilium is complete; a further period of refinement occurs until the developing bundles achieve their ultimate orientation by P10. The kinocilium begins to retract from approximately P3 onward. Inset shows lateral view of hair cell with kinocilium and stereociliary bundle. (D) Scanning electron micrograph of the stereociliary bundles on four outer hair cells at P0 (basal turn of the cochlea). The crescent-shaped stereociliary bundles (red arrowhead) with the single kinocilium (black arrow) located at the vertex are indicated. Microvilli on the supporting cells between the hair cells are still present at this stage. Scale bar is 2.5 μm.

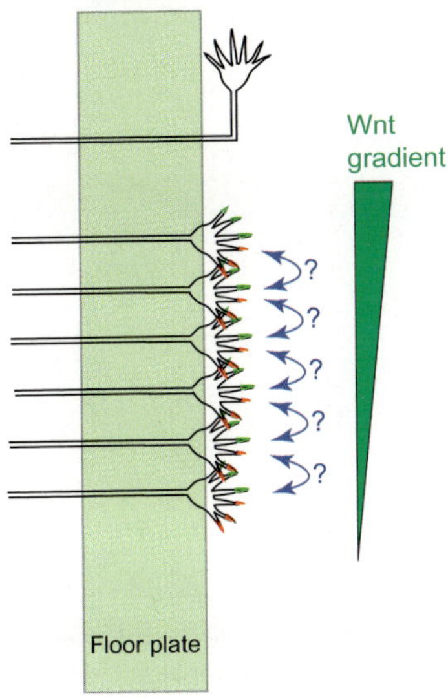

Yimin Zou, Figure 6.4 Possible growth cone–growth cone interactions during midline crossing and anterior turning of commissural axons.

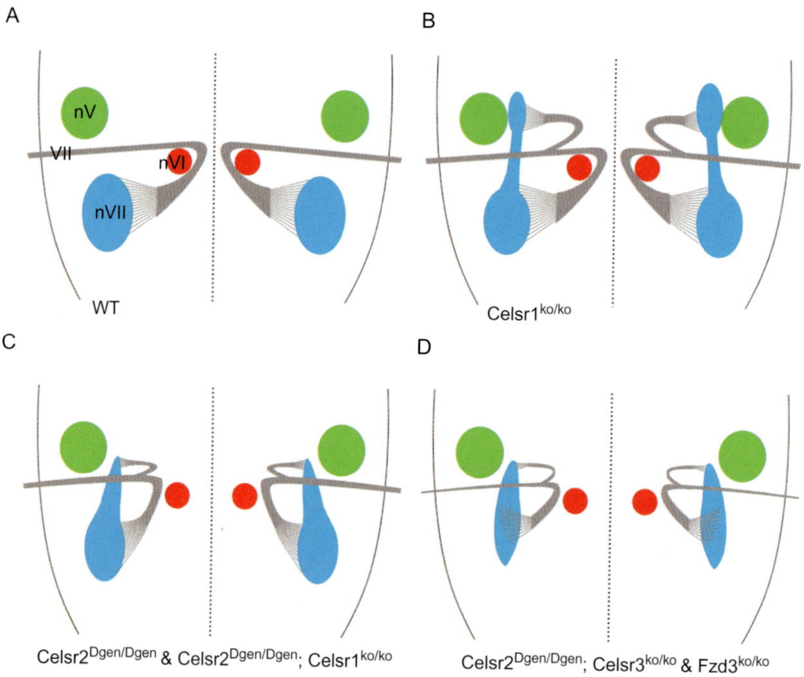

Camille Boutin *et al.*, Figure 7.4 Migration of facial branchiomotor (FBM) neurons in normal and *Celsr1–3* knockout mice. Drawings summarize the phenotype at P0. In wild-type animals (A), FBM neurons form a single nucleus (nVII) in a region that derives from r6. Their axons loop around the abducens nucleus (nVI) before exiting the rhombencephalon dorsally, caudal to the trigeminal nucleus (nV). In *Celsr1$^{ko/ko}$* mutants (B), in addition to the normal nVII, FBM neurons form another ectopic nucleus at the level of nV, but their axons leave the rhombencephalon at the right position. This is due to aberrant migration of FBM neurons in lateral r2 during embryogenesis. In *Celsr2$^{Dgen/Dgen}$* and *Celsr2$^{Dgen/Dgen}$; Celsr1$^{ko/ko}$* (C), the facial nerve *genu* is completely abnormal and axons do not loop around nVI because FBM neurons migrate prematurely in lateral r4–r5, forming lateral heterotopias. In *Celsr2$^{Dgen/Dgen}$; Celsr3$^{ko/ko}$* and in *Fzd3$^{ko/ko}$*, in addition to the absence of the *genu* of the facial nerve, the size of nVII is reduced because of embryonic cell death (D). "r," rhombomere; nV, motor trigeminal nucleus; nVI, abducens nucleus; nVII, facial nerve nucleus.

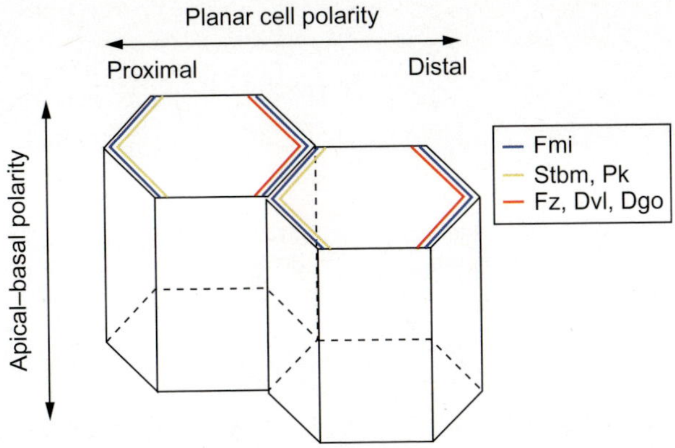

Thomas J. Carroll and Jing Yu, Figure 8.1 Schematic illustration of planar cell polarity (PCP). PCP is perpendicular to the apico-basal polarity of the epithelial tissue. PCP proteins are asymmetrically localized at the cell junctions between neighboring cells. This asymmetry defines the tissue polarity.

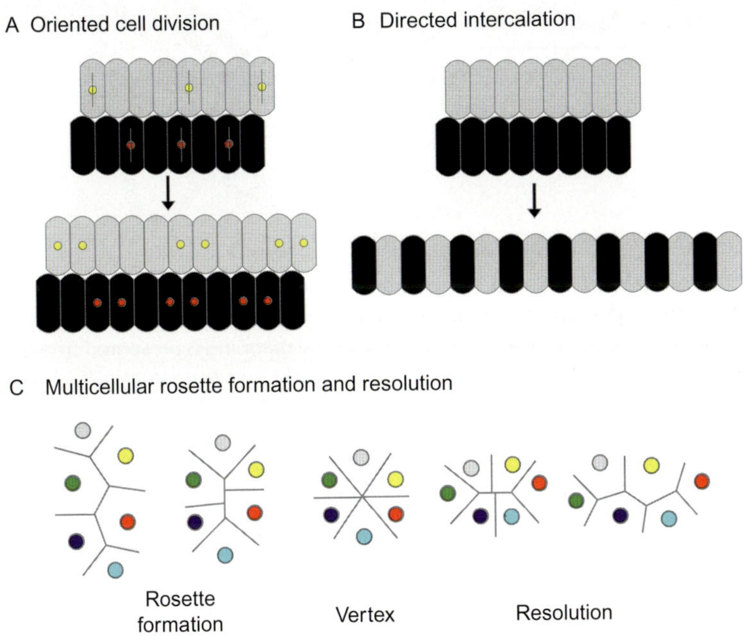

Thomas J. Carroll and Jing Yu, Figure 8.2 Processes regulating tubule elongation. (A) Oriented cell division leads to lengthening of the tubule, while the diameter stays the same. (B) Directed intercalation leads to the tubule elongating while it also becomes thinner. (C) Multicellular rosette formation and resolution (adapted from Zallen & Blankenship, 2008) describe the changes in cell morphology and adhesiveness that result in a transient rosette with a vertex forming. The rosette resolves resulting in the tubule becoming longer and thinner. Only oriented cell division has been demonstrated in the kidney. Panels (B) and (C) require live imaging to confirm.

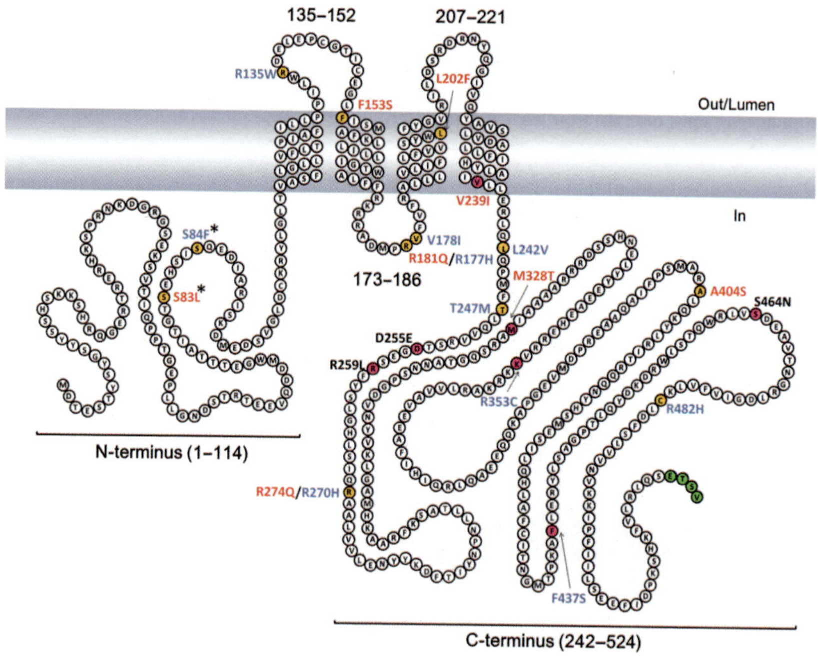

Elena Torban *et al.*, Figure 10.3 Topological model, structural features, and mutant variants of the Vangl protein family. The primary sequence and secondary structure of human VANGL1 protein as established by epitope mapping is presented. The nomenclature adopted in the figure has been standardized to residues in hVANGL1, which is used as molecular backbone. Sequence motifs and individual amino acid residues defining sequence landmarks are depicted in different colors. This includes Vangl protein variants (both human hVANGL1 and hVANGL2 and mouse mVangl2) associated with neural tube defects in mice and humans (orange, pink) and where loss of function has already been demonstrated (pink) or has yet to be characterized (orange). Residues altered in mVangl2 (D255E, R259L, S464N) are depicted in black, hVANGL1 (S83L, F153S, R181Q, L202F, V239I, R274Q, M328T, A404S) are depicted in red, and hVANGL2 (S84F, R135W, R177H, V178I, L242V, T247M, R270H, R353C, F437S, R482H) are depicted in blue. Also depicted is the ETSV sequence corresponding to a PDZ-binding motif (green) and two serine phosphorylation sites for which variants have been detected in human NTD patients (*; S83L in hVANGL1 and S84F in hVANGL2).

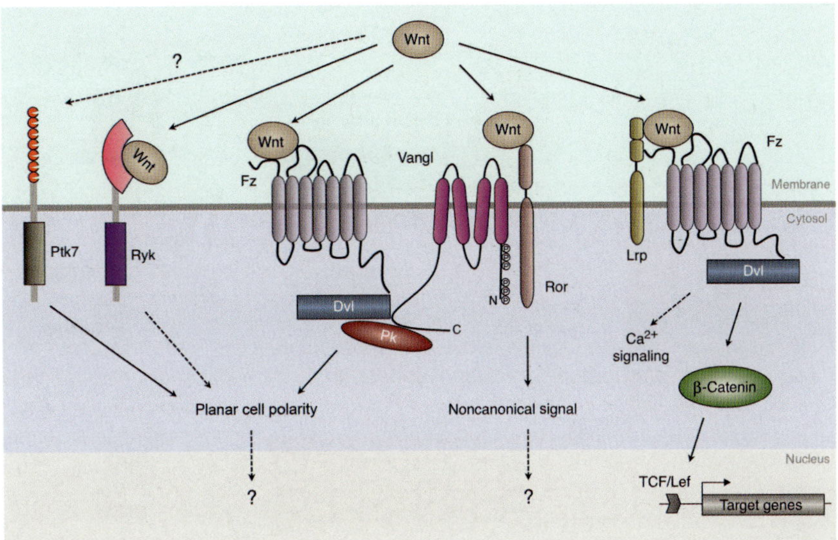

Bo Gao, Figure 11.2 *Wnt signaling pathways.* In the canonical Wnt/β-catenin pathway (right panel), secreted Wnt proteins bind the Frizzled (Fz) and low-density lipoprotein receptor-related proteins 5 and 6 (LRP) coreceptors to activate Disheveled (Dvl) leading to the stabilization of β-catenin, which then enters into the nucleus, where it binds T cell factor/the lymphoid enhancer factor (TCF/Lef) to transactivate target genes expression. In the Wnt/PCP pathway (middle panel), Wnt proteins also bind Fz, which recruits Dvl. On the other hand, Wnt proteins bind receptor tyrosine kinase orphan receptor (Ror) to form a complex with Vangl, which recruits Prickle (PK). As a result, Vangl is phosphorylated in a Wnt dose-dependent manner. It is proposed that PK and Dvl bind and antagonize each other to generate asymmetrical protein localization, a critical regulatory event of PCP. Other receptor tyrosine kinases including PTK7 and Ryk are also involved in PCP (left panel).

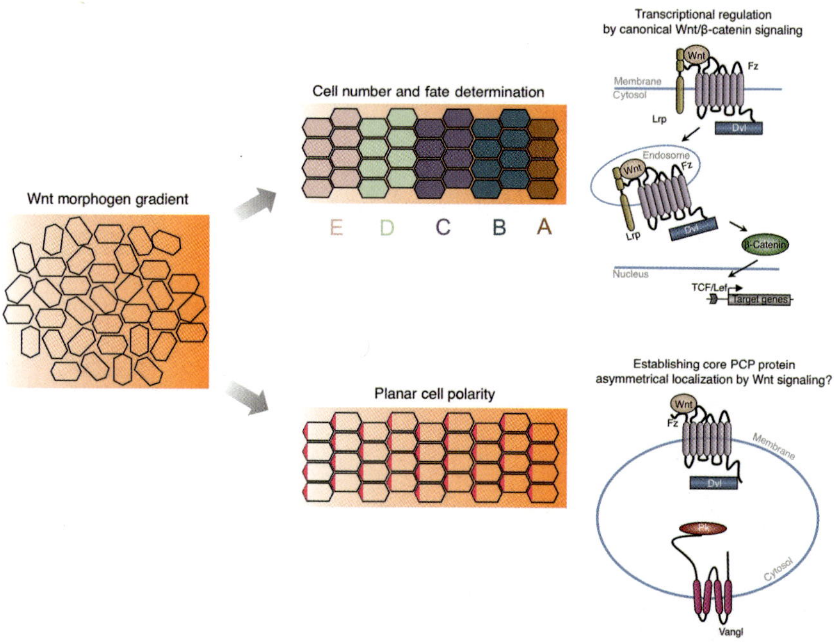

Bo Gao, Figure 11.3 *Dual function of Wnt morphogen gradient*. First, Wnt signaling gradient provides quantitative information to control cell proliferation and cell differentiation through canonical Wnt/β-catenin pathway, which is achieved via endocytosis of key regulatory components, β-catenin stabilization and nuclear translocation, and, finally, transactivation of target genes. Second, Wnt morphogen acts as a global cue to establish PCP to convey directional information, which is evidenced by asymmetrical localization of core PCP proteins through an unknown mechanism.